Introduction to

Diagnostic Electron Microscopy

Introduction to

Diagnostic Electron Microscopy

Bruce Mackay, M.D., Ph.D.

Associate Pathologist
Associate Professor of Pathology
Department of Pathology
The University of Texas M.D. Anderson Hospital
and Tumor Institute
Houston, Texas

ACC

APPLETON-CENTURY-CROFTS/New York

81 82 83 84 85/10 9 8 7 6 5 4 3 2 1

Prentice-Hall International, Inc., London
Prentice-Hall of Australia, Pty. Ltd., Sydney
Prentice-Hall of India Private Limited, New Delhi
Prentice-Hall of Japan, Inc., Tokyo
Prentice-Hall of Southeast Asia (Pte.) Ltd., Singapore
Whitehall Books Ltd., Wellington, New Zealand

Library of Congress Cataloging in Publication Data
Main entry under title:

Introduction to diagnostic electron microscopy.

Bibliography
Includes index.
1. Diagnosis, Electron microscopic. I. Mackay, Bruce, 1930– . [DNLM: 1. Microscopy, Electron. 2. Cytodiagnosis—Methods. QY 95 M153i]
RB43.5.I57 616.07′58 81-8109
ISBN 0-8385-4315-4 AACR2

Cover design: Jean M. Sabato
Text design: Faith Gertner
Production editor: Holly A. Reid

PRINTED IN THE UNITED STATES OF AMERICA

CONTRIBUTORS

Paul Schuh Baur, Jr., Ph.D.
Research Associate Professor
Division of Cell Biology
University of Texas Medical Branch

Chief, Division of Cell Biology
Shriners-Burns Institute

Galveston, Texas

Willard A. Burns, M.D.
Chief, Laboratory Service
Veterans Administration Medical Center
Lyons, New Jersey

Professor of Pathology
Department of Pathology
College of Medicine and Dentistry of
New Jersey-Rutgers Medical School
Piscataway, New Jersey

Richard H. Conklin, M.D., Ph.D.
Assistant Professor
Department of Pathology and Laboratory Medicine
University of Texas Medical School
Houston, Texas

Bruce Mackay, M.D., Ph.D.
Associate Pathologist
Associate Professor of Pathology
Department of Pathology
The University of Texas M.D. Anderson Hospital
and Tumor Institute
Houston, Texas

Nelson G. Ordóñez, M.D.
Assistant Pathologist
Assistant Professor of Pathology
Department of Pathology
The University of Texas M.D. Anderson Hospital
and Tumor Institute
Houston, Texas

Barbara M. Osborne, M.D.
Assistant Pathologist
Assistant Professor of Pathology
Department of Pathology
The University of Texas M.D. Anderson Hospital
and Tumor Institute
Houston, Texas

Sydney S. Schochet, Jr., M.D.
Professor of Pathology
Department of Pathology (Neuropathology)
West Virginia University College of Medicine
Morgantown, West Virginia

CONTENTS

INTRODUCTION

The transmission electron microscope (TEM) has become accepted in surgical pathology. Most large institutions have at least one electron microscope (EM) that is used for diagnostic problem-solving, complementary to routine light microscopy. Many pathologists in smaller hospitals and private pathology laboratories who do not possess their own ultrastructural facility periodically send specimens to an electron microscopist for interpretation. Since the cost of setting up an ultrastructural laboratory can be considerably less than that of many of the analytical instruments in a clinical laboratory, it is reasonable to anticipate that the number of EMs in pathology will continue to increase, and some manufacturers, cognizant of this situation, are attempting to make available an instrument that will satisfy the requirements of diagnostic pathology without the costly refinements and embellishments of a high-resolution research microscope.

Many older pathologists have not received much exposure to electron microscopy or cell fine structure during their medical school years, and have had to acquire a working knowledge through postgraduate courses to appreciate the range of applications of diagnostic electron microscopy, interpret reports, and comprehend publications that include ultrastructural observations. The opportunity for residents in training to obtain instruction in practical electron miscroscopy is still relatively limited. Even some of the larger medical centers devote much of their ultrastructural laboratory time and effort to research rather than to clinical application. The need for publications on diagnostic electron microscopy is gradually being met by authors and publishers, and multiple volume treatises are becoming available for detailed instruction and reference.

The object of this book is to provide a starting point for the pathologist or resident who wishes to obtain an introduction to the field. It makes no attempt to cover every situation where electron microscopy may contribute information of diagnostic significance, but focuses instead on the areas of medicine where ultrastructural studies are well established and their contribution is widely recognized. The opening chapter provides basic information about the electron microscope, particularly the TEM, since scanning and analytical electron microscopes are still mainly research instruments. Considerable emphasis is placed on effective specimen procurement and preparation, because the results of an ultrastructural diagnostic study are directly proportional to the quality of material with which the electron miscroscopist has to work. In electron microscopy, a picture is often worth if not the proverbial thousand words, at least a paragraph of text, and while selection in a book of this size is necessarily somewhat arbitrary, the publishers have allowed a generous number of figures. No commercial instrument or product is specifically endorsed by the illustrations or text.

If this book stimulates an interest in applying transmission electron microscopy in diagnostic medicine, or in acquiring more information on any aspect of its increasingly broad range of applications, it will have succeeded in its purpose. I sincerely thank the authors of the various chapters for their contributions. Some of the information included in the chapter on tumors was obtained in the course of my work with the Cancer Clinical Research Center at the M.D. Anderson Hospital, supported by NCI Grant 05831. I also wish to thank my senior technician, Ms. Laura Keyes and Ms. Elizabeth Stueck of Appleton-Century-Crofts, for their assistance.

1

THE ELECTRON MICROSCOPE LABORATORY

PAUL S. BAUR

The electron microscope (EM) was first developed by Ernst Ruska and his associates in Germany between 1931 and 1933. The first commercially available EM was offered by Siemens (Germany) in 1939. Since that time, thousands of instruments have been placed into service throughout the world, and there have been vast improvements in the design and performance of the commercial instruments (Fig. 1-1). The introduction of the EM opened up vast new frontiers in morphology, and in a few decades the EM has revolutionized much of our understanding of the structure of cells and tissues. It has been used exhaustively for 30 years in cellular and molecular biology, but its clinical application has proceeded more slowly. Even today, the use of ultrastructural studies in the diagnostic evaluation of human biopsy material is principally in the fields of nephrology and oncology.

The EM allows the pathologist to scrutinize specimens more closely than is possible by conventional light microscopy and to compare the observations of diseased tissues with similar studies of normal tissues and cells. Subtle nuances of the diseased state, not visible in the light microscope (LM), can be detected, and more precise diagnoses become possible, leading in turn to greater precision in therapy.

Ultimately, the EM offers its user a closer look at his samples. It amplifies measurably the meaning of the microscopist's terms, "contrast," "resolution," and "magnification." The EM can give the microscopist more than 1000 times improvement in resolution and magnification over the LM, and the instrument's capabilities are best appreciated when compared to those of the LM (Table 1-1). However, despite the impressive advantages of the EM over the LM, the EM has its limitations. For example, despite its sophistication, the EM can produce only black-and-white images of inanimate material. Color images are not even theoretically possible.

The transmission electron microscope (TEM) is analogous to the light microscope in nomenclature, operation, and design. The TEM passes or transmits (hence "transmission") a flux or stream of electrons instead of a beam of light to generate the final image. The electrons are preferable to light in that their mass—infinitely larger than a photon of light, which lacks mass—imparts to each electron a characteristic wavelength ideally suited for high-resolution microscopy. The wavelength of the electrons, after some mathematical adjustments, can readily be calculated using the formula $\lambda = 12.3/\sqrt{V}$ (de Broglie's relationship), where λ is the electrons' wavelength in Angstroms, 12.3 is Planck's constant, and V is the voltage used to accelerate the electrons within the microscope. If all the electrons in the beam have the same

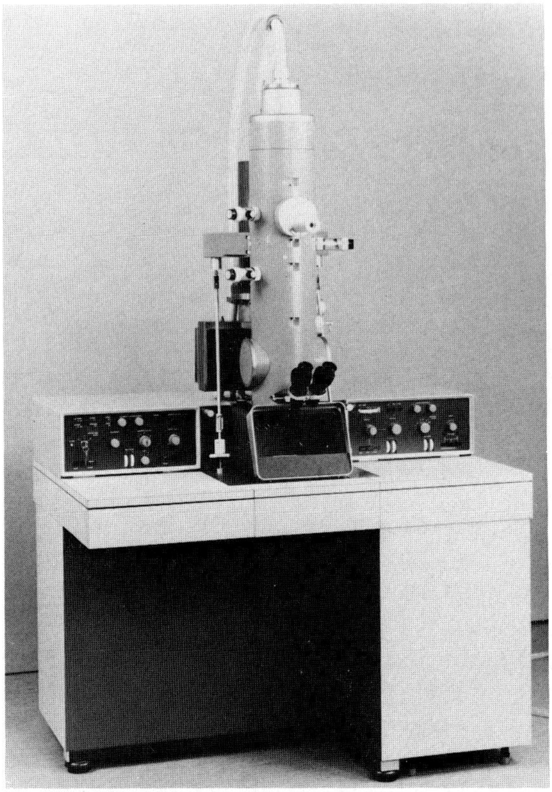

FIG. 1-1. A modern transmission electron microscope. (*Courtesy Carl Zeiss, Inc.*)

mass and speed (±1%), they also will have an identical wavelength. Thus, the electron beam is, by definition, monochromatic and is unable to produce a color image.

Derivation of the electrons' wavelength is of little value unless one recognizes the role that wavelength plays in determining the theoretical limit of resolution (LR) of the instrument. Generally speaking, LR = 0.61λ/NA. In the case of the electron microscope, the numerical aperture (NA) is very small and can be approximated by the value 0.015. Thus, the theoretical resolution of a 100 kV electron microscope would be determined as follows:

$$\lambda = \frac{12.3}{\sqrt{100{,}000}} = \frac{12.3}{316} = 0.04\text{ A}$$

$$\text{LR} = \frac{0.61 \times 0.04\text{ A}}{0.015} = \frac{0.024\text{ A}}{0.015} = 1.63\text{ A}$$

A light microscope's LR is about 1700 A for violet light (λ = 4000 A) and 2500 A for white light. The NA of a light microscope is approximately 1.4. With those two formulae well in hand, the basic mathematical aspects of electron microscopy have been completed.

TABLE 1-1
Comparison of microscope capabilities

	Light Microscope	*Electron Microscope*
Resolution (greatest)	0.2 μm	1.4 A
Magnification (greatest)	x1250	x1.5 x 10^6
Size of red blood cell at greatest magnification	8.75 cm	10.5 m
Smallest resolvable cellular structure	Mitochondria (very little detail)	Portions of membranes; single DNA strands; molecules
Contrast of specimens	The natural contrast can be observed using bright field, phase, interference contrast microscopy. Mordant stains increase versatility.	None can be seen naturally. EM requires the use of heavy metal stains.
Section thickness	7 μm	700 A
Sections/cell (RBC)	1	100
Capacity to view viable tissue	Yes	No
Embedding medium	Paraffin-fresh Frozen	Usually epoxy Resin

One does not need to know the electronics, physics, theory, design, etc., of the EM to become a competent operator. However, a general understanding of the basic theory of the EM enhances appreciation of the microscope's capabilities.

An electron microscope is commonly comprised of a column, console, and, in some instances, remote power supplies. The uppermost section of the EM column is known as the gun chamber (Fig. 1-2), for this is where the electron gun is located. Here, a constant stream of electrons is produced by heating a small filament (commonly a V-shaped tungsten wire) nearly to its melting point. The filament is encased within a cup-like chamber called the cathode cap. The cap has a small (1 mm) hole in its tip and is positioned several millimeters from an anode, which has a hole through its axial center. The cathode cap is negatively charged with respect to the filament, thus forcing the electrons through the aperture. This serves to focus the electrons into a beam. The anode is highly charged (positive with respect to the cathode) and attracts the electrons, most of which pass through its center and on down the column.

The difference in electric potential (volts) between the anode and the cathode cap is called the accelerating voltage and is measured in kilovolts (kV). Most microscopes offer a range of kVs (40, 60, 80, 100) and thus allow the operator to quickly choose the accelerating voltage best suited to the sample. The greater the accelerating voltage the higher the electron's speed, the shorter its wavelength, and the more collimated (coherent) the beam.

After the beam of electrons has been emitted from the gun chamber, it passes through and is deflected by a series of electromagnetic fields, each produced by a

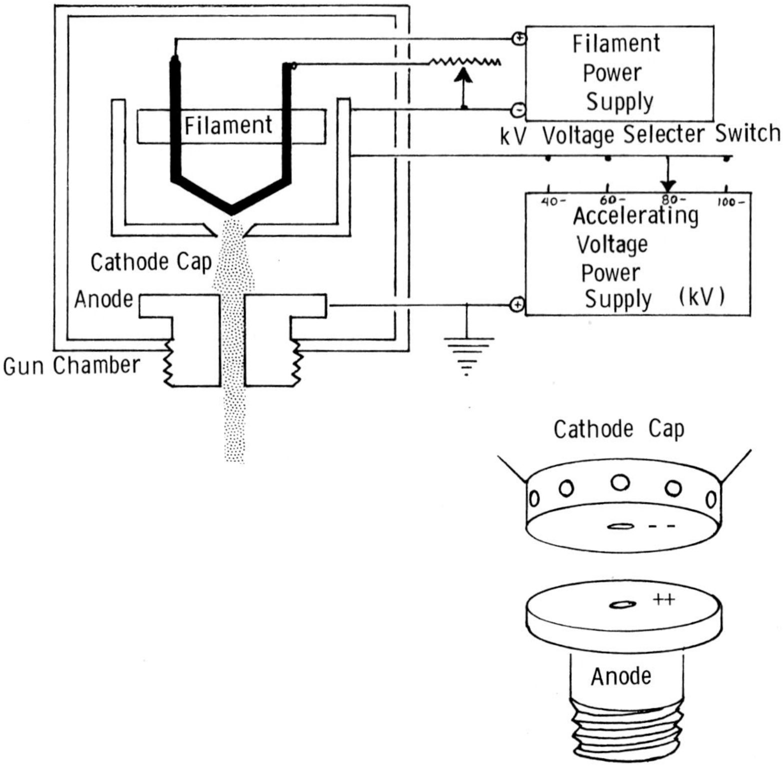

FIG. 1-2. The basic circuit of the EM gun.

magnetic lens. The electrons are affected by each of these fields (lenses) in a fashion analogous to the refraction of a beam of light through a glass or quartz lens. However, in an electron microscope all lenses are positive and thus cause the electrons to converge toward a single point (Fig. 1-3). The first two lenses, which are occasionally combined into a single unit, are the condenser lenses 1 and 2. The first condenser lens, with its aperture, serves to purify the beam of electrons by removing the slow, fast, erratic, and/or nonaligned electrons, and converges (demagnifies) the beam to a small point. The second condenser lens focuses this small point on the specimen below (Fig. 1-4), while the second condenser lens aperture restricts the diameter of the beam.

Most beam electrons passing through the specimen, which is usually very thin, remain unaffected. However, some electrons may react with nuclei or orbital electrons of atoms in, on, or combined with the specimen. If an electron interacts or collides with an orbital electron, an inelastic scattering occurs, in which the beam electrons lose energy. Elastic scattering (no energy loss) results when an electron passes close to, but does not collide or interact with, the positively charged nucleus (Fig. 1-5).

When the beam of electrons passes through a thin (700 A) plastic section, the image within the beam is formed mainly by the scattering of electrons interacting with the atoms of the heavy metals used to "stain" the specimen (Fig. 1-5). The scat-

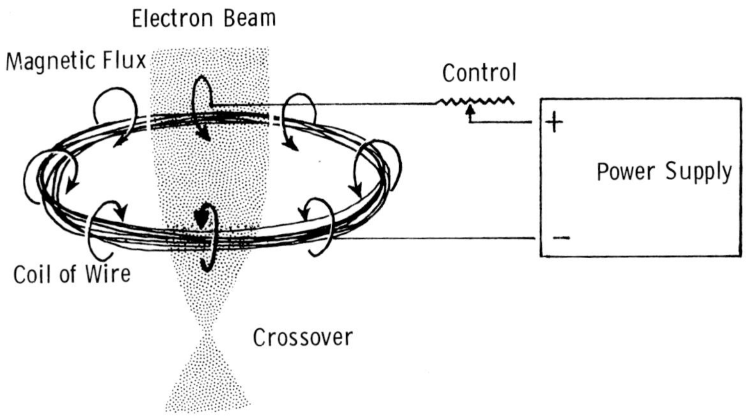

FIG. 1-3. The magnetic field produced by current through a lens coil. The field produces a positive lens effect on the stream of electrons that passes through its center.

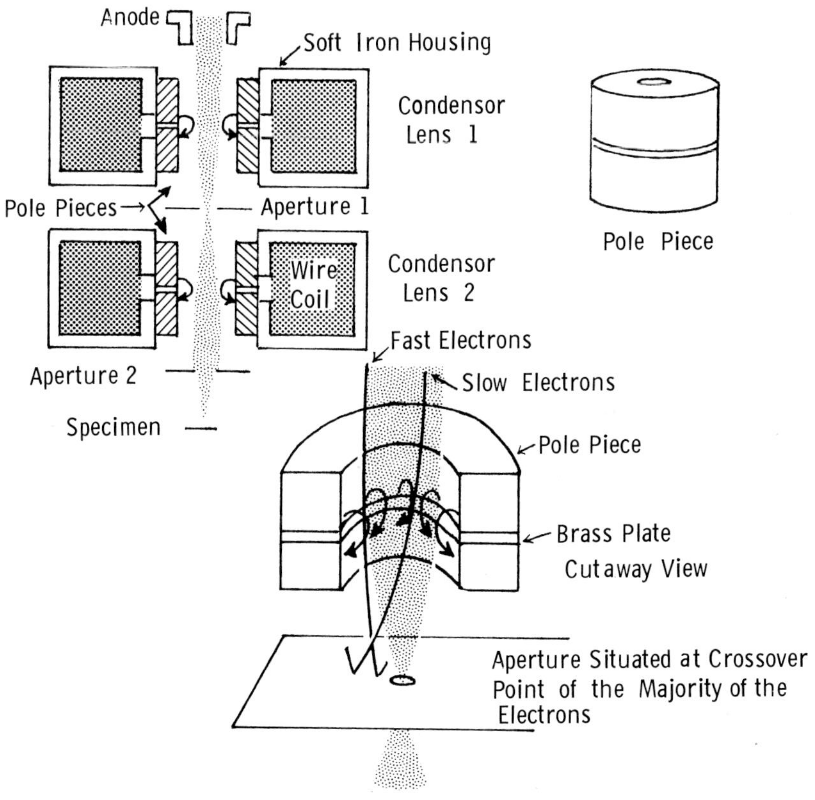

FIG. 1-4. This view shows the physical arrangement of the two condenser lenses and their effect on the electron beam. Slow and fast electrons are removed by the aperture plate, which is situated at the point of beam crossover.

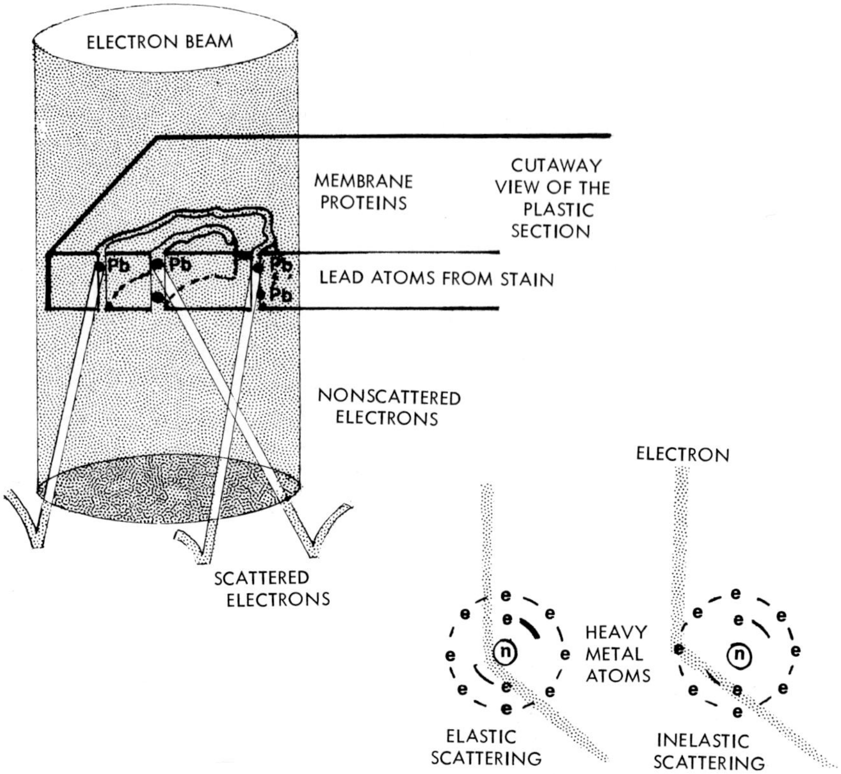

FIG. 1-5. The electron beam interacts with heavy metal atoms that are attached to the embedded tissue proteins during the fixation (Os) or staining (U or Pb) procedure. The electron's direction is altered by the elastic or inelastic scattering.

tered electrons are removed from the beam by the objective aperture. This separation of informational and scattered electrons is one of the most important aspects of the TEM. X-rays and low-energy electrons (secondary electrons) are also emitted as a result of an inelastic collision.

As one might expect, the thicker the specimen, the greater the probability of electron interactions as the beam traverses the specimen. The converse is also true, with the thinnest sections of the specimen having little effect on most of the electrons. While there are other interactions between primary beam electrons and the specimen's atomic structure that contribute to image formation (i.e., absorption, interference, and diffraction), they are of less importance than electron scattering and will not be discussed here. It is important to note that the larger the atom, the greater the probability of interaction. The electron beam cannot "see" or interact with most natural biologic atoms, for they are of the same low atomic weight elements that comprise the embedding plastic. Most biologic thin sections are therefore stained with a heavy metal solution to enhance interaction, and beam electrons are then deflected by the metal atoms (lead, osmium, uranium) that have become attached to molecules in the specimen. The indirect nature of the image formation lends itself to significant loss in resolution.

The specimen and the objective aperture are both located within the center of the objective lens. The beam of electrons is focused by the objective lens and then passes to the intermediate lens, where it is magnified. The final lens, the projector lens, projects the image onto a fluorescent screen at the base of the column (Fig. 1-6). In those areas of the beam where scattering did not occur in the specimen, the electrons strike the screen. The electron-screen interaction emits visible light, allowing visualization of the bright areas. Regions of the sample that block, scatter, or diffract electrons are represented by the dark areas on the screen.

The fluorescent screen can be removed from the path of the beam and the image recorded on film situated beneath the screen. The electrons react with the photographic emulsion in a somewhat similar manner to light. Silver grains are ultimately produced after development and fixing by an electron-silver halide interaction. As the TEM has a very long depth of focus (measured in meters), the location of the screen and film is not critical. However, due to the nature of magnetic lenses, the depth of field of the objective lens is very small (measured in microns), necessitating precise placement of the focal plane (electronic focus) on or slightly below the specimen plane.

The electron microscope's console consists of a series of highly regulated power supplies and controls that are needed for lens operation and/or beam production (high voltage or accelerating voltage). To maintain stability in these power supplies, the microscope requires a dependable source of electricity that is not subject to drastic alterations in line voltage, amperage, and frequency.

The pumps, pipes, hoses, and valves that compose the vacuum system are also situated within the console of the TEM. The interior of the column is evacuated to a high vacuum. (Vacuum is inverse to pressure, so the term "high" refers to a very low

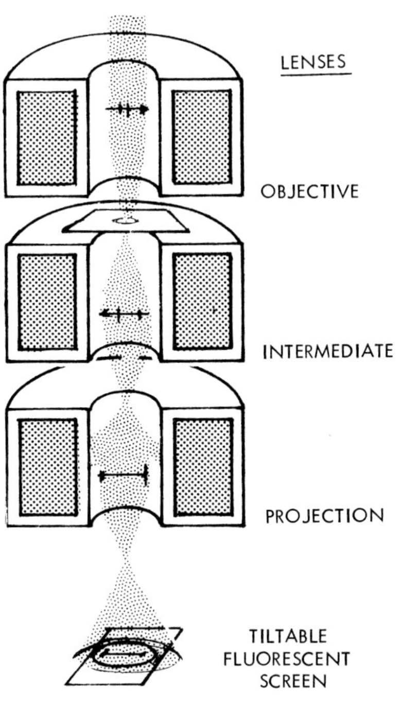

FIG. 1-6. This view shows the physical arrangement of the lower three lenses in the TEM.

pressure.) The initial vacuum within the system is reached using a mechanical roughing pump (also called a rotary pump or backing pump), which produces a vacuum of the order of 10^{-2} torr (1 torr = 1 mm mercury). When some microscopes are switched on, this pump is expected to run continuously. The roughing pump is first used to evacuate the column, and later to exhaust another type of pump known as a diffusion pump. This ingenious pump utilizes a stream of hot oil or mercury to create the high vacuum that removes most of the remaining gasses from the column (Fig. 1-7). The diffusion pump can produce a vacuum of 10^{-6} or 10^{-7} torr, depending on the fluid used, the pump's design, and the manner in which it is connected to the microscope and roughing pump. The better the vacuum within the microscope, the less the chance that the beam electrons will collide with a gas molecule. Such collisions scatter the beam and adversely affect the quality of the image. The life of the filament is also related to the vacuum level within the microscope. A sudden inrush of air into the column will cause oxidation of the hot tungsten filament, with subsequent loss of the metal's integrity.

Because of the minuscule diameter of the electron beam (2 μm or less prior to entering the specimen and as little as 20 μm after passing through the objective aperture), the very small size of the specimen (grid 3.05 mm, thickness 700 A, field of view 4.0 μm at 250,000 magnification), and the precision required for specimen

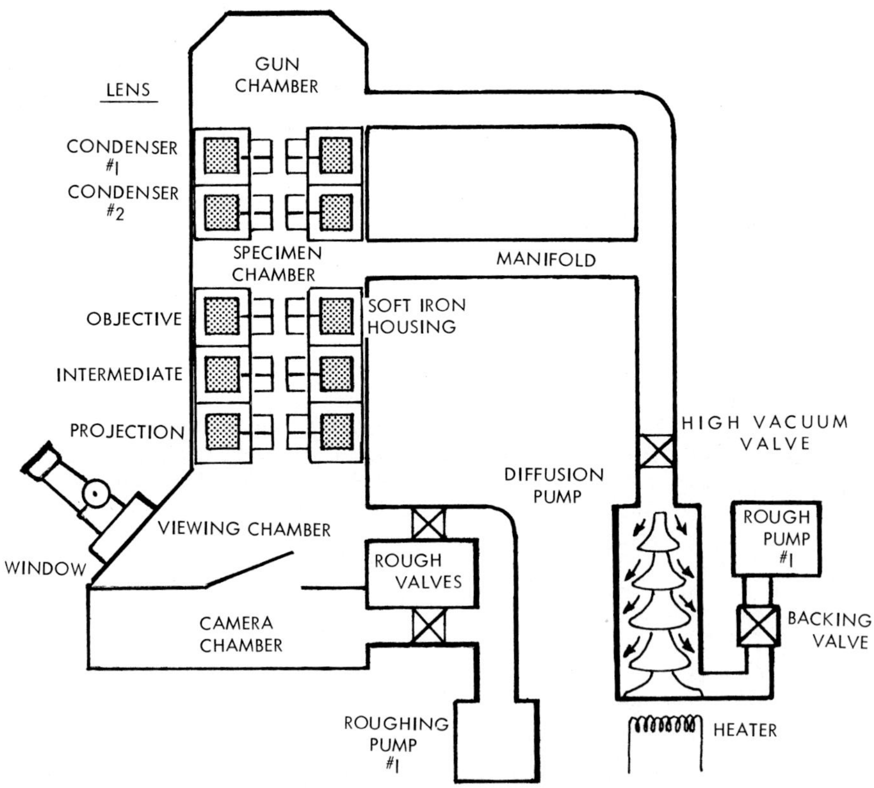

FIG. 1-7. This figure illustrates the vacuum system in a TEM.

movement, there is an absolute requirement for precise and finished machining in the manufacture of the microscope. Imperfections and irregularities in the mechanical aspects of the microscope are often more bothersome than the vacuum and/or electronic system eccentricities.

CHOOSING THE RIGHT MICROSCOPE

Before an instrument is selected, the purchaser and the operator (not always the same person) must first decide on the microscope's ultimate mission. If it is to be used solely for diagnostic studies, the instrument need not be a highly flexible, high-resolution microscope. However, for some research purposes, a more refined, high-resolution microscope will be required. As one might expect, there is a direct relationship between price and the complexity of the instrument. Also, the more sophisticated the instrument, the more that can and will go wrong with it. When instrument growth is planned (a microscope to which additional components will later be added), it is often the case that the planned additions will become obsolete long before their acquisition. Similarly, there is no assurance that components planned for later addition to the instrument will remain available.

Buying an electron microscope is similar to buying a car. The basic EM consists of a gun, column, viewing chamber, and camera system. Embellishments, accessories, and options, which always cost more but provide flexibility, are usually offered by the manufacturer. A brief discussion of some of these may prove useful.

Accelerating Voltage. Some instruments offer ranges from 40 to 200 kV in 20- or 40-kV increments. Routine transmission electron microscopy (TEM) uses 60 or 80 kV most of the time. Other voltages are only attractive when thicker (100 to 200 kV) or thinner (40 kV) specimens are to be examined or when the instrument is to be operated in either the scanning transmission (STEM) or scanning (SEM) modes of operation.

Resolving Power. Biologic samples, because of the nature of their preparation (indirect staining, embedded in plastic, chemically fixed), seldom, if ever, permit a resolution greater than 15 A. An instrument guaranteeing 1 A resolution is, therefore, contraindicated, unless the higher resolving power is required for an optional accessory. Usually a 3-to 4-A resolution is more than adequate for all intended uses.

High Contrast Option. The contrast of the specimen can be changed by positioning it at different levels (heights) within the objective lens. Some instruments offer dual placement of the specimen as a standard feature. A similar effect can also be achieved by reducing the accelerating voltage and thus increasing the incidence of electron scattering within the plane of the specimen, and to some degree by reducing the size of the objective aperture.

High Vacuum. The higher the vacuum, the better the resolution and the lower the rate of contamination (removal of vapors and gaseous spoils from the column). Contaminants progressively coat the specimen surface, apertures, lens components, and viewing screen, limiting resolution and contributing to the development of astigmatism. A truism thus exists: The cleaner the column, the better the operation and the results. Ion-getter pumps, cryopumping devices, series-coupled diffusion pumps, and cold traps are all efficient devices used to achieve or support high vacuum in the EM.

Indicator Lights. Many newer instruments provide panel lights to indicate "go" or "no go" situations. Though operation is simplified, diagnosis of instrument trouble is impaired. For instance, if a diffusion pump burns out a heater filament, the lack of a "go" light only indicates that there is a problem somewhere in the microscope. A meter reading or digital display of the vacuum level showing a column and roughing pump vacuum level of 10^{-2} suggests that the diffusion pump is not functioning. Similarly, if one can accurately monitor the lens currents, filament current, etc., then aberrant performance in those circuits will immediately be obvious. The faster trouble is spotted, the more quickly the instrument can be repaired, and money is often saved by catching problems early.

Digital versus Meter Dials. The trend in electron microscopes is toward the digital presentation of data (magnification, vacuum level, negative number, etc.) with alpha numerics (letter/numbers) printed out on cathode ray tubes, nixie tubes, or light-emitting diode displays. Meters may remain dependable, but they are often difficult to read because of parallax or inadequate illumination. Digital displays prevent such errors, and in the next decade they will undoubtedly replace meters on all microscopes.

Panel Controls versus Computer Assist. Until recently, the illumination, beam current, focus, and magnification of microscopes have primarily been controlled by the operator's altering the position of variable resistors that are coupled to panel knobs. A new technology involving computer-based instrument control is now being marketed. The operator simply types on a keyboard the electron microscope set-up (magnification, condenser setting, etc.) that he wishes, and the computer electronically sets up and regulates the instrument accordingly. The computer is even able to memorize the settings and retrieve them on command. Thus, routine or special microscope set-ups can be recalled instantly, with the confidence that a microscope can be operated on any given day exactly as it was on a previous day. Changes in focus, magnification, and illumination are achieved in a stepwise fashion by push-button switches. Thus, few, if any, "knobs" are to be found on the instrument panel. The question of flexibility and ease of operation can be argued, but most future electron microscopes will be computer controlled and/or assisted. This does not mean, however, that the current knob/potentiometer technology, which at the present time is readily available and a great deal less expensive than the computer-based units, is significantly less accurate or reliable than these units.

Film Options. Most microscopes collect their data on single sheets of cut film or glass plates. A cassette holding as many as 60 sheets or plates, depending on the model, can be loaded into the microscope. A 70-mm camera is often an available option. Thirty-five–millimeter cameras are less suitable for detailed EM work. As the gelatin-based film is placed inside the EM, there is always an effect on the vacuum system. The gelatin molecule is hydrophilic, and therefore each piece of film accumulates a great number of water molecules that ultimately escape as the microscope attains high vacuum. Thus, the length of time required for the removal of water vapor from the camera chamber becomes a factor when changing film. Usually, a long delay is avoided by prepumping the newly loaded film cassettes in a holding vacuum chamber to remove the water molecules.

New technology on at least one commercial model offers the transmission of a light-based image from within the microscope to film located outside the vacuum system. The transferral of imagery is achieved by means of fiber optics. The advantage in this arrangement is that the vacuum is never affected, and loading and retrieval of the film is greatly simplified. In this instrument 70-mm roll film is used in lieu of sheets or plates.

Future microscopes may incorporate the electronic recording of data whereby the image can be displayed on a high-resolution videocom tube. The data could then be stored on video tape, and observation of the images could be accomplished by means of television monitors. Routine image enhancement and/or analyses by computer-based technology would then be within the grasp of the microscopist. Additionally, the electronically recorded image could be manipulated (rotated, reversed, inverted, magnified, or contrast-enhanced) to the benefit of the operator.

Simple versus Complex. If the instrument is to be used only by a small number of skilled individuals, a unit having considerable complexity and flexibility is feasible. However, if many users are anticipated, and particularly if some are inexperienced (as is the case in the training of pathology residents), a microscope that is simpler and less vulnerable to operator-caused damage should be selected. In general, the smaller and less expensive the instrument, the fewer things there are to go wrong, and a neophyte will be less afraid of an array of 12 knobs and switches than of a series of panels with 40 or more controls. It should be borne in mind that the training and supervision of an operator takes a considerable amount of time, and a damaged instrument resulting from an operator's inadequate training can cost a department or an institution time and money.

Instrument Size. It is not possible to install an electron microscope in a room the size of a janitor's closet. A certain amount of space around the instrument is necessary for maintenance and servicing. Manufacturers will often cite the minimum room dimensions required for their microscopes, although salesmen, in their zeal to sell an instrument, may suggest that a smaller room would be adequate. Remote components of the microscope, which often take up much space, may not be shown in the brochure views. They include roughing pumps, power supplies, water chillers, air compressors, and transformers, and are often attached to the instrument by tubing or cable assemblies of specific lengths with molded connectors at each end preventing their modification or extension. Thus, an estimate of 4 to 6 square meters of floor area for the main console may have to be augmented by additional space for the components.

Maintenance Contracts. Electron microscopes break down occasionally. Repairs are expensive, and, after the first year, which is usually covered by warranty, a hedge against unforeseen repair costs is the service contract. In this agreement, the company (usually the seller) provides all repair service, replacement parts, and scheduled maintenance to the microscope owner in return for an annual fee.

A commonly asked question concerns the worth of service or maintenance contracts. At the end of a year that has not seen a single breakdown, it is not easy to say that the contract was worthwhile. However, without a service contract, the field engineer on a service call may charge the microscopist for air fare, car rental, hotel accommodations, and portal to portal (from the time he leaves his front door until the time he returns) labor charges at $40 to $60 per hour. Additionally, the owner cannot hurry the engineer; agrees to pay for the service call when he requests it; may not be able to foresee if the defect will take one hour or one week to find, much less to repair; has no control over the retail price of any replacement parts, which are usually very expensive; and does not have priority with respect to when the call for service will draw a response. Thus, a small transformer failure may cost as much as a yearly service contract, which, of course, would make the annual charge a good bargain. Service contracts are of little value if there is not a service center or field engineer located nearby (within 200 to 300 miles), if the company has a bad reputation with respect to response time and/or quality of repair, if adequate transportation facilities are not available, or if the company lacks a complete inventory of parts and supplies. The best way to avoid these service contract problems is to investigate the company's reputation. If a particular company's reputation is poor, it is best not to purchase the electron microscope from that company.

OPERATION OF THE UNIT

The electron microscope is usually simple to operate in the regular transmission mode. The procedure follows the same pattern on all the different models and includes the following steps:

1. The main power to the microscope is switched on, and a button is depressed to activate the sequence of pumping that will evacuate the column and backing system. Generally, the roughing pump is initially connected to the column and gun chamber for approximately 10 minutes while the diffusion pump warms up. After the diffusion pump has reached its operational temperature, the pump is valved in line with the column and gun chamber, and evacuates these areas to a high vacuum. This entire sequence usually takes 20 to 30 min.
2. A "high voltage" switch is depressed, together with one of several high voltage selector switches. This turns on the high voltage transformer and places it in a configuration to produce the chosen accelerating voltage (i.e., 40, 60, 80, or 100 kV). The microscope will immediately respond to these settings and produce a precisely regulated potential difference across the anode and cathode.
3. A "filament-current" knob is rotated while observing either a meter reading or the filament image on the viewing screen. The latter requires correct magnification and condenser lens settings as well as correct aperture configurations. The knob control is connected to a potentiometer that permits a

change in resistance. A change in current production is regulated by a "gun-bias" circuit, which alters the potential difference between the filament and the cathode cap.

4. A few alignment steps are required. The various parts of the column (electron gun, pole pieces, and apertures) are adjusted relative to one another, generally by centering them with reference to the objective lens. When this has been completed, any astigmatism, which usually arises from contamination of the objective aperture, is corrected. Most microscopes have as many as three different alignment procedures differing in complexity and precision. An in-depth alignment is necessary after the column has been dismantled; a moderately detailed alignment is usually necessary after filament replacement or after changing from one mode to another (TEM to STEM to SEM); and simple adjustments are needed on an occasional basis during the day.
5. The filament is turned off momentarily, and the tissue sample is inserted into the column. The plastic sections (approximately 700 A thick) of the tissue are usually mounted on thin copper grids (3.05 mm in diameter), which are placed in the specimen holder using a jeweler's forceps. A tiny clamping device holds the grid securely in place. The specimen holder is situated within the geometric center of the objective lens (Fig. 1-8). The holder may penetrate the side of the lens ("side entry") or be inserted from above ("top-loading"). Both configurations have inherent advantages. The specimen holder is usually prepumped in a small chamber, which allows its insertion into the column without high vacuum loss.
6. The filament current is reapplied, and the specimen's detail becomes apparent. Changes in magnification are effected instantly using a single selector switch. Focus corrections are easily made using the "coarse," "medium," and "fine" focus controls. The specimen is moved about within the column

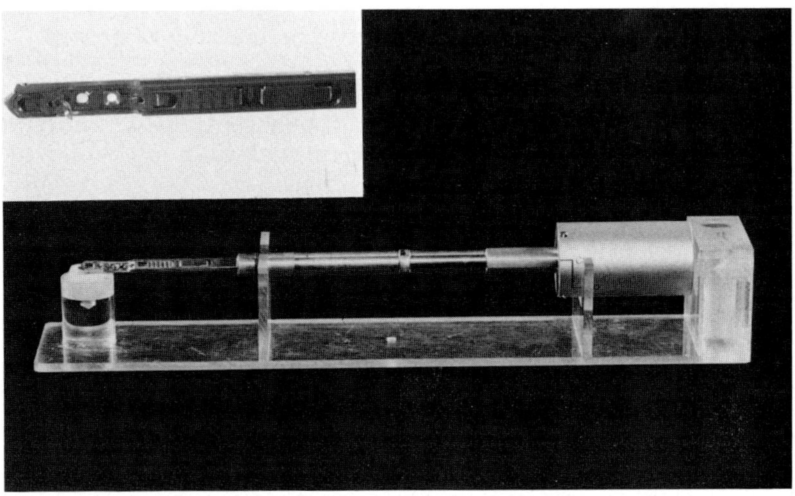

FIG. 1-8. A specimen holder of a modern TEM. The inset shows a close-up of the grid holders (2) located at the end of the holder. (*Courtesy JEOL-USA.*)

by means of specimen control knobs mounted on or near the panel face. These controls are mechanically linked to the specimen holder and thus allow the operator to traverse the grid and observe the sections.

7. *Visualization.* The images are viewed on the surface of the fluorescent screen, which in some models can be tilted at approximately 45 degrees to facilitate observation. A binocular viewing microscope (x10 to x20) assists the operator in focusing the image, and an accessory screen for focusing may be present or available. The screens emit green light, which is the color most sensitive to the dark-adapted eye.
8. *Photography.* When the fluorescent screen is lifted out of the way, the electrons can then strike the underlying photographic emulsion. Usually the exposure time and film transport are automatic features, with mechanical overrides possible to allow flexibility. Nearly all new microscopes now print data such as date, I. D. numbers, and the magnification on the negative border as the exposure takes place. Thus, negatives do not have to be labeled prior to loading or after unloading. This particular feature is timesaving, ensures accurate data retrieval, and, if offered as an option, should be given high priority.

FACTORS THAT AFFECT THE PERFORMANCE OF THE ELECTRON MICROSCOPE

Environmental aspects of the area in which the microscope will be placed must be considered. These factors can and will affect the performance of the instrument and could negate a sizable investment in time and money. A high-resolution instrument (2 to 3 A) should not be placed permanently in a room that robs it of 20 A of resolution. Conversely, there is little need to spend large sums of money in building costs to provide a perfect environment for an electron microscope having nominal performance specifications.

The main environmental factors to consider when selecting a site are:

1. *Vibrations.* Excessive floor vibrations caused by machinery, elevators, trains, construction, air conditioners, or venting fans can seriously impair the quality of performance of the instrument, and will result in blurred photographic images.
2. *Temperature.* Electron microscopes generate a considerable amount of heat, so the temperature of the room must be regulated to provide adequate heat removal by convection from the nonwater-cooled electronic circuits and to provide comfort to the operator. It is distracting for an operator to sit in front of a microscope with an occasional drip of perspiration falling off the tip of his nose. Most of the heat from the power supplies, lenses, and diffusion pumps is removed by an independent chilled-water circulator, which should be placed in a remote area that allows adequate heat transfer. Preferably, the chiller should not be in the same room as the EM.
3. *Humidity.* Humidity affects modern electronic circuits and can eventually cause problems. If the humidity is high, some parts of the column may rust. Although microscopes are less susceptible to moisture today than they were 10 years ago, they should nevertheless be protected from excessive humidity changes. An air-conditioned environment is usually adequate. If the air

conditioner fails and the water chiller continues to cool the column (lenses) and power supplies, condensation will form on the various parts and run into cracks, crevices, and seams.

4. *Magnetic Fields.* Excessive magnetic fields can deflect the beam so much that alignment of the EM becomes impossible. Similarly, the fields can produce an astigmatism that defies correction. Transient magnetic fields are difficult to detect and measure and are thus difficult to cope with, but modern microscopes have enough magnetic shielding so that magnetic fields of small to moderate intensity have little or no effect on operation or performance. Transformers, electric motors, compressors, and fluorescent lights are the most common sources of magnetic disturbances.
5. *Dust, Dirt, and Debris.* A particle of dust 1/5 of a micrometer (0.2 μm) in diameter contaminating the surface of the thin section will appear at x50,000 magnification as a 1-cm black object on the image. Particulate matter can affect o-ring seals on the microscope and cause vacuum leaks. The viewing windows are also sensitive to such contamination. Charged dust particles within the gun chamber will lead to high voltage discharges, rendering the microscope so unstable that it becomes unusable. Discharges also lead to filament failure. Dirt in the column leads to astigmatism and image distortions. The EM suite should, therefore, be kept very clean.
6. *Isolation.* Because EM work requires isolation, concentration, and dark adaptation, the instrument must be situated in an area that affords all three. The room should be of adequate size, light-proof, and free of such distractions as people traffic, noises, and odors. The microscopist should be vigilant with respect to these aspects of the EM suite and take appropriate action when changes in building design threaten to subject him to any such disruptions. Nonmicroscopists seldom appreciate these considerations.
7. *Maintenance.* A single individual should be responsible for all in-house maintenance. This usually includes filament changes, aperture cleaning or replacement, fuse replacement, cleaning of specimen holders, etc. The assigned individual thus becomes knowledgeable about the nuances of the instrument and can hopefully prevent cross-threaded screws, chipped glass windows, insufficiently lubricated moving parts, dirt or dust in the column, defective o-ring seals, inadequate vacuum system performance, and other problems. Similarly, he can keep track of support equipment such as oil levels in roughing pumps, water levels in water chillers, motor lubrication, pulley belt wear, and dust filter and desiccant replacement. The supervisor should also keep an operational log (users, dates, times, number of photos) and ensure that all users operate the microscope correctly. He should interface directly with the field service engineer and establish a good rapport with him and other company personnel. The supervisor should be responsible for notifying the company of a microscope breakdown; should aid the company's service department in diagnosing the problem and in dispatching parts when needed; and in some cases should be able to repair the instrument while being "talked through" the procedure over the telephone. Some aspects of microscope maintenance (filament replacement, aperture cleaning, specimen holder cleaning, etc.) are not covered by the service contract and thus become the assignee's responsibility. He should ensure that service contracts do not lapse, that adequate numbers of spare filaments and apertures are on hand, that lubricants and greases are available, that film stocks are adequate and rotated, and that all gauges, meters, lights, and indicators on the microscope are working. An average of two to three hours per week

will usually suffice for this responsibility. A good EM technician should be able to handle these duties.

SETTING UP THE ELECTRON MICROSCOPE LABORATORY

As has been mentioned, the suite will require a designated EM room of adequate size and isolation. Additionally, areas devoted to tissue preparation (120 to 200 sq ft), ultramicrotomy (50 to 100 sq ft), a darkroom (120 to 200 sq ft), and a storage area suitable for film, chemicals, shipping cartons, supplies, etc. are essential. Usually, existing laboratory floor plans can be altered without great difficulty to provide the required areas (Fig. 1-9).

Utilities

Electricity. A totally devoted 220-V circuit with sufficient amperage (40 amp maximum) is required for the EM. The building supply should be of sufficient capacity to ensure that current and voltage fluctuations due to excessive load by other instruments, equipment, or machinery do not occur. A ground line (from earth ground) should be provided at the tie-in outlet box. All circuits (supply) should be hard wired (conduit). The preparative laboratory, darkroom, and microtomy suite all need a generous supply of 110-V outlets. Duplex outlets at 4- to 6-foot intervals around each room should be sufficient.

Cooling. A cooling device (water recirculator and chiller) is required for the microscope, and it should be located in a separate room. Distance traveled and elevation differences between microscope and chiller can, if excessive, prevent the unit from providing adequate water volume and/or pressure. Sensor switches on the microscope should prevent or stop its operation if water pressure fails or is inadequate. Never consider hooking the EM to an existing cold water supply: It is expensive to use and puts one at the mercy of the supplier for water purity (filters are always necessary) as well as for unannounced fluctuations in flow and pressure.

Air. Some microscopes require an air pressure supply; other models have an air compressor as a component part. The compressors can be located in areas far distant from the unit if hoses and electrical cabling (hook-ups) allow. Sensors on the microscope monitor the pressure and will terminate operation if adequate pressure is lost. An air outlet separate from the microscope's air compressor system will also be useful in the preparative laboratory.

Sewer. Many of the solutions and solvents used in electron microscopy are extremely toxic or corrosive, and require proper disposal after use. A corrosion-resistant (glass) sewer line of adequate size should be provided in the preparative laboratory and in the darkroom. However, some chemicals cannot even be discharged through a corrosion-resistant sewer system and should be discarded using practices suggested or mandated by federal pollution standards.

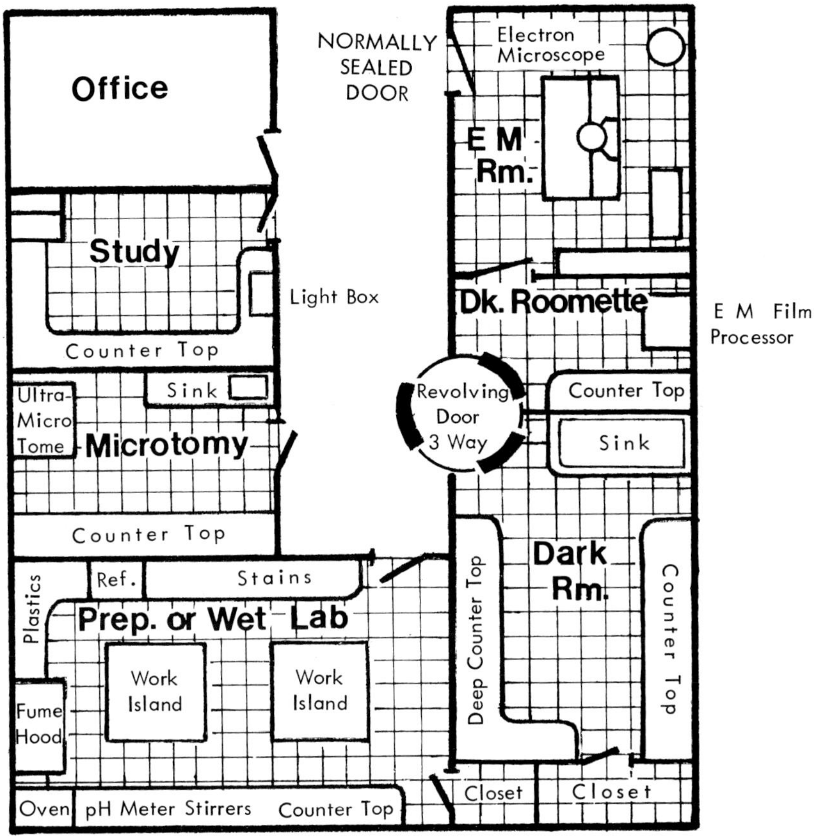

FIG. 1-9. An idealized EM laboratory.

Telephone. It is exasperating to lose one's dark adaptation to answer a telephone call outside the microscope room. However, missing an important call may mean long delays in service, supplies, or acquisition of samples. Thus, a telephone in the EM room is essential. Telephone "talk throughs" and diagnostic evaluations would otherwise be impossible.

ANCILLARY EQUIPMENT FOR THE EM LABORATORY

Ultramicrotomes. Ultramicrotomes work in a similar fashion to that of rotary microtomes in light microscopy. The ultramicrotome, however, cuts much thinner sections (as thin as 500 to 600 A) and holds plastic-embedded specimen blocks of small dimensions. The specimen holder is mounted on a movable mechanical arm that advances by a precise increment, passes the knife edge, and then recycles (Fig. 1-10). The various models differ primarily in how the advance is accomplished—thermally or mechanically. In a thermal advance instrument, the mechanical arm is slowly warmed. The thermal expansion of the metal arm causes the specimen block to ad-

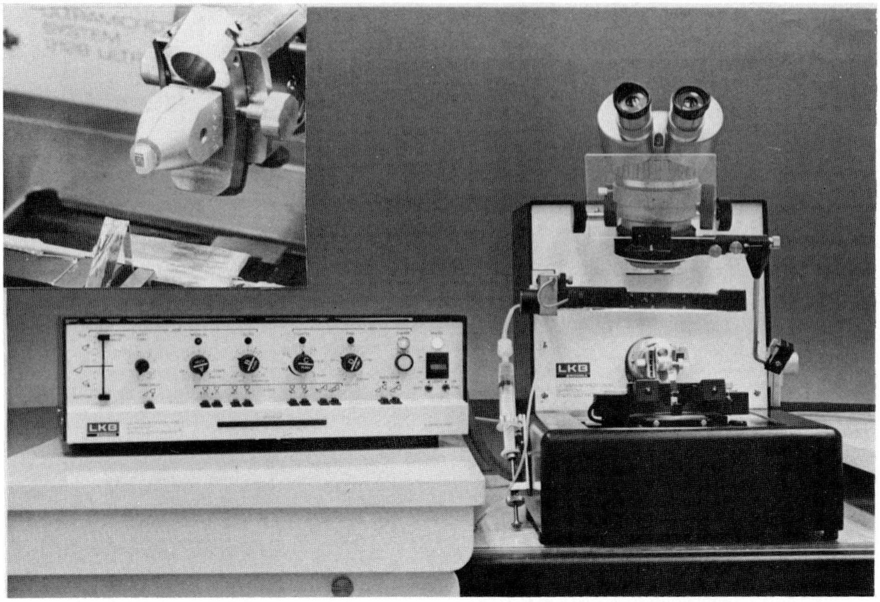

FIG. 1-10. A modern ultramicrotome. Inset: Close-up view of a plastic embedded specimen mounted in the microtome chuck. (*Courtesy LKB, Inc.*)

vance towards the knife edge. The mechanical advance units achieve this same effect by means of mechanical linkages, lever ratios, and lead screw escapement.

Diamond Knives. After a few months of experience with glass knives, most operators turn to a diamond knife to cut thin sections. Although quite expensive (currently approaching $2500 for a knife with a 4-mm edge), a carefully used diamond knife can last for many months and produce sections of high quality that are less likely to be marred by the nicks, scratches, tears, and chatter that are prone to occur with glass knives. The person cutting the sections must be familiar with the idiosyncracies of the particular diamond knife being used, since minor adjustments in the angle of the knife edge relative to the section, and variations in quality from one part of a knife edge to another, can markedly alter the quality of the thin sections. Diamond knives are undoubtedly a worthwhile investment despite their cost, and a manufacturer will usually guarantee at least one, and probably several, resharpenings, at a cost that is at least half the initial cost of the knife. A diamond knife usually has a warranty period of up to 90 days, during which the operator can assess the quality of the knife and if necessary return it for refurbishment or replacement. Prior to 1979, only a minority of American-made knives were worth keeping. There has recently been a dramatic improvement in manufacturing techniques, and the quality of new knives is consistently high as a result. In ordering a diamond knife, the type of tissue to be sectioned should be specified, such as, for example, soft biologic tissues, or calcified material. It is also worth checking on the anticipated delivery time, since in the past this has tended to be lengthy.

For the continued production of high quality sections by a diamond knife, careful maintenance is necessary. Physical damage to the knife edge must be avoided, and it is important to clean the knife carefully after use to prevent the adherence and drying of sections on the cutting edge.

Light Microscope. A binocular light microscope of sufficient quality and versatility to allow low (x30) and high magnification (x500 hi-dry), bright field, and possibly phase contrast operation is essential in the EM lab. Plastic thick sections (1 μm) of each embedded specimen should be routinely made and stained for viewing with the light microscope. In this way it can be confirmed that the material is representative, and with experience the quality of preservation can be determined from these thick sections. If only part of the selected tissue block is representative, this information must be conveyed to the technician who will be carrying out the thin-sectioning, so that he can trim the block without sacrificing the desired area.

Fume Hood. Nearly all biologic tissue that is examined will have been fixed or postfixed in a 1 or 2% solution of osmium tetroxide (OsO_4). Its vapor is extremely toxic and can blind. Therefore, all tissue processing should be carried out in a fume hood or fume sink (a stainless steel sink with a down-draft flow of air). The organic plastics and solvents used in the EM procedures are also harmful and present a definite fire hazard. These vapors can also be removed by the fume hood.

Explosion-Proof Refrigerator. A refrigerator with its lights and switches removed and its thermostat exteriorized becomes an explosion-proof device. Such refrigerators are commercially available. Solvents used in electron microscopy, fixative and postfixative solutions, specimens (either stored or being processed), resin mixtures, and heavy metal stains are a few of the items that will be placed within the cold box. An additional refrigerator (regular type) is also useful in the lab for the storage of film and biochemicals.

Incubators or Ovens. The liquid plastic into which the specimens will be placed is polymerized using heat (48 to 60 C) for various periods of time. As the temperature level is not high, the laboratory can use a modest incubator device instead of a more costly oven. A model having a controllable temperature ranging from 35 to 65 C, with a maximum deviation of $\pm$0.5 C, is sufficient. Interior lights, gas supply, or moisture control are not required, but a removable and washable shelf of aluminum or stainless steel is a handy feature.

Glass Knife-Maker. The embedded biologic specimens are commonly trimmed and thick-sectioned using a glass knife mounted in an ultramicrotome. Although these knives can be made by hand, their production is slow and of inconsistent quality. Commercially available knife-makers produce 45 or 50 degree knife edges on standardized glass with amazing rapidity and quality (Fig. 1-11). The knife-maker quickly pays for itself by saving valuable technical time. The manufacturers also sell expensive precut glass strips for use in their knife-makers.

Analytical Balance. An indispensable instrument in the EM suite is the analytical balance. The formulation of buffer and fixative solutions requires careful weighing

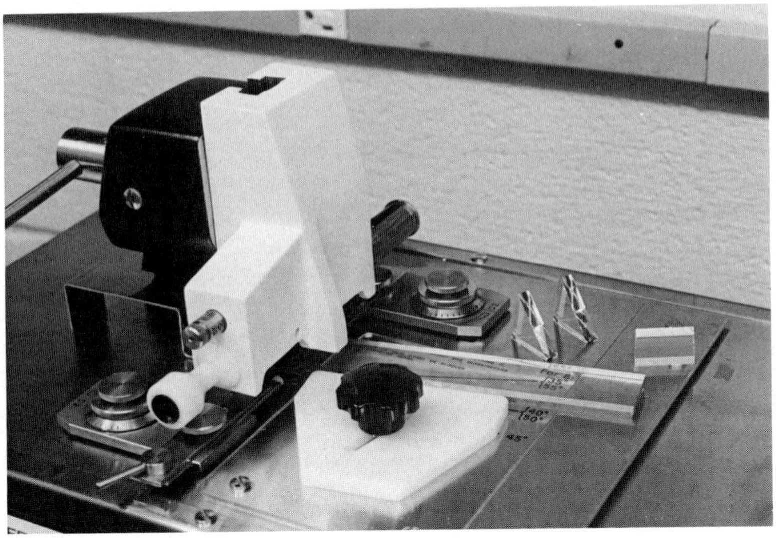

FIG. 1-11. The stage area of a glass knife-maker. (*Courtesy LKB, Inc.*)

of the various components. The balance should be of sufficient range (0.0001 to 200 gm) and accuracy to allow determination to at least tenths of a milligram. A sturdy balance table (granite or cement) located in a vibration-free area of the laboratory facilitates accurate weight determinations. A second balance, capable of weight determinations ranging from 0.1 to 2000 gm, is also useful in the operation of an EM facility.

Dissecting Microscope. It is useful to be able to view samples, tissues, and small microscope parts at low magnifications. A good dissecting microscope having a variable range in magnification (x3 to x70) aids in the microdissection of tissue (prior to fixation), orientation of specimens, and the cleaning of microscope parts. A rack-mounted unit with a separate epi-illumination light source is very handy. The models offering the zoom option are the most popular and provide the greatest flexibility.

Viewbox. A back-lighted viewbox (fluorescent light-milk glass cover) of sufficient size (16″ x 20″) to permit the simultaneous study of 10 to 20 EM negatives is an absolute requirement. Often a negative's quality will not warrant its printing, enlargement, or study, and the quality can best be determined by using a small (x5 to x8) magnifying glass or jeweler's loupe and the viewbox.

Infiltration Device. The penetration of the fixatives, solvents, and plastics into the tissue blocks is facilitated and expedited by agitation. This can be accomplished by placing the processing vials in a rotating infiltration device during the various processing steps. The device consists of a wheel-like arrangement holding the vials and rotating at approximately 10 rpm (Fig. 1-12). The wheel surface is tilted to control the degree of agitation.

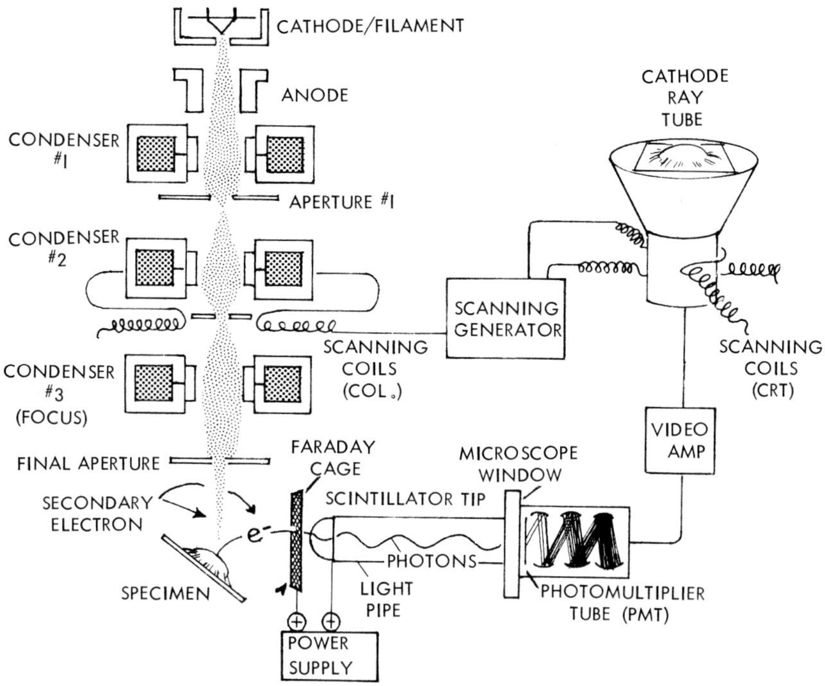

FIG. 1-13. The basic scheme of the scanning electron microscope.

Just before the electron beam passes through the magnetic field of the final lens, it is deflected by a separate set of magnetic fields produced by a pair of scanning coils. These two coils are positioned at 90 degrees from each other. As each set of scanning coils is energized, the beam is deflected according to the strength and direction of the resultant magnetic field. The beam, after deflection, is then forced back through the center of the final lens by its magnetic field. Thus, the scanning coils act in concert with the final lens to cause the beam to rock back and forth, or forward and backward, with a beam fulcrum existing in the electronic center of the final lens (Fig. 1-14). If the coils are selectively energized in a uniform and repetitive fashion, the beam can be synchronously swept both from left to right and from front to back across the specimen surface. The specimen has to be located within 40 mm of and immediately beneath the final aperture, and be situated within the center of the scanning electron beam. The SEM scanning coils are electronically connected to a similar set of scanning coils located on the cathode ray tube (CRT). As the beam of electrons is deflected within the microscope across the specimen, a second beam of electrons is deflected in a coordinated fashion onto the screen of the CRT.

When the primary beam strikes its target, a variety of particulate and nonparticulate emissions are produced (Fig. 1-15). Photons, x-rays, back-scattered electrons, and secondary electrons are generated at the point of beam-specimen interaction. The SEM most commonly uses the secondary electrons to generate its image, but in certain research situations the photons or back-scattered electrons may be used. The

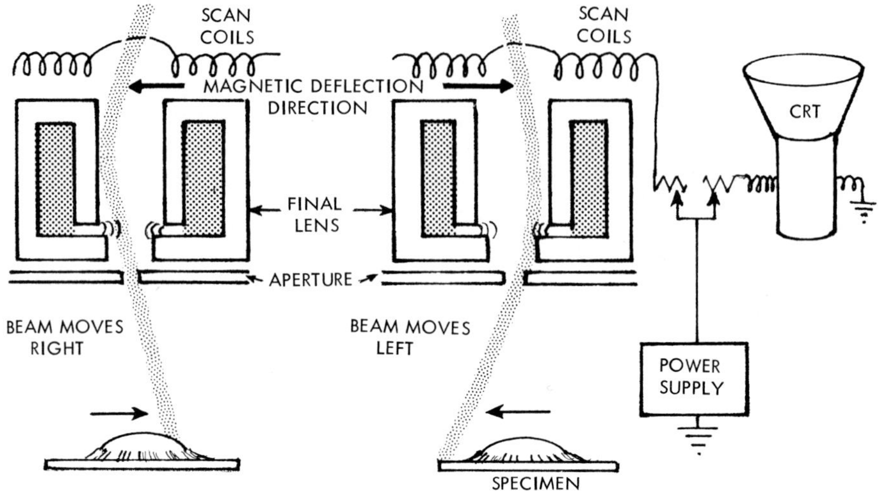

FIG. 1-14. The effect of the final lens and scanning coil on the electron beam in the SEM.

low-energy secondary electrons (up to 40 eV) are ejected from atoms in the specimen's upper surface (500 A), attracted to a collector or Faraday cage, and then drawn further to a scintillator tip (Fig. 1-13). The attractive forces on the collector and scintillator tip are positive charges. As an electron strikes the scintillator tip, it interacts with phosphor to emit a photon of light that is then conducted down an

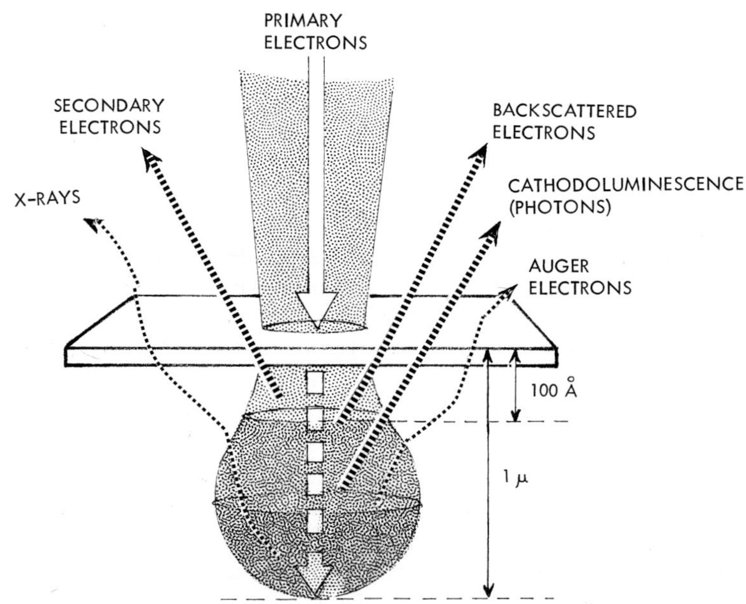

FIG. 1-15. The various particulate and nonparticulate emissions produced when the primary beam strikes the specimen surface in the SEM.

acrylic plastic light pipe, passed through a window in the wall of the microscope, and finally directed to the surface of a photomultiplier tube (PMT). Within the PMT the photon is converted into a photo electron. This electron is then amplified within the PMT to produce an electronic signal, which is further amplified and/or processed by a video amplifier unit and passed to the grid circuit of the cathode ray tube. The SEM works because the scanning events in the CRT and microscope are perfectly coordinated and because there is an insignificant delay occurring between secondary electron production, their detection, signal generation and amplification, and finally the CRT image display. The image on the CRT screen is produced simultaneously with the sweep of the primary beam across the specimen surface.

Magnification is produced on the SEM by keeping the CRT image fixed in size (i.e., 10 × 10 cm) and by varying the size of the sweep of the primary beam. For instance, if the primary beam sweeps over the specimen surface 5 cm from left to right and 5 cm from top to bottom, the image on the CRT will be magnified ×2. However, if the primary beam sweeps a pattern over the specimen 0.1 × 0.1 mm, then the CRT image will show a magnification of ×1000 (10 cm/0.1 mm or 100 mm/0.1 mm or ×1000). Thus, the magnification of the SEM is determined by Size of the image/Size of the object, and is achieved by altering the dimensions of the sweep.

The vacuum systems of both instruments are virtually identical. A series of pipes, manifolds, and valves connects the column to one or more diffusion pumps and a roughing pump. The vacuum level is usually displayed by a meter, while indicator lights indicate the status (go/no go) of the vacuum. In the SEM, the specimen chamber is large enough to facilitate the study of bulky specimens. However, most SEM biologic studies involve specimens no larger than 1 × 1 cm. Exotic specimen stages are offered by most manufacturers to allow the examination of warmed, frozen, multiple, stretched, or compressed specimens, and these are principally applicable in metallographic studies.

Critical Point Dryer (CPD)

The fixed tissue must be dried before it can be introduced into the evacuated column of the SEM, and there is general agreement that the most effective method is what is known as critical point drying. The physics of the process are complex, involving the placement of tissue samples, which have been dehydrated to absolute alcohol, in a cooled chamber into which carbon dioxide gas is introduced. The gas condenses to fill the chamber with liquid. The temperature of the chamber is then raised, and at a certain combination of temperature and pressure (roughly 1300 psi and 42 C for CO_2) the densities of the liquid and gaseous CO_2 become equal and the liquid is replaced by gas without significant perturbations in the morphology of the tissues. The gas can then be vented from the chamber, leaving dry tissue. If feasible budget-wise, it is worth investing in a CPD that has a chamber of adequate size with a viewing window and reliable temperature controls.

Vacuum Evaporator

SEM samples are usually coated with a thin layer (50 to 300 A) of a heavy metal, such as gold or platinum, prior to being placed in the microscope. The image is actually produced by interaction of the beam with this metal coating, and consequently its deposition must be of uniform distribution and thickness. The metal is deposited by means of a vacuum evaporator or a sputter coater. Many different models of each are available. The basic principle is to place the dried specimen, mounted on its stub, within a chamber (usually a bell jar), which is then evacuated. Within the chamber is a wire filament that can be heated. A length of gold wire is looped round the filament, and when the metal is heated it vaporizes and is deposited on everything inside the bell jar, including the surface of the specimen. To facilitate uniform coating, the specimen support is usually rotated and variably inclined. If the coating is inadequate, a charging effect will be apparent as the specimen is viewed in the SEM. An SEM view of two Chinese hamster ovary cells grown in culture, fixed and dehydrated by critical point drying, and coated with 200 A of gold can be seen in Figure 1-16.

THE ANALYTICAL ELECTRON MICROSCOPE

The analytical electron microscope (AEM) combines unique features of the transmission and scanning microscopes in the same instrument. The AEM may be operated as a regular TEM, as a high-resolution SEM (20 A or less), or as a scanning transmission electron microscope (STEM).

In the STEM mode the primary beam is demagnified by the condenser lenses to produce an electron probe (beam) with a diameter of 10 A or less. The beam is scanned over and through a plastic thin section located in the specimen chamber. The rastered beam in the STEM has sufficient energy (100 kV accelerating voltage) to penetrate the entirety of the section on a point-by-point basis. As in the case of the TEM, the electrons are scattered by their interactions with large atoms in the section (specimen). The scattered electrons (both elastic and inelastic) are removed from the beam by the objective aperture. The beam electrons then strike a detector located beneath the specimen that converts the beam to an electronic signal that is displayed on a CRT in the same manner as in an SEM. The STEM has an advantage over conventional TEM in that it can penetrate much thicker sections. It accomplishes this by virtue of the large number of high energy electrons that are able to pass through the section at each point in time. However, the STEM has a distinct disadvantage in that there is a sizable loss of resolution in this mode of operation (15 A) when compared to that possible in the conventional TEM (3 to 4 A).

What makes an analytical electron microscope "analytical" is the addition of an x-ray detector to the microscope column. The detector is coupled to an x-ray spectrometer located close to the microscope console. The detector and spectrometer detect the x-rays produced by the beam-specimen interaction, sort them out according to their energy or wavelengths, collate the information, and then display the resultant spectrum. The energy dispersive system (EDS) is the most popular x-ray system on AEMs. It provides the operator an energy spectrum of x-rays unique to the atoms within the specimen under study. For instance, calcium atoms, when bombarded with a stream of electrons, primarily emit x-rays with an energy of 3.617 keV. By careful analysis of generated spectra, the operator can determine the qualitative and semiquantitative microchemical composition of targets within the sample. If, for ex-

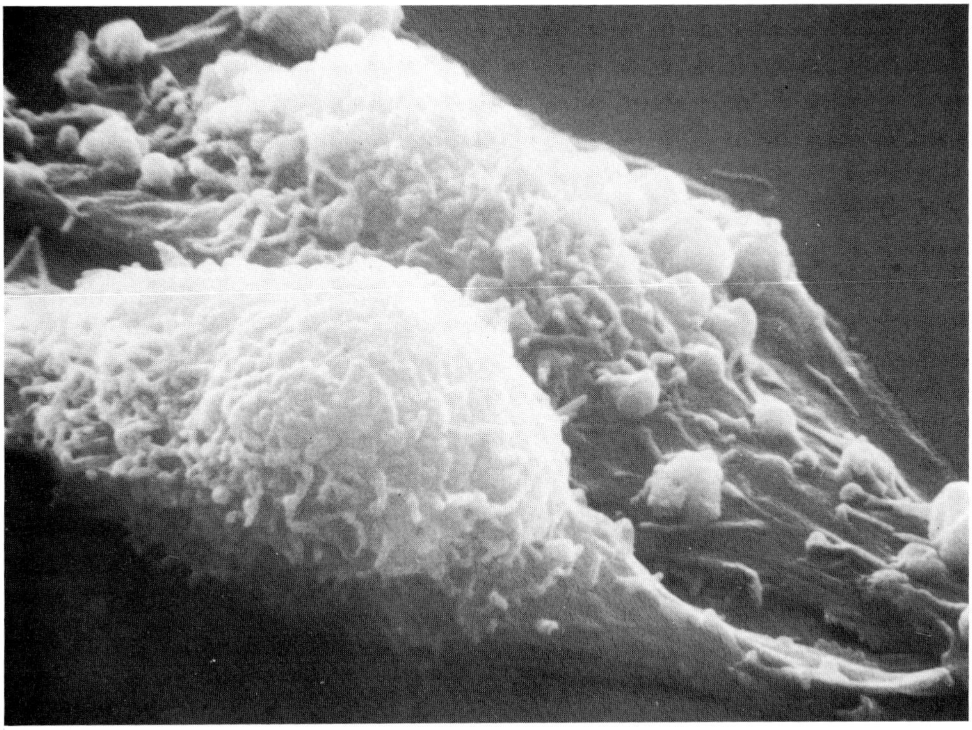

FIG. 1-16. Two Chinese hamster ovary cells, grown in vitro, as seen by the SEM. One cell is covered with blebs while the other's surface has microvilli. X3750.

ample, microcrystals observed in mitochrondria are selectively bombarded by the beam for an adequate length of time, a spectrum equivalent to the one observed in Figure 1-17 could be produced by the x-ray analyzer. The location of the spectral peaks and their heights (ordinate = energy; abscissa = counts) would suggest that most atoms within the crystals are calcium and phosporus and that the material is

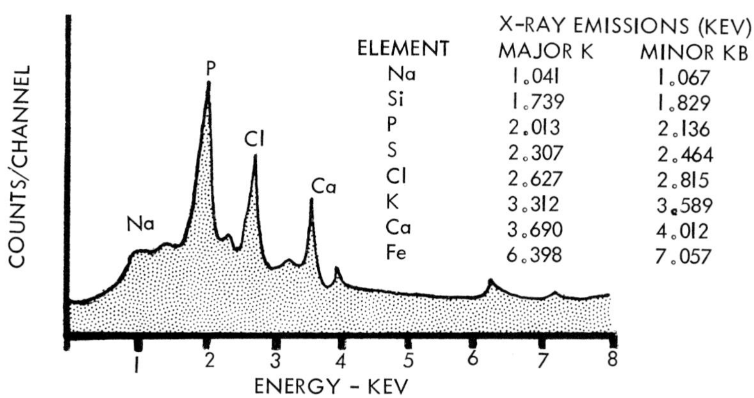

FIG. 1-17. A stylized energy-dispersive x-ray spectrum produced by analysis of a crystalline deposit in a mitochondrion.

either a calcium phosphate salt or calcium hydroxyapatite. Elements lower than carbon on the atomic chart are usually not analyzable because of their weak x-rays.

Analytical electron microscopes are very expensive and require a great deal of technical skill for their operation. Thus, they are commonly used in research facilities. Most pathology laboratories would find little use for an AEM.

ACKNOWLEDGEMENT

The author would like to thank Mr. Edmond Stevenson, Director of Medical Communications, Shriners Burns Institutes, for his help in the preparation of the illustrations in this chapter.

REFERENCES

1. **Hayat MA:** Principles and Techniques of Electron Microscopy. New York, Van Nostrand-Reinhold, 1970, vol 2
2. **Hren JJ, Goldstein JI, Joy DC (eds):** Introduction to Analytical Electron Microscopy. New York, Plenum Press, 1979
3. **Koehler JK (ed):** Advanced Techniques in Biological Electron Microscopy. New York, Springer-Verlag, 1973
4. **Meek GA:** Practical Electron Microscopy for Biologists, ed. 2. New York, John Wiley, 1977
5. **Sjostrand FS:** Electron Microscopy of Cells and Tissues: Instrumentation and Techniques. New York, Academic Press, 1967, vol 1

2

TECHNICAL PROCEDURES

PAUL S. BAUR AND BRUCE MACKAY

The results from a diagnostic study by electron microscopy are primarily determined by the tissue specimens that are provided for the study. The tissue must be representative and must be properly handled and processed. The procurement and preparation of tissue specimens for transmission electron microscopy are discussed in this chapter. The basic sequence of steps is similar to that for light microscopy. Small pieces of the tissue are immersed in a fixative solution, commonly glutaraldehyde, then rinsed in buffer and postfixed in osmium tetroxide. After dehydration, tissue blocks are embedded in an epoxy resin. Once they are hard, sections approximately 1 μm thick are cut and examined by light microscopy. Much thinner sections are then cut from selected blocks and placed on small copper grids. The sections are usually soaked in a solution of a heavy metal to improve tissue contrast in the electron beam. They are then ready for study with the electron microscope.

FIXATIVES

The choice of fixatives for diagnostic electron microscopy is not as extensive as one might suspect from some manuals on technique. Only two or three fixatives are widely used. Most laboratories carry out their primary fixation using glutaraldehyde. This five-carbon dialdehyde (Fig. 2-1) effectively cross-links protein molecules, and the glutaraldehyde molecules become incorporated into the molecular matrix of the tissue. Fixatives that become part of the structure of a tissue are known as additive fixatives, and if they do not cause alterations in the secondary and tertiary protein configurations, they are also referred to as noncoagulative fixatives. The range of concentrations through which glutaraldehyde is effective is relatively broad: It may be used as a 0.5 to 6% solution, the different concentrations being more a personal preference rather than dependent on any overriding physical or chemical considerations. Although glutaraldehyde causes shrinkage of some tissues (up to 6%), it is nevertheless the most effective fixative in preserving fine structure for morphologic studies and cytochemistry. It not only cross-links proteins and prevents cytologic distortion during the subsequent steps of the preparative procedure, but in some way also increases the permeability of the tissues to plastic embedding media. Glutaraldehyde penetrates tissues slowly. The rate of penetration is temperature-dependent, ranging from 2.5 mm per day at 4 C to 4.5 mm per day at 27 C. This slow infiltration by glutaraldehyde is a primary reason that tissue specimens for electron microscopy must be extremely small or thin. The chemical reactivity of glutaraldehyde suggests that it combines primarily with proteins, mucopolysaccharides, and polyalcohols. It shows little affinity toward lipids or carbohydrates. It is also pH-dependent; an optimum pH of 7.5 is often suggested. Wherever possible, the osmolarity of the solution should be relatively hypertonic (approximately 450 mOsm) to aid penetration. The buffer component of the fixative also contributes significantly to the total osmolarity of the solution. Effective fixation can be achieved by using 2% glutaraldehyde for up to 24 hours at 1 to 27 C as the primary fixative. The composition and size of the tissue and the objective of the study may influence the selection of time, temperature, and concentration. In diagnostic work, much briefer (eg. 2 hours) fixation is feasible, but the tissue must be sliced very thin (see Tissue Procurement section). The fixative should be made up from a purified vial of concentrated glutaraldehyde. It may be stored at room temperature for a few days.

Osmium tetroxide can be used as either a primary fixative or a postfixative. It is both an additive and a noncoagulative fixative, and, in addition to preserving tissues, it is an electron stain. The molecule has at its center a heavy metal atom (atomic number 76, atomic weight 190), which can effectively scatter incident electrons in the electron beam (Fig. 2-1). Osmium tetroxide primarily combines with lipids and shows its highest reactivity with the unsaturated forms. Osmium shows little affinity toward proteins, although there is a slight amount of reactivity with sulfur-containing amino acids. Although osmium does not react with proteins with a sufficient in-

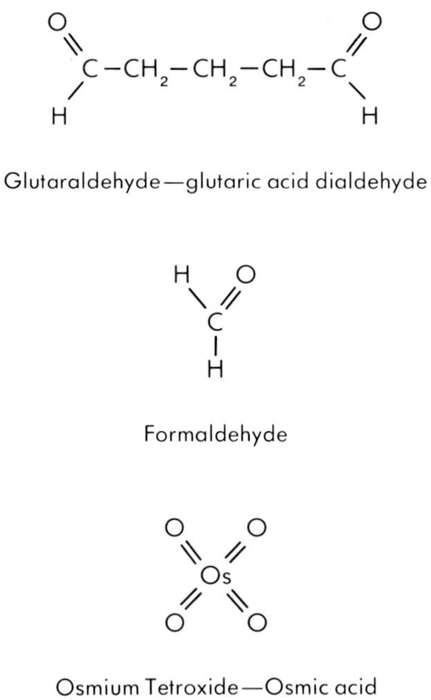

FIG. 2-1. Glutaraldehyde, formaldehyde, and osmium.

tensity to stain them, it undoubtedly contributes to their preservation by virtue of sol-gel transformations. Osmium does not react with nucleic acids or carbohydrates.

Because of its affinity for lipids, osmium is an ideal membrane fixative. It penetrates at approximately the same speed as glutaraldehyde, and is also temperature-dependent. Primary fixation with osmium tetroxide can be carried out using a 1 to 2% solution for not more than two hours at 4 C. For postfixation, a 1% solution for one hour is suitable. Prolonged exposure to solutions of osmium causes extraction of some lipids and proteins, and unlike glutaraldehyde, osmium tetroxide causes tissue swelling. The solution is highly volatile. Its vapors are extremely reactive and combine with lipids in the mucous membranes and conjunctiva. It must therefore be used with an efficient fume hood, and care should be taken to prevent its coming into contact with exposed skin.

Formaldehyde is the simplest of the aldehydes (Fig. 2-1). It is a noncoagulative and additive fixative that reacts readily with molecules containing active hydrogen atoms. The products of these reactions cross-link among themselves by the formation of methylene bridges. Formaldehyde penetrates tissue rapidly and allows relatively good retention of enzymatic activity. Unfortunately, it is at best a fair preserver of cell fine structure. When formaldehyde is used for electron microscopy, it should be prepared from paraformaldehyde, which is simply a large polymer of formaldehyde molecules. Some commercial forms of formaldehyde contain so many impurities that they are useless as EM fixatives. Formalin is a 37% (weight/volume) solution of formaldehyde gas in water, which also contains 10 to 15% methanol to prevent polymerization; the methanol limits its value as an EM fixative. Formalin-fixed material should thus be used only as a last resort for ultrastructural studies. Formaldehyde reacts with proteins, lipids, nucleic acids, mucoproteins, and unsaturated fatty acids, but not with polysaccharides. When nothing better is available, a well-buffered solution of formaldehyde is acceptable for diagnostic electron microscopy.

BUFFERS

Nearly all the aqueous solutions used in typical EM preparative routines contain a buffer component. The type ranges from the common diphosphate buffer to the zwitterionic organic buffers such as PIPES. Probably the most common buffer used in ultrastructural studies is sodium cacodylate, although phosphate buffers are also popular and collidine is widely used for osmium tetroxide solutions.

The function of the buffer is to maintain the pH of the cell and its components as close to the natural level as possible. While there may be considerable variation in pH within specific compartments of a cell, the cell average is on the alkaline side of the isoelectric pH of the proteins. Therefore, most fixative solutions are buffered in the pH range 7.2 to 7.4.

Fixatives react with the basic amino acids of protein molecules, and an adequate buffer must therefore be used to compensate for shifts in pH that would otherwise alter the fine structure of the cells. Buffers must never be viewed as inert agents in the fixation process, because several types have been shown to cause excessive protein extraction, organelle dissociation, alterations in enzyme activity, or crystal formation within tissues. Some buffers complex with divalent cations, and phophate buffers are particularly detrimental by forming insoluble calcium phosphate (di- and tri-basic) precipitates in the cytoplasm. These precipitates can be so extensive as to ruin the value of the tissue for electron microscopy. All buffers are temperature-sensitive with respect to their ionization constant. The choice of buffers is large (S-collidine, veronal acetate, Sorenson's phophate, Millonig's phosphate, cacodylate, and the organic buffers), and no one buffer has a clear advantage over another.

The speed of penetration of a fixative is directly proportional to its tonicity, and rapid penetration is necessary for good ultrastructural preservation. Some alterations in cell ultrastructure inevitably occur toward the center of a tissue block before the fixative arrives on the scene to arrest catabolism. A hypertonic solution shows the highest rate of penetration but causes some shrinkage. Hypotonic solutions infiltrate more slowly and cause tissue swelling. An isotonic fixative solution does not always prevent swelling or shrinkage, but its use can undoubtedly reduce cell surface artifacts caused by osmotic pressure imbalance. However, the deleterious effects of slow penetration of an isotonic fixative may offset any advantage in enhancement of structural detail.

Temperature affects the rate of fixation and can also alter the pH of the buffer. Advantages of lowering the temperature of the fixative to 0 to 4 C are that it slows enzyme-mediated cytoplasmic degradation, and suppresses respiration of the cells while the fixative is penetrating. Theoretically, use of a low temperature should result in better preservation of cytologic detail. However, certain vital cellular structures such as microtubules are dissociated at low temperatures, and their loss can drastically alter the shape and internal morphology of a cell long before the fixative can exert its effect. When the primary consideration is retention of enzymatic activity, some sacrifice of cytologic detail may be necessary and fixation can then be carried out at 4 C. A good compromise on temperature is approximtely 22 C, which reduces enzymatic processes and enhances the reactivity of the fixative. Fortunately for the microscopist, this is normal room temperature.

TISSUE PROCUREMENT

Most specimens for diagnostic electron microscopy are of solid tissue, but it should be possible, through processing, to harvest and retain cells in any type of biologic material. Some specimens require special methods of handling. A fundamental requirement for effective electron microscopy is prompt immersion of the excised tissue in the fixative solution. The specimen must be sufficiently small to allow rapid penetration of the fixative. As has been noted, autolytic change can be retarded by cooling the fixative, but as a general rule a solid tissue specimen should not be greater than 1 mm in size or thickness. In the case of a solid tumor, an effective procedure is to expose a freshly cut surface of the neoplasm, and, with a clean, sharp scalpel or razor blade, to slice a thin wedge from the cut surface, avoiding areas of hemorrhage and necrosis. A tapering wedge of tissue 1 mm in maximum thickness is ideal. Without handling it with forceps, the tissue is lifted on the blade and transferred immediately to the fixative solution. The specimen can be diced prior to immersion in the fixative, but this introduces the likelihood of squashing artifact. A thin slice of tissue will fix adequately provided the fixative can penetrate freely from both sides. Once it has fixed, dicing into 1-mm cubes is easy. The surface of the tissue must not be allowed to dry at any time during processing.

There is no hard and fast rule for the amount of tissue to be taken for diagnostic study. A single EM tissue block may be adequate, provided it is representative, but in some situations many blocks must be sectioned before a suitable area is found. Fixatives must be on hand at the time the biopsy is performed. An intermediate solution, such as saline or even a physiologic salt solution, should not be used, as important soluble proteins and carbohydrates can be leached from the cells and interstitial spaces in a short period of time. Logistic problems may have to be overcome in a large institution to have a readily accessible supply of fixative. Small specimen bottles containing approximately 10 cc of glutaraldehyde should always be available in the frozen section laboratory, and also in the radiology department, wards, and clinics if biopsies are performed frequently in these locations. It is useful to have a label on each bottle conveying brief instructions on how to handle the biopsy specimen for electron microscopy. Otherwise, there is the danger of receiving a lymph node the size of a walnut, stuffed into a small container, with the tip of the tissue barely nestling in the glutaraldehyde.

Education of the clinicians who are carrying out the biopsy procedures is often necessary to obtain suitably preserved EM specimens. It is time consuming and rarely feasible for a representative from the EM service to be present when a biopsy is being performed. Prior discussion between clinician and pathologist concerning the clinical situation and the most desirable method for handling the biopsy specimen is often extremely helpful. A decision can be made on the most effective method for apportioning the specimen, as for example between light microscopy and electron microscopy. Percutaneous renal biopsy specimens provide a good example. Generally, a 1-mm core from 1 to 2 cm in length is obtained and tissue should, when facilities are available, be submitted for light microscopy, immunofluorescence studies, and electron microscopy. The pathology is generally located within the glomeruli, and since cortex may only be present in part of the biopsy, it is customary to cut 1 or 2 mm from both ends of the specimen, placing these pieces in glutaraldehyde for electron microscopy. The remainder of the specimen can then be sliced longitudinally (requiring some manual dexterity in a 1 mm-specimen) and half can be placed in a suitable medium for immunofluorescence studies and the rest in formalin for light microscopy.

OTHER SPECIMENS

It should be possible to process any type of tissue for electron microscopy. However, special methods of handling are necessary in some instances, as described below.

Needle biopsies are a convenient and often fruitful source of material for diagnostic electron microscopy. The biopsy procedure is easy on the patient, does not require surgical scheduling, and provides an acceptable quantity of tissue. Disadvantages are the risk of missing a small lesion, and susceptibility to squashing and drying artifact.

Needle biopsy specimens are usually 1 or 2 mm in diameter. They should be placed directly into the fixative solution and diced only after the tissue has been hardened by fixation.

The use of fine needle aspirates for cytologic diagnosis has burgeoned in recent years, and the tissue, provided it is representative, should be adequate for electron microscopy. Localization of the needle tip within a small lesion can be achieved by a competent radiologist, but there is a tendency to aspirate necrotic cells rather than viable tissue, and the specimen is often heavily contaminated with blood cells. The specimen should be handled like a bone marrow aspirate. Otherwise, searching for representative cells can be like looking for a needle in a haystack.

Bone marrow aspirates are collected in a heparinized syringe and centrifuged in hematrocrit tubes. If desired, marrow granules can instead be picked from the aspirate over a chilled paraffin slab in a petri dish, and processed as a spun-down pellet. Granules are not always present in significant numbers in the aspirate, and centrifuging in a hematocrit tube is therefore preferable. The buffy coat can be withdrawn in a Pasteur pipette and transferred to glutaraldehyde. If performed gently, it will form a worm-like cylinder that can be sliced, once it is fixed, and processed as solid tissue. An alternative method is to centrifuge the specimen, remove the supernatant plasma, and layer glutaraldehyde on the surface of the buffy coat. If the latter is of a suitable amount, it will fix as a disc that can be gently dislodged with a fine wire, or by filing and breaking the tube. However, when the buffy coat is very thin, it is unmanageable and if it is thick, it does not fix uniformly.

A core biopsy of bone for ultrastructural studies should yield a specimen not more than 2 mm in thickness. The tissue can be decalcified in 10% buffered ethylenediamine-tetra-acetic acid (EDTA), but this procedure invariably damages the ultrastructure of the cells to some degree. A preferable method is to fix the core, and then, under a dissecting microscope, delicately pry the soft tissue from the marrow spaces with a fine needle. The procedure requires some practice.

Cells in effusions can be spun down to form a pellet, which, after fixation, is diced and processed as for solid tissue. Where the effusion is frankly bloody, it is desirable to eliminate as many of the red cells as possible by hemolysis prior to centrifugation.

Specimens of cerebrospinal fluid generally contain very few cells. Even when the specimen is centrifuged, a pellet is not visible. It is, therefore, preferable to spin the specimens in a container with a pointed capsule at its base so that the pellet forms at the tip of the capsule. Processing can then be carried out without dislodging the pellet from the tip of the capsule. The results are often disappointing.

In some situations, tissue has already been fixed in formalin before the need for electron microscopy is appreciated. The tissue can be taken from formalin and processed in the usual way for electron microscopy, but the results depend on the quality of the initial fixation. There have been attempts to devise multipurpose fixatives suitable for both light and electron microscopy, combining the preservation of glutaraldehyde with the penetration of formaldehyde, but these fixatives are less than effective in practice since good preservation is only obtained if the tissue slice is thin, in which case it is preferable to use glutaraldehyde for the EM and formalin for the LM.

When nothing else is available, it is possible to take tissue from a paraffin block or even a microscope slide, and process it for electron microscopy. The results are determined primarily by the quality of initial fixation and are generally disappointing, though occasionally signficant features such as cell junctions, secretory granules, or organisms can be identified.

POSTFIXATION

After fixation and prior to postfixation, the tissue is diced into small cubes no larger than 1 mm. Larger pieces will not process adequately. The tissue is rinsed to remove residual fixative. After glutaraldehyde fixation, the glutaraldehyde residues can be removed by two brief buffer rinses, or, if more convenient, the tissue can be left in buffer overnight. The concentration of buffer in the rinsing solution should be at the same level as that used in the fixative and postfixative solutions (e.g., 0.1 M used throughout). Postfixation is carried out by immersing the tissue in a small quantity of osmium tetroxide, generally a 1% buffered solution, in a fume hood for one hour. Some buffers may form a precipitate when the tissue is transferred to the first dehydrating solution, and, in this event, a brief water rinse can be employed prior to dehydration.

DEHYDRATION

Dehydration of the tissues can be accomplished by using any one of a variety of organic solvents. Ethyl alcohol (ethanol), isopropyl alcohol, methanol, acetone, polyvinyl alcohol, polyethylene glycol, and water-miscible embedding resins have all been used to remove the free water from the tissue. During the process of dehydration, some tissue shrinkage and extraction of cellular components undoubtedly occur. The two best dehydrants are probably ethanol and acetone. Ethanol is the more popular and has the advantage of causing less extraction, but since it is not readily miscible with most embedding resins, a transitional solvent, such as propylene oxide, must be used. Propylene oxide can extract both osmium-fixed and unfixed lipids, and propylene oxide residues remaining in the tissue after dehydration can be incorporated in the epoxy resin. If excessive contamination by propylene occurs, the plastic will fail to polymerize properly, and blocks will not section well. Acetone is an excellent dehydrant and is readily miscible with the resins, but has the disadvantage of excessively hardening some tissues to the point that sectioning may be difficult or impossible. It is therefore not often used in the routine processing of tissues for electron microscopy.

INFILTRATION

Following dehydration, the tissue is introduced into the liquid resins through a series of infiltration steps, which gradually replace the dehydrant or transitional solvent with increasing concentrations of the plastic medium. Three or four steps can be used with proportions of solvent to resin such as 3:1, 1:1, 1:3, and 100% (unpolymerized resin). Each change of solution may take from one-half to one hour, depending on the nature of the tissue, the type of plastic used, and whether or not agitation is employed during the infiltration process. For small pieces of tissue, it is possible to reduce the number of steps and to pass the tissue from propylene oxide into a 1:1 solvent-to-resin mixture for one hour, and then into unpolymerized resin (100%) for a further two hours. Gentle agitation facilitates penetration of the resin mixture.

EMBEDDING

The fixed and dehydrated tissue must be embedded in a plastic resin for thin-sectioning, and to enable the tissue to resist the adversities of the electron beam. Two groups of resins are useful in EM work: the nonwater-miscible resins and the water-miscible resins. The nonwater-miscible resins are commonly used. This group includes the resins known as methacrylates (acrylics), the polyesters, and the epoxies. The methacrylates are soluble in alcohol and acetone, are colorless, penetrate the tissue easily, and polymerize spontaneously. However, they tend to polymerize unevenly and cause a large degree of shrinkage; the polymer is soluble in organic solvents, and the sections are volatilized by the electron beam.

The polyester resins are colorless, cause only moderate tissue shrinkage, are resistant to the electron beam, penetrate tissues slowly but polymerize evenly, are soluble only in acetone, and have relatively high viscosities. Polyester resins are difficult to store, are insoluble in ethyl alcohol, produce hard tissue blocks, and penetrate tissues very slowly, and for these reasons are not widely used.

Epoxy resins are widely used in electron microscopy. They are readily polymerized by cross-linking; they polymerize evenly, cause little tissue shrinkage, are stable in the electron beam, and facilitate high-resolution electron microscopy. A disadvantage is that they cause contact dermatitis on coming in contact with exposed skin. They are used in conjunction with other chemicals that affect the physical qualities of the final polymer. These additives are hardeners that act as linking agents, plasticizers that add flexibility and strength, and catalysts that promote cross-linking but are not incorporated into the polymer. Variations in the concentration of the additives and monomer significantly alter the viscosity and penetrability of the liquid resin mixture and the hardness and sectioning quality of the polymerized plastic block. Nonwater-miscible resins are usually polymerized by heat. Sixty degrees Centigrade for at least 12 hours is necessary. Longer polymerization is desirable, but in the interest of expediting the examination of diagnostic material, the blocks can be sectioned after overnight curing.

Water-miscible resins are excellent for cytochemical studies, since they preserve biochemical activities within fixed tissues and do not require dehydration of the tissue by organic solvents or heat for their polymerization. Dehydration is carried out by passing the tissue through a graded series of aqueous resin solutions. Polymerization is brought about by the use of ultraviolet light at low temperatures. The thin-sectioned material is easily treated with, or stained by, cytochemical reagents in an aqueous phase. Disadvantages of the water-miscible resins include extraction of lipids, nucleoproteins, and lipoproteins; contact dermatitis; section swelling on the water in the knife trough; and a ghost-like (negatively stained) appearance of membranes.

Automatic processing machines are available for electron microscopy, but before purchasing one it is advisable to look carefully into the specifications. These instruments are expensive and tend to use larger volumes of solutions than are used in processing by hand. Because some may not be used for infiltration, this often must be done manually. They take a certain amount of time to set up, dismantle, and clean, and are probably not justified for a laboratory with a small volume of specimens.

After the tissue blocks have passed through the solvents and liquid resin mixtures, they are placed in embedding capsules filled with liquid resin. When tissue orientation is important, a flat embedding mold can be used. The blocks can be moved about in the liquid resin using a slender needle. If necessary, a dissecting microscope may be used to observe the orientation of the tissue blocks. Care must be taken to ensure that the position of the blocks is not altered once it has been established. The number of pieces of tissue that are embedded will vary with the type of specimen.

When the tissue is likely to be homogeneous, as is the case with many solid tumors, a satisfactory practice is to embed five tissue blocks in capsules and to place the remainder in a plastic receptacle, such as a specimen bottle cap filled with liquid resin. These extra pieces of tissue can, when necessary, be cut out and reembedded, or, if speed is of the essence, can be mounted on machined slugs with a rapid epoxy. Commercial capsules and molds are designed for ultramicrotomy purposes and are the correct size and shape for easy mounting in an ultramicrotome, for proper plastic polymerization, and for eventual sectioning (Fig. 2-2). The tip design facilitates tissue orientation and minimizes the amount of trimming required to remove excess plastic from around the tissue. Capsules with conical tips are particularly suitable for

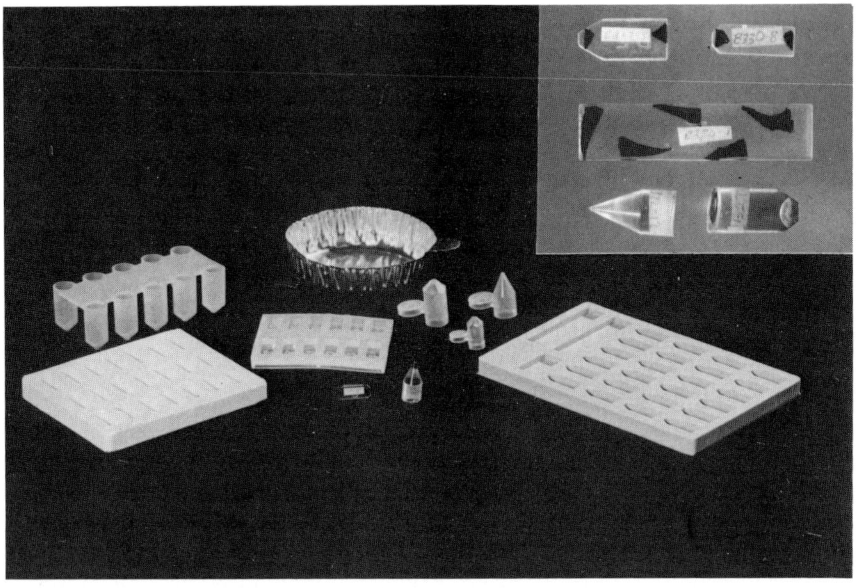

FIG. 2-2. Embedding paraphernalia used in electron microscopy. Inset shows the variety of embedded tissue blocks used in transmission electron microscopy.

embedding specimens in which the cells are in suspension. Most capsules are constructed of polyethylene or gelatin and are discarded after use. Flat molds are made of polyethylene or latex rubber and can be reused repeatedly. Small paper tapes bearing the specimen number and block number are placed in the liquid resin. They become incorporated in the polymer and are a permanent record of the specimen and block. The information can be printed using a No. 3 lead pencil or a typewriter whose ribbon ink will not run.

Once the plastic has polymerized, the plastic block is removed from the capsule or mold. Flat embedding blocks separate from the molds readily when the latter are flexed. A plastic block can be removed from a polyethylene capsule by slicing the capsule off with a razor blade (not recommended), squeezing it out with pliers, or by using a convenient commercial press. Care must be taken to avoid damaging the tip of the capsule where the specimen is located. Capsules with conical tips are particularly vulnerable to pressure damage. In every instance the block should be allowed to cool after removal from the oven before an attempt is made to remove it from the mold or capsule.

ULTRAMICROTOMY

Next comes the most exacting and often the most tedious phase of preparing tissues for study with the electron microscope. Ultramicrotomy entails placing the block in an ultramicrotome and cutting sections with a suitable knife. The sections may be of the order of 1 μm in thickness (called semithin or thick sections) for light microscopy, or thin sections, approximately 700 A, for ultrastructural study.

The 1-μm thick sections are an essential preliminary to thin-sectioning, since they are necessary for determining that a tissue block is representative and that the material is acceptably preserved. The number of tissue blocks on any one specimen varies considerably and is determined in part by the type of specimen. For example, a renal biopsy specimen will generally provide from two to six tissue blocks, all of which should be embedded and thick-sectioned. In contrast, a homogeneous tumor can usually be adequately assessed from a study of a single block. It is our practice to cut and evaluate thick sections on two EM blocks of tumors, selecting one of these for thin-sectioning if it is representative and adequately preserved, and contains a sufficient number of tumor cells for study. If neither block is satisfactory, the remaining three blocks in capsules are then thick-sectioned.

Thick sections are cut using a glass knife. The knife is triangular, and can be broken with hand pliers, or is preferably made using a commercial knife-maker and commercially available strips of glass. The tissue block is mounted in an ultramicrotome, and the surface of the tissue is exposed either with a razor blade or more conveniently by filing away the resin. The knife is advanced to the surface of the block, and a few sections are cut to smooth the block face. The knife is then moved laterally so that one part of the knife is used for rough-facing and another part for cutting the better sections, and two or three sections are cut and allowed to accumulate on the surface of the glass knife. They are lifted off using a finely tapered camel's-hair brush or a hypodermic needle, and are transferred to the surface of a single drop of water centered on a clean glass slide. The water is allowed to evaporate on a hot plate (approximately 150 C). The sections flatten out on the slide and adhere tenaciously.

For rapid diagnostic work the procedure outlined in the preceding paragraph is convenient and efficient. It is facilitated by having a Porter-Blum MTI ultramicrotome that is reserved for cutting the 1-μm sections. In this way, a more elaborate ultramicrotome can be reserved for thin-sectioning, avoiding unnecessary wear and tear on its bearings. In many laboratories, a single ultramicrotome is used for both thick- and thin-sectioning, and the microscopist may prefer to shape the tissue block prior to thick-sectioning. The following procedure can then be followed (Fig. 2-3).

The tissue block is mounted in the sectioning arm of the ultramicrotome, and a glass knife is positioned in the knife holder. A series of sections 2 to 5 μm thick is

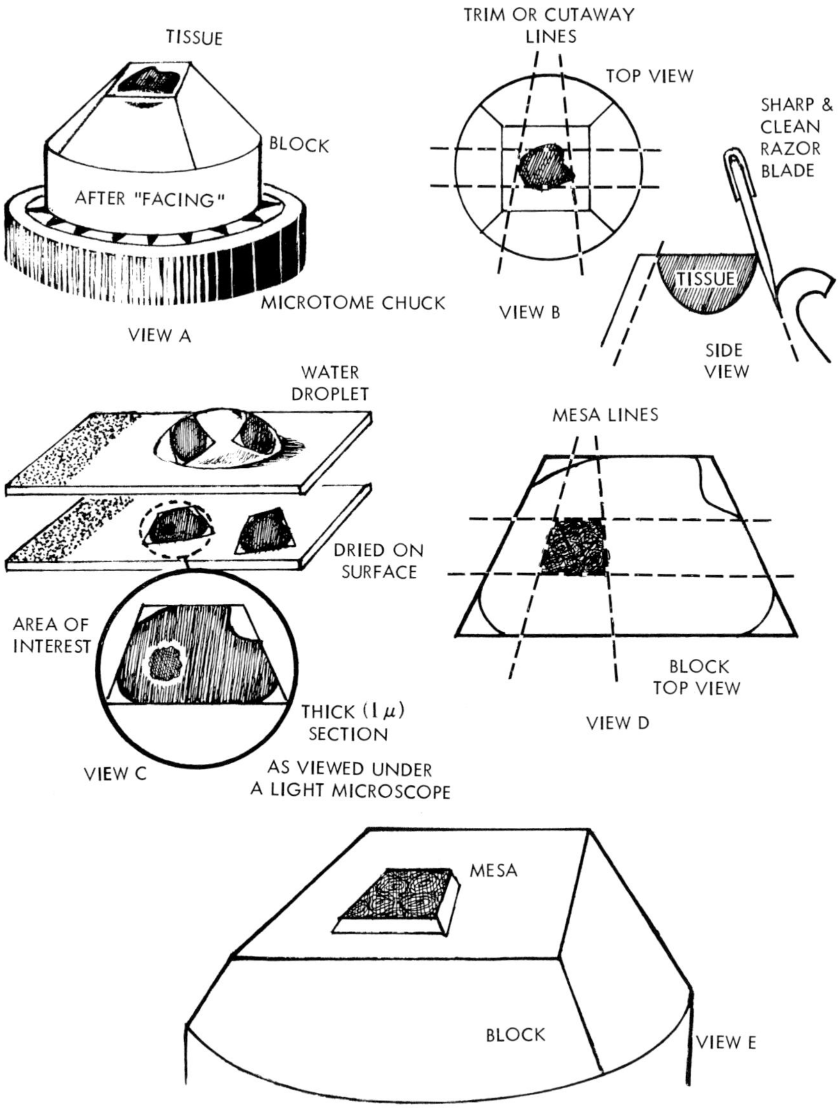

FIG. 2-3. Steps used to trim the block for thin-sectioning.

taken until a reasonable amount of the tissue is exposed on the block face. The block is mounted in a special trimming block holder (in some ultramicrotome models, the block, or chuck, must be transferred from the specimen arm to the trimming holder), and the block face is observed through the viewing microscope. The stage holder allows the chuck to be rotated through a 360-degree arc for ease of operation. A clean (organic solvent removal of protective oils) razor blade (double-edged blade broken in half) is used to cut away excess plastic at the sides of the tissue. The plastic is removed in such a manner as to produce a truncated trapezoidal block face with sides approximately 45 degrees to the face. The base of the trapezoid should be located at the bottom of the block face and becomes the part of the specimen that first makes contact with the knife edge. The size of this trapezoidal face is normally around 1 mm in greatest dimension, but larger block faces can be thick-sectioned with relative ease.

The next phase in the operation is to return the block (with chuck) to the cutting arm of the microtome, and to place a new, sharp, clean glass knife in the holder. Several thick sections (½ to 1 μm) are then taken using a moderately slow sectioning speed (less than 1 mm per second). The sections should have a shiny surface and appear rainbow-colored (mostly pinks and greens) when observed through the viewing microscope. The sections accumulate on the back side of the glass knife edge and can be lifted off this surface using a small finely tapered camel's-hair brush. Each section is transferred to the surface of a single drop of water (double-distilled and filtered) centered on a clean glass slide. The droplet is allowed to evaporate on a hot plate (about 150 C), and the section flattens out on the slide and adheres tenaciously. After two to five minutes of heat treatment, the slide is removed and stained with a methylene blue/azure II aqueous stain for approximately one minute. The slide is then rinsed, dried, and cover-slipped using an appropriate mounting medium. The section is examined by light microscopy and the field of interest selected. Using the section size, shape, and tissue features for reference, the field of choice can be located in the block face. That area of the block face is where the embedded tissue must be thin-sectioned (Fig. 2-3).

To further reduce the block face to a size that will permit ultrathin-sectioning, excess plastic and tissue must be removed from the margins of the field. The block is again transferred to the trimming holder, and a sharp, clean razor blade is used to produce a "mesa" (0.5 mm x 1 mm) with a trapezoidal shape. The top and bottom edges of the small trapezoid should be parallel so as to produce a straight ribbon of thin sections. The edge angle of the mesa should be at least 45 degrees, and its surface should be no higher than ½ mm above the surrounding supportive plastic (which would allow production of 700 thin sections).

The block is then returned to the sectioning arm of the ultramicrotome, and the knife to be used for cutting thin sections is positioned. Orientation is greatly facilitated by a light, or reflecting foil, below the knife edge to reflect its shadow onto the block face. The trough on the knife is filled with absolutely pure water, the microtome controls are set to provide a 700-A advance increment per cycle, and sectioning is begun. Thickness of the thin sections is judged from their interference color. Silver or straw-colored (700 to 800 A) sections are usually best for routine electron microscopy. The sections accumulate on the water surface and can, if necessary, be herded together using a single eyelash mounted on a stick. Excess, partial, thick, or folded sections can be removed from the water surface by stabbing them with a tweezer tine. The sections quickly attach to the tine and can subsequently be wiped off using clean lens paper. A clean EM grid (dilute acid solution followed by acetone rinses in conjunction with an ultrasonic device) is held in fine forceps and can be either laid down on top of the sections or lowered through the water and lifted out with the sections on its surface (Fig. 2-4). Some microscopists are "dippers" and others "lifters."

The grid is dried on the filter paper and then stained using solutions of uranyl acetate and lead citrate. Thorough rinsing with water must follow each staining step. After the final water rinse, the sections are air dried, section side up, on a clean piece of filter paper in a petri dish.

Selection of the proper tweezers is important. A No. 4 jeweler's tweezer made of stainless steel is widely used. The tips should occasionally be cleaned using a dilute soap solution and rinsed with water or solvent to prevent grid/section/stain contaminations. The points should be protected when not in use with a tip shield, which is usually provided with the tweezer at purchase.

Fresh, filtered, deionized, and distilled water should be used during all steps in the fixation, dehydration, sectioning, and staining processes. Bacterial contaminants commonly occur in large, plastic storage containers and wash bottles in spite of the low elemental content of distilled water. The mucopolysaccharides produced and shed by these microorganisms can ruin section surfaces, contaminate diamond knife edges and troughs, and cause precipitates to form in preparative solutions.

There is an extensive variety of grid designs commercially available for EM work. Most grids come in the standard 3.05-mm diameter (Fig. 2-5). The most popular grids are made of copper by means of an electroplating process and come in a wide range of mesh sizes (50, 100, 200, 300, and 400 bars per inch). The 200- to 400-mesh

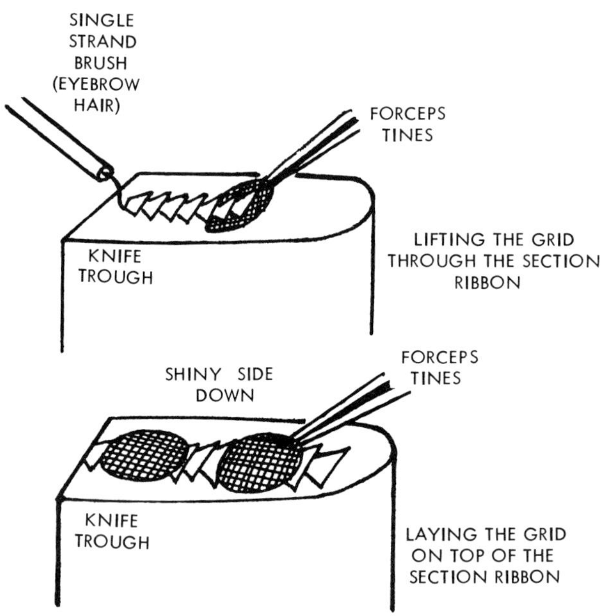

FIG. 2-4. Picking up the sections off the water surface in the trough. The grid can be lifted through the ribbon or laid on its surface.

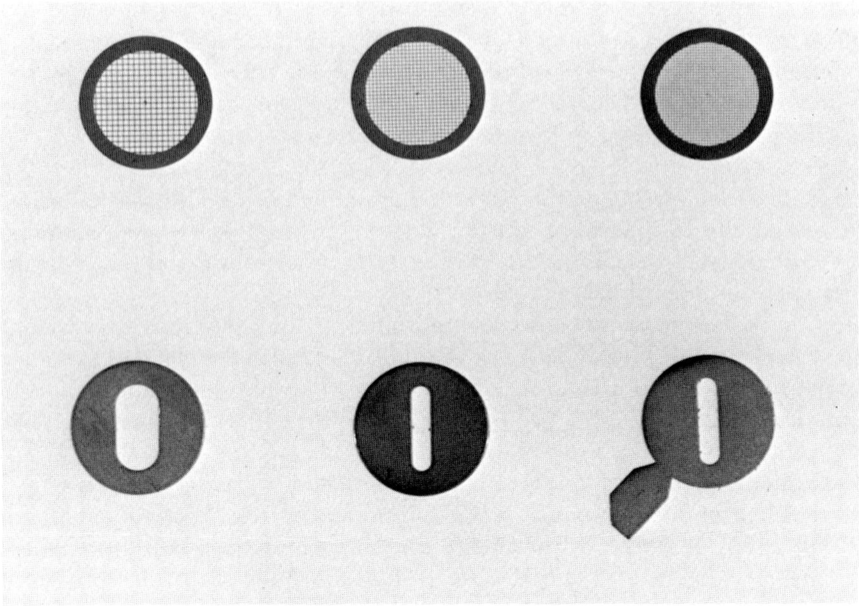

FIG. 2-5. Several grid designs used with the TEM.

sizes are convenient because they do not require an additional film for support of the thin sections. Supporting films (formvar or collodion, strengthened if necessary by carbon coating) will be necessary for the study of particulate suspensions, such as in the demonstration of viruses by negative staining.

Electron microscopy has a penchant for Murphy's law. The hedge against failure is the employment of good technical help, and a trained histology technician frequently makes a good electron microscopy technician. In setting up a new EM lab, it may be necessary to arrange for a new technician to spend some weeks in an established laboratory learning the techniques. Learning the art of thin-sectioning generally requires a minimum of two weeks of concentrated, supervised practice.

The embedding routine presented in Table 2-1 will produce exemplary preservation of fine structure provided the tissue is properly taken and properly fixed. With experience it will be found that some steps can be omitted or time reduced without detrimental effect. Rapid processing procedures taking as little as four hours are in fact feasible.

STUDY OF THIN SECTIONS

The importance of examining the appropriate 1-μm section before doing the electron microscopy has been emphasized. It is useful to have a mental picture of the topography of the particular area to be examined, and to know whether the entire specimen is representative of the pathologic process, or whether other entities such as inflammation, necrosis, or normal tissues are also present. A scan of the grid at low power

TABLE 2-1
A Typical Fixation, Dehydration, and Embedding Procedure

Step	*Solution*	*Temperature*	*Time**
1.	Fixation (0.1 M buffer—3% glutaraldehyde)	24 C†	1–4 hr
2.	Buffer rinse (0.1 M buffer)	24 C	15–30 min
3.	Buffer rinse (0.1 M buffer)	24 C	15–30 min
4.	Buffer rinse (0.1 M buffer)	24 C	15–30 min
5.	Postfixation (1% osmium tetroxide in 0.1 M buffer)	24 C	1 hr‡
6.	H_2O rinse	24 C	Several times
7.	75% ETOH	24 C	10–15 min
8.	95% ETOH	24 C	10–15 min
9.	100% ETOH (absolute)	24 C	10–15 min
10. §	100% ETOH (absolute)	24 C	10–15 min.
11.	Propylene oxide:ETOH (1:1)	24 C	10–15 min
12.	Propylene oxide (100%)	24 C	30 min
13.	Propylene oxide (100%)	24 C	30 min
14.	Propylene oxide:Epoxy resin mixture (3:1)	24 C	1 hr
15.	Propylene oxide:Epoxy resin mixture (1:1)	24 C	1 hr
16.‖	Propylene oxide:Epoxy resin mixture (1:3)	24 C	1 hr
17.	Epoxy resin mixture (100%)	24 C	1 hr
18.	Epoxy resin mixture—in capsules	45–60 C	24–48 hr

* All times may be shortened considerably if infiltrator device is used.
† Often fixation at 0–4 C will be desired for studies involving cytochemical surveys (enzymatic activity). Times of fixation must then be altered accordingly.
‡ Postfixation may require longer times if the tissue blocks are large. The tissue should turn black during this step.
§ A stopping place.
‖ A special stopping place. If catalyst is not in resin, then the propylene oxide may be allowed to evaporate overnight.

with the electron microscope is a useful preliminary, providing orientation and bridging the gap between the 1-μm section and more detailed ultrastructural study. Specific areas can then be selected for examination. In many laboratories, it is common practice for a technician to scan the grid and take photographs that are subsequently interpreted by a pathologist. While acknowledging the considerable abilities of many technicians, there are advantages to having a pathologist perform the study. Significant features are less likely to be missed, areas most representative of the pathologic process can be selected, and a better evaluation of morphologic features that may enter into a differential diagnosis is possible. In addition, a more rapid diagnosis is likely, since in many instances a verbal report can be given immediately without waiting until photographs are processed. The total number of photographs taken will also be reduced, since only those areas necessary for documentation of the disease process need be recorded.

When photographs are taken for publication, a record of a recent calibration of the microscope should be available. It is a good practice to calibrate the instrument on a regular basis, using a commercially available line grid. The practice of calculating magnifications for every routine print is inordinately time consuming and quite unnecessary. A skilled microscopist can estimate the approximate magnification of a print, and calculation of magnifications can be reserved for prints for which a precise determination of the size of structures, such as viral particles, is required.

The information from a diagnostic study must be communicated to the patient's medical record. The report may be sent by the electron microscopist performing the diagnostic study, but in many instances the tissue is also examined by light microscopy, and, in the case of renal biopsies, also by immunofluorescence. Separate reports on these different facets of the study introduce duplication, and in controversial cases it is better to coordinate the findings in a single report, or at least make it the responsibility of one pathologist to evaluate the entire case. This can be effected by submitting the report of the ultrastructural study to the pathologist carrying out the light microscopy, who can then at his discretion incorporate the findings in a single report or provide a supplemental report with the results of the ultrastructural study. The amount to be charged for the pathology service can also be more appropriately coordinated.

DOCUMENTATION IN THE ELECTRON MICROSCOPY LABORATORY

Regardless of the size of an electron microscopy facility, an efficient routine for the documentation and storage of specimens must be devised and strictly followed. This is mandatory in a laboratory where large numbers of specimens are received, where material comes from several different sources, and where human tissue is concerned. The procedure for numbering specimens varies, but it is convenient to have sequential accession numbers with cross-referencing information (patient's hospital number and name, surgical pathology number) in a logbook. It is also useful to have a small printed card with basic information on the case for laboratory use while the specimen is being processed and examined. This card should subsequently be filed under the EM accession number, with the diagnosis typed on the card. A second card, color-coded for convenience, should be prepared for a classified file, which can greatly facilitate retrospective study of groups of cases.

The EM tissue blocks can be stored in various ways. The most convenient storage container is a small plastic or cardboard box with the specimen number written on it. Before EM negatives are filed, relevant data must be written on the glassine or plastic envelope containing the negative. Prints and negatives are generally kept in separate files, and a cumulative EM numbering system for accessions makes filing and retrieval a simple procedure.

REFERENCES

1. **Glauert AM:** Fixation, dehydration, and embedding of biological specimens. In Glauert A (ed): Practical Methods in Electron Microscopy. New York, North-Holland-Elsevier, 1974

2. **Hayat MA:** Principles and Techniques of Electron Microscopy: Biological Applications. New York, Van Nostrand-Reinhold, 1974, vol. 1
3. **Hayat MA:** Positive Staining for Electron Microscopy. New York, Van Nostrand-Reinhold, 1975
4. **Meek GA:** Practical Electron Microscopy for Biologists, ed 2. New York, John Wiley, 1977
5. **Sjostrand FS:** Electron Microscopy of Cells and Tissues: Instrumentation and Techniques. New York, Academic Press, 1967

3

PREPARATION AND INTERPRETATION OF SEMITHIN SECTIONS

WILLARD A. BURNS

This chapter covers the preparation of semithin sections as used in traditional electron microscopy, variations of these techniques to increase the options for the microscopist, and adaptations for light and/or electron microscopy. These are outlined in Figure 3-1.

Semithin sections are cut from tissue embedded in plastic and viewed under a light or sometimes a phase microscope. These sections, also referred to as thick sections, are approximately 0.5 to 1.0 μm in thickness and traditionally are used for orientation of what has been embedded in the plastic block. These semithin sections are much thicker than the sections viewed under the electron microscope but much thinner than sections of paraffin-embedded tissue. Since the terms "semithin sections" or "thick sections" create a problem in terminology, many individuals prefer the phrase "light microscopy of plastic-embedded tissue."

The thick-section technique has been a step in the processing of tissue for electron microscopy since the inception of ultrastructural studies. However, its use has increased over the years, has become routine in most laboratories, and must be considered obligatory when one is processing tissues for diagnosis. The thick-section technique is most helpful in permitting rapid identification of organ structures in a particular block and in assuring that appropriate samples have been embedded for examination under the electron microscope. This enables the electron microscopist to better predict what will be available in the thin sections taken from the same epoxy block. The technique serves as a pathway or bridge between a light microscopy image and a low-power electron micrograph. If the bridge is traveled in both directions, the benefit for both the light and electron microscopist is enhanced. These sections can complement paraffin sections in diagnosis, even if a laboratory has no electron microscope facility. The semithin section is, therefore, becoming adopted by light microscopists because of the superior preservation of the tissues and the better resolution available in plastic sections.

PREPARATION OF SEMITHIN SECTIONS

Following the traditional processing of tissue for electron microscopy, the plastic blocks are ready for sectioning. If paraffin-embedded tissue is also processed, it can be useful to submit adjacent "mirror-image" portions representing samples as similar as possible in the paraffin and plastic blocks.[1]

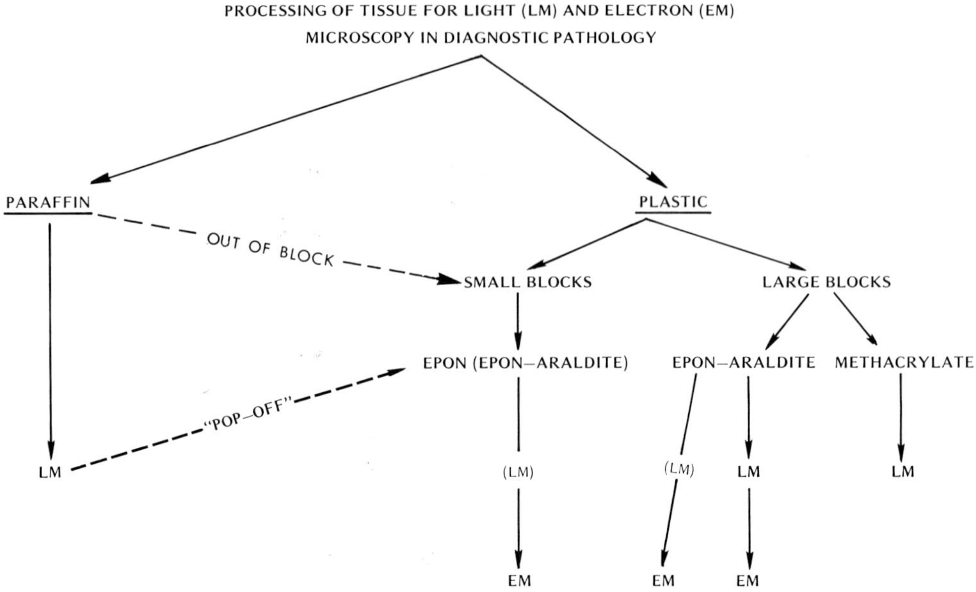

FIG. 3-1. Microscopy options for plastic embedding of tissue.

Since many tissue samples are usually embedded from a single specimen, random blocks are sectioned until the desired structure is found. The blocks are trimmed of excess epoxy with a razor blade until the tissue is exposed. During the trimming, which is done either on an ultramicrotome or on a chuck, viewed through a dissecting scope, a truncated pyramid is formed (Fig. 3-2). Blocks can be sectioned on the same ultramicrotome that will subsequently be used for preparing the thin sections to be viewed under the electron microscope. The block is cut with a glass knife made either by a knife-maker or, less frequently, by hand. Metal knives will not produce suitable sections of plastic blocks. The sections are floated into a water-filled trough constructed on the glass knife from electrical tape or fashioned from metal. One or more of the sections, which vary between 0.5 and 2.0 μm in thickness, are picked up by a shaved applicator stick and placed on a large drop of water on a clear glass slide. The sections are slowly heated on a 60 C hot plate to remove wrinkles, aided by gentle bursts of air from the mouth or from a spray can. The sections are then fixed to the slide by gentle heating. Staining is carried out on a low-temperature hot plate with a limited number of stains. Although toluidine blue is probably the most frequently used stain applied to plastic-embedded tissue,[2] we often employ the Paragon stain or its recent replacements because it differentiates acidophilic and basophilic structures in a manner similar to that of the hematoxylin and eosin stains.[3] The Paragon stain is no longer marketed, but companies that manufacture products for electron microscopy sell stains similar to Paragon. By following their instructions an acceptable stain can be produced for most tissues.

The procedure for staining tissues embedded in epoxy resins, and destined for electron microscopy, has its limitations. For example, there are difficulties in trying to apply many biologic stains to sections from these tissue blocks.[4–8] If a variety of

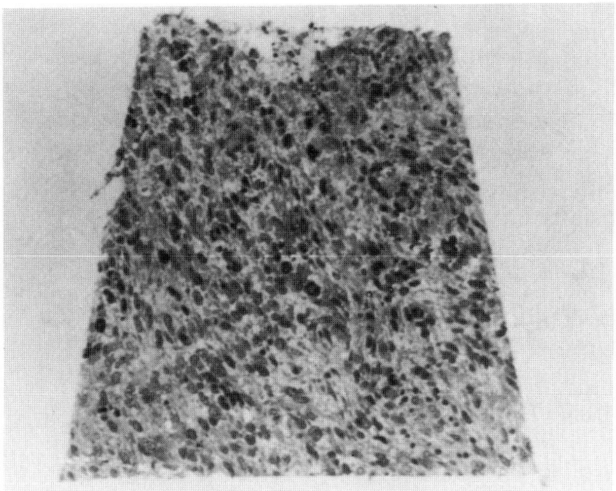

FIG. 3-2. Truncated pyramid of plastic block as visualized under low power in the light microscope (Epon-embedded tissue).

staining procedures are desired for light microscopy, another medium, such as methacrylate, must be employed. Alternately, the embedding resins may be removed, but the methods are often destructive, impeding subsequent ultrastructural studies, or are too inconvenient and time consuming to be acceptable as a routine procedure.[7,8]

The stained sections are cover-slipped with a drop of commercial mounting media or by using catalyzed unpolymerized epoxy resins prepared as the embedding medium. The permanence of the stains and the brilliance of the sections prepared by this technique makes it the method of choice for preserving stained sections.[9,10]

Semithin sections prepared by these techniques for light microscopy will provide a preview of what can be expected in the thin sections. They will serve as a bridge if paraffin sections are available and they can yield added diagnostic information not available in the usual paraffin sections.

VARIATIONS OF THE TRADITIONAL PREPARATION OF SEMITHIN SECTIONS

Although the superiority of plastic-embedded sections for viewing under the light microscope has been recognized for years,[11–13] the usual approach to processing tissues, especially for diagnosis, using two separate pathways is still followed in most laboratories. However, there are variations of the traditional techniques of preparation, which give both the light and electron microscopist options previously unavailable. The principal limitations that precluded the blending of light and electron microscopy techniques were the use of different fixatives; the small block size, which made sampling a tremendous problem, especially in diagnostic pathology; and the inability to stain epoxy resins with a variety of stains.

Tissue destined for light microscopy is usually fixed in formalin, while tissue to be viewed under the electron microscope is traditionally fixed in glutaraldehyde. Fixatives that give the microscopist the option of doing either light or electron microscopy have been available for the last few years. Carson et al described a fixative that may be routinely employed as the fixative for all surgical, autopsy, or animal tissue.[14] McDowell and Trump have also developed an all-purpopse fixative that gives equally good results at either the light or ultrastructural level.[15] The small size of tissue in plastic-embedded blocks has always been a major problem in integrating the dual system. The fixatives described above [14,15] allow larger samples to be fixed, and techniques are now available to embed specimens comparable in size to these which are embedded in paraffin (Fig. 3-3). Microtomes such as the JB-4, and the availability of large knives, now make it possible to routinely section at 0.5 to 2.0 μm a variety of sizes of tissues. By utilizing larger blocks and different embedding media, the following options become available to the microscopist.

Processing for Light and Potential Electron Microscopy

Specimens can be fixed in one fixative, embedded in Epon-araldite, surveyed by light microscopy, and, if so desired, have an area of the same block sectioned for electron microscopy.[16] For example, renal biopsies can be embedded directly in plastic and the entire specimen viewed under the light microscope (Fig. 3-4). Areas with glomeruli are then selected for ultrastructural study and that portion of the block processed for electron microscopy. A further advantage of this technique is that a small needle biopsy specimen need only be divided longitudinally, with one portion set aside for combined light and electron microscopy and the other for immunofluorescence, instead of the usual three divisions. The main disadvantage of this technique is that special stains are limited by the osmication and the epoxy resin. However, the

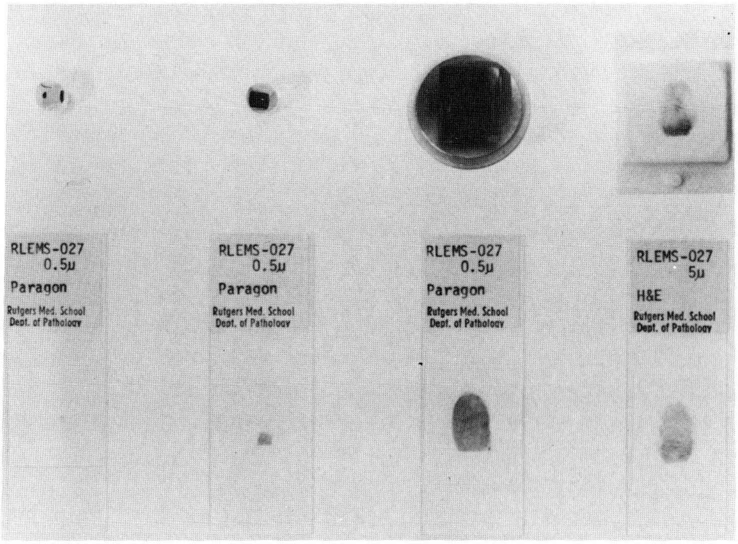

FIG. 3-3. (Left to right) BEEM capsule, inverted BEEM capsule, large block (all plastic blocks), paraffin block. Slides with sections from blocks are below the blocks.

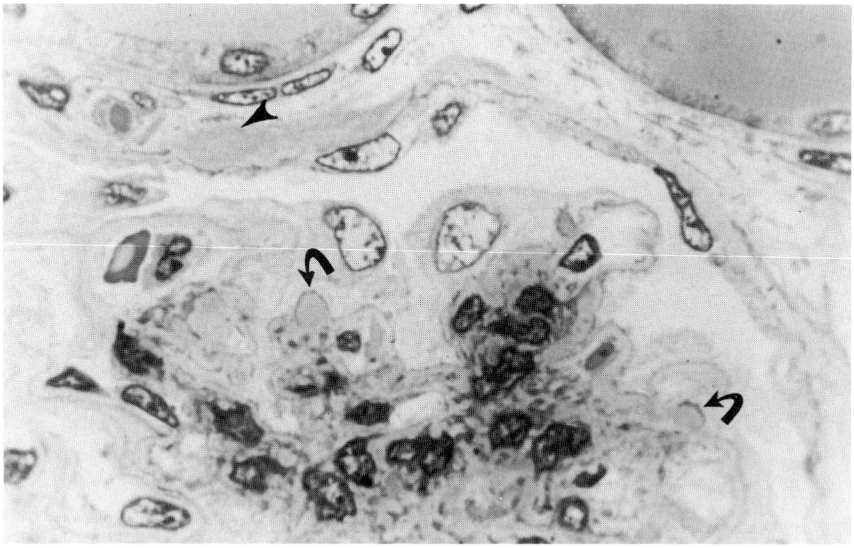

FIG. 3-4. Humps (dark arrows) and similar material in Bowman's capsule (arrow head) from renal biopsy specimen of a case of acute glomerulonephritis.

need for special stains is debatable when one is able to view a biopsy specimen at 0.5 to 1 μm by light microscopy and do subsequent electron microscopy. In addition to renal biopsy specimens, this technique is most suitable for liver specimens, hematopathology specimens, and a variety of tumors.

Light Microscopy of Large Plastic Sections

The goal of some microscopists is to obtain superior light microscopy and still have available a variety of special stains. If subsequent electron microscopy is not desired, either prospectively or retrospectively, then the techniques for processing large blocks can be applied to obtain semithin sections, with special stains, to be viewed under a light microscope. Over the years, plastic, especially methacrylate, has been recognized as being far superior to paraffin as an embedding medium.[4,11,12,17,18] Finer features could be resolved, because there is less overlapping of structures, and significantly lower distortion. And because of the chemical polymerization of plastic, there is much less shrinkage of tissue, which remains firmly embedded and bonded to the plastic block. Methacrylates were originally used by electron microscopists, but most tissue now destined for ultrastructural study is embedded in epoxy resins, since they are harder and more stable under the electron beam. A number of methacrylates have been employed over the years, but glycol methacrylate is now probably the most frequently used plastic for embedding tissues for light microscopy. The subject has recently been reviewed in depth by Bennett et al.[19] The hydrophilic properties of the polymer and the water solubility of glycol methacrylate permit easy dehydration, and a wide range of simple, quick, reliable staining procedures is possible (Fig. 3-5).

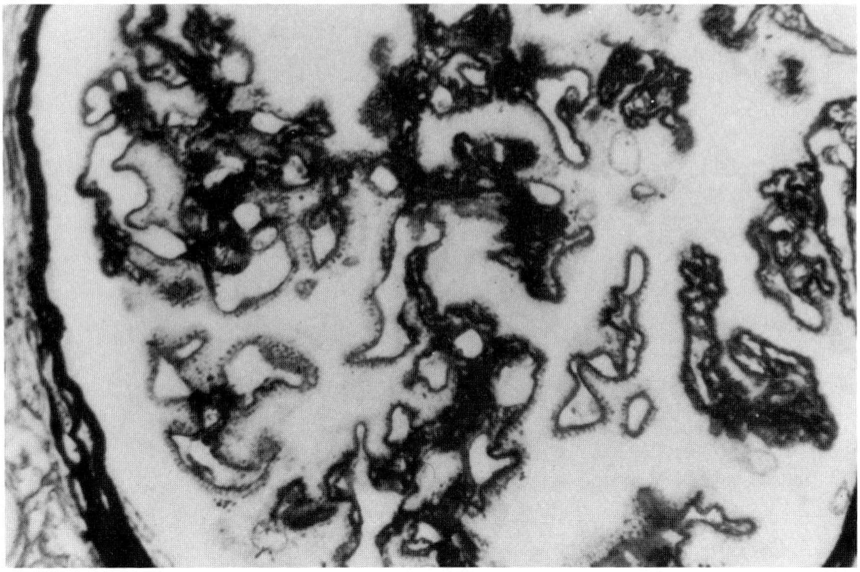

FIG. 3-5. "Spikes" are easily seen in a portion of a glomerulus from a case of idiopathic membraneous glomerulonephritis (methacrylate-embedded autopsy specimen stained with methenamine silver).

Reembedding of Paraffin Embedded Tissues into Plastic

Occasionally, tissue that one wishes to view under the electron microscope is already embedded in paraffin. Reembedding of such tissue is now an accepted procedure in diagnostic pathology.[2,20] If a dual-purpose fixative, as described previously, is utilized, paraffin-embedded tissue can often yield valuable information if reembedded into plastic. The area of interest in the paraffin block is cut out with a scalpel or a single-edged razor blade. The paraffin is dissolved in xylene for a few hours or melted gently and slowly in an oven. Alternately, the entire block can be melted and the area of interest cut out. The selected tissue is hydrated through decreasing concentrations of alcohol and then processed by the usual procedures for electron microscopy. Semithin sections of the now plastic-embedded tissue should reveal the desired area that originally was seen in the H&E stained section, which then may be studied under the electron microscope (Figs. 3-6 and 3-7). If the semithin section does not contain the desired area, the procedure must be repeated.

"Pop-Off": Reembedding of Stained Paraffin Sections into Plastic

Sometimes a specific area in a stained paraffin-embedded section warrants further cytologic evaluation. Techniques have been devised for reembedding the section into plastic so as not to lose the precise area by reembedding the entire specimen into plastic.[21]

After removing the cover-slip, the slide is dipped in xylol. It is then placed for five

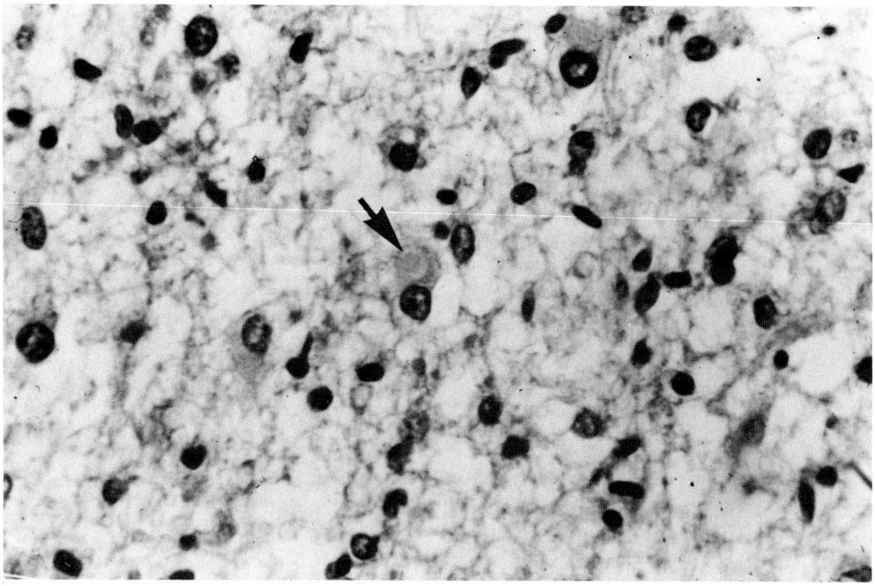

FIG. 3-6. Section of cerebrum showing "inclusion" in cytoplasm (arrow) of cell (section of paraffin-embedded tissue).

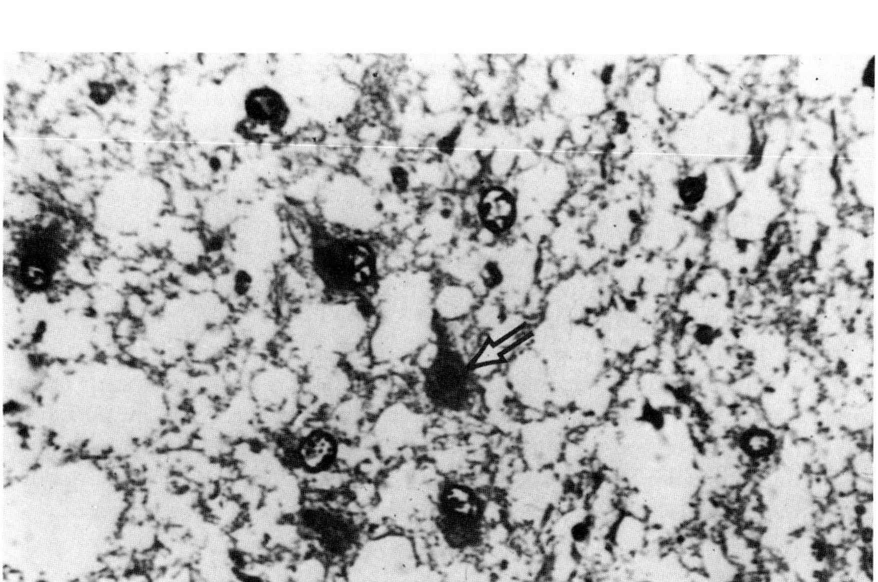

FIG. 3-7. Plastic section of area seen in Fig. 3-6 removed from paraffin block and reembedded into Epon. Similar inclusion is seen at arrow.

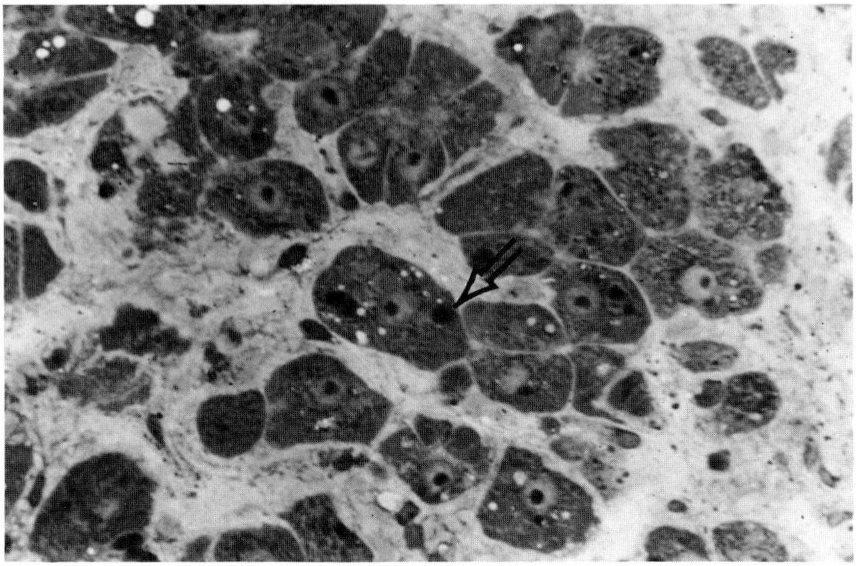

FIG. 3-11. Same specimen as in Fig. 3-10, with arrow pointing to similar structure (LM of Epon-embedded section).

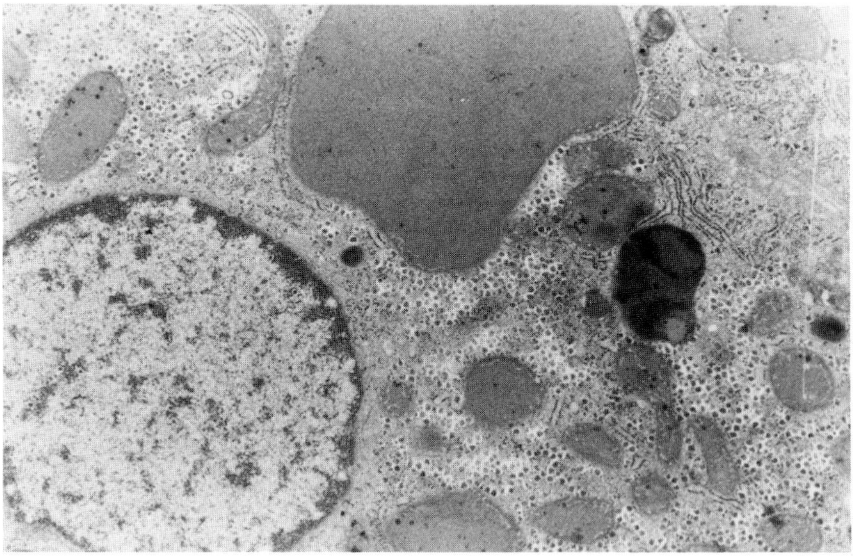

FIG. 3-12. Electron micrograph of area from Fig. 3-11, demonstrating that granules seen in Figs. 3-10 and 3-11 are giant mitochondria.

pathology. Their continued use for these purposes is both informative and instructional. What is even more exciting is the use of these sections for light microscopic evaluation, with the possibility of subsequent electron microscopy, if so desired.

Although usually not appreciated, the morphologic details present in a variety of tissues can readily become "visible" in plastic sections (Figs. 3-13 and 3-14). In renal tissue the immune deposits usually visible only in electron micrographs can be seen in semithin sections (Figs. 3-4 and 3-15). Cytoplasmic granules may be easily elucidated (Fig. 3-16). Bone marrow aspirates, which so often show artifacts, are crisp and easy to interpret. Bone need not be decalcified since undecalcified plastic-embedded bone can be sectioned with glass knives. Microorganisms are easily located and identified (Fig. 3-17). Liver biopsy specimens often show added details in plastic sections (Figs. 3-9 and 3-18). It will probably be in the area of tumors that the most value will be derived from the use of plastic-embedded tissue. Cells in lymphoid tissue can be examined in a single layer (Fig. 3-19), and cellular morphology is superior in most sections of tumors embedded in plastic (Fig. 3-20).

SUMMARY

Plastic-embedded tissue offers the microscopist a choice: specimens may be examined by electron microscopy only, by both light and electron microscopy, by light microscopy with the option of subsequent electron microscopy, or by light microscopy only. The age of plastic is here.

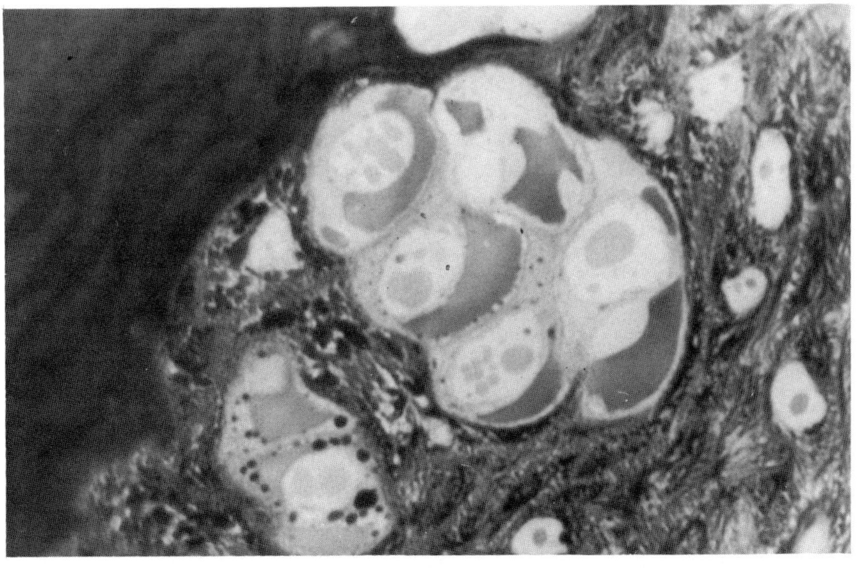

FIG. 3-13. Skin biopsy specimen from a case of verruca vulgaris. Note tonofilaments and intracellular bridges in right half of photo (methacrylate-embedded tissue).

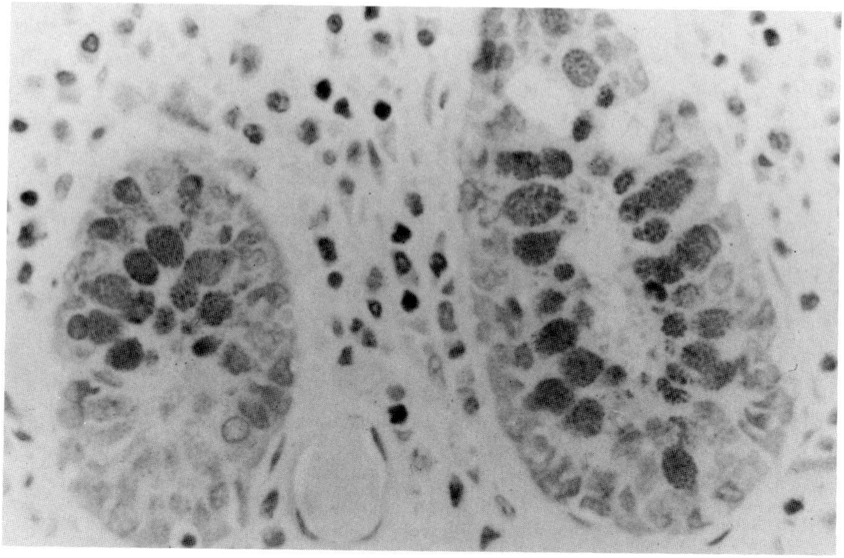

FIG. 3-14. Glands from ascending colon. Note granularity of mucous secretion.

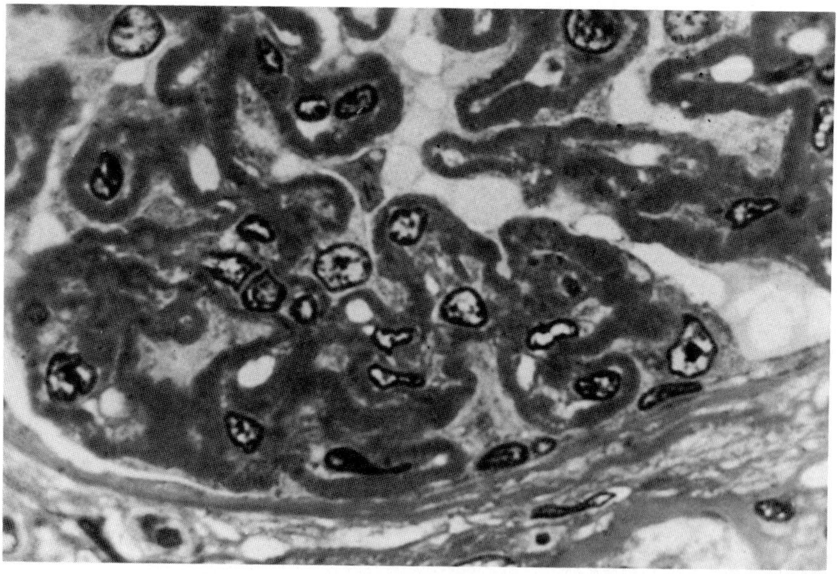

FIG. 3-15. Same as Fig. 3-5, showing dark immune deposits throughout capillary loops (Epon-araldite; Paragon stained).

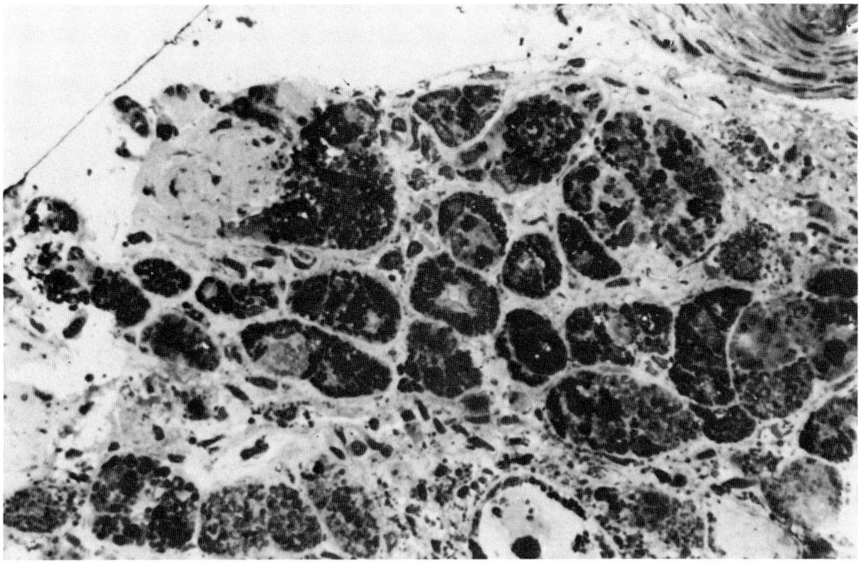

FIG. 3-16. Renal tubules from a patient with Fabry's disease. Tubules contain dark accumulations of storage product. Note edge of block in upper left corner (Epon; toluidine blue stain).

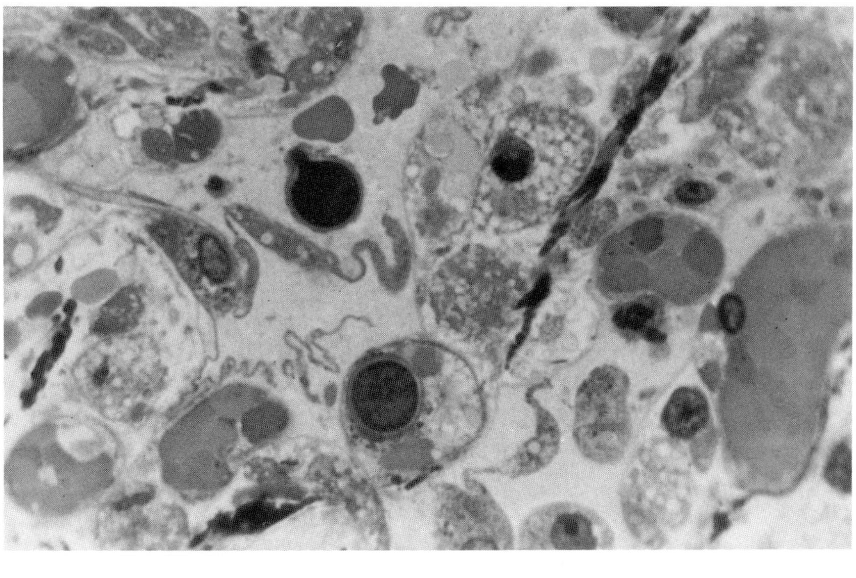

FIG. 3-17. Lung autopsy specimen from a patient with blastomycosis, with budding organism seen in alveolar space.

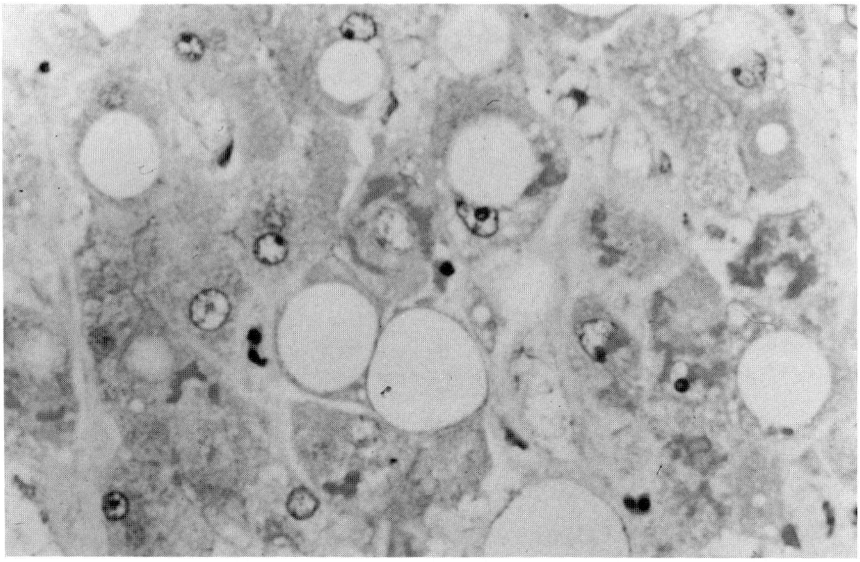

FIG. 3-18. Liver autopsy specimen from patient with acute alcoholic hepatitis (methacrylate-embedded specimen).

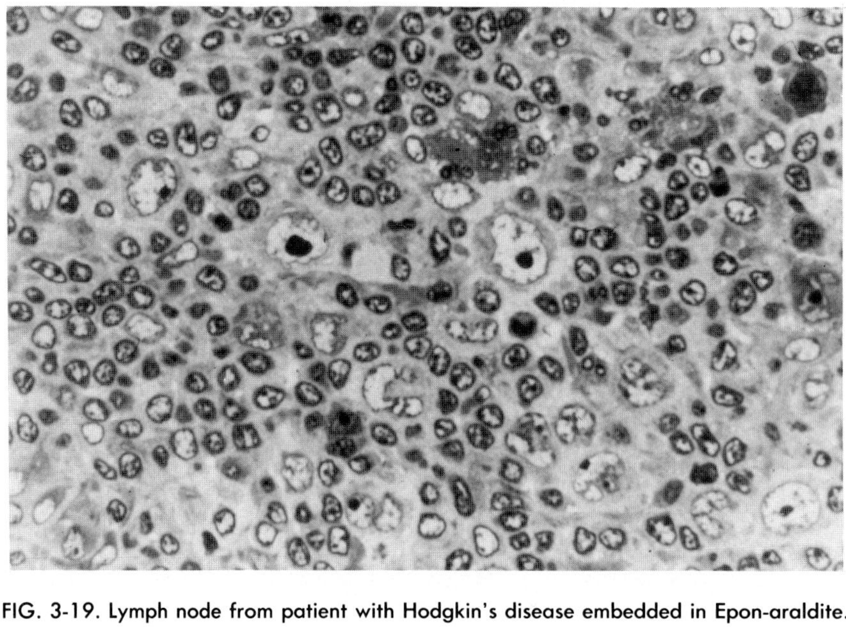

FIG. 3-19. Lymph node from patient with Hodgkin's disease embedded in Epon-araldite.

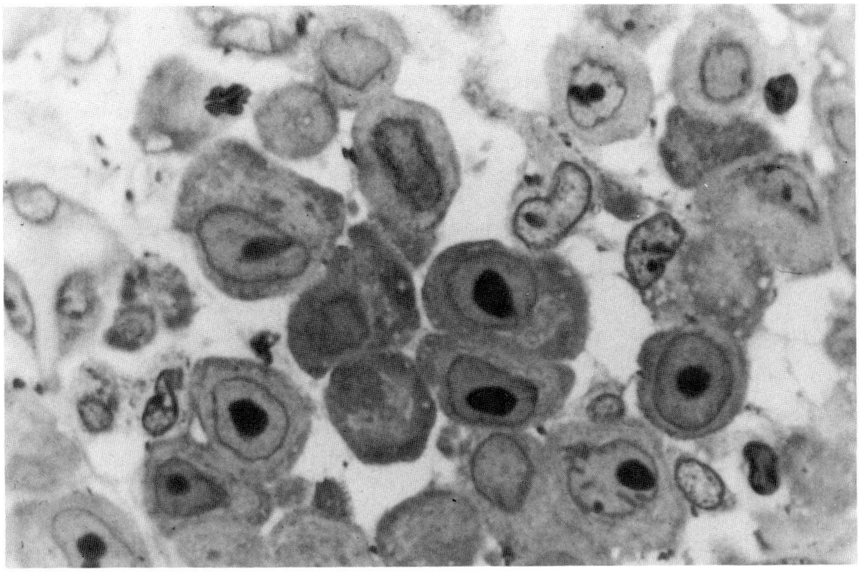

FIG. 3-20. Cells with prominent nucleoli, thought possibly to be melanocytes, from a case of adenocarcinoma of the breast (specimen embedded in Epon-araldite).

REFERENCES

1. **Lynn JA:** 'Adjacent' sections: A bridge in the gap between light and electron microscopy. Hum Pathol 6(4):400–402, 1975
2. **Burns WA:** Thick sections: Techniques and applications. In Trump B, Jones R (eds): Diagnostic Electron Miscroscopy. New York, Wiley, 1978, vol 1, pp 141–166
3. **Martin JH, Lynn JA, Nickey WM:** A rapid polychrome stain for epoxy-embedded tissue. Am J Clin Pathol 46:250–251, 1966
4. **Munger BL:** Staining methods applicable to sections of osmium-fixed tissue for light microscopy. J Biophys Biochem Cytol 11:502–506, 1961
5. **Grimley PM:** A tribasic stain for thin sections of plastic-embedded, osmium tetroxide-fixed tissues. Stain Technol 39:229–233, 1964
6. **Shires TK, Johnson M, Richter KM:** Hematoxylin staining of tissue embedded in epoxy resins. Stain Technol 44:21–25, 1969
7. **Snodgress AB, Dorsey CH, Bailey GW, et al:** Conventional histopathologic staining methods compatible with Epon-embedded, osmicated tissue. Lab Invest 26:329–337, 1972
8. **Chang SC:** Hematoxylin-eosin staining of plastic-embedded tissue sections. Arch Pathol 93:344–351, 1972
9. **Lynn JA, Martin JH, Race GJ:** Recent improvements of histologic technics for the combined light and electron microscopic examination of surgical specimens. Am J Clin Pathol 45:704–713, 1966
10. **Lynn JA, Martin JH, Matthews LJ:** Electron microscopy techniques. In Race GJ (ed): Laboratory Medicine. Baltimore, Harper & Row, 1973, pp 16–20
11. **Houck CE, Dempsey EW:** Cytological staining procedures applicable to methacrylate-embedded tissues. Stain Technol 29:207–211, 1954

12. **Jennings BM, Farquhar MG, Moon HD:** Staining methods for osmium-methacrylate sections. Am J Pathol 35:991–997, 1959
13. **Lane BP, Europa DL:** Differential staining of ultrathin sections of Epon-embedded tissues for light microscopy. J Histochem Cytochem 13:579–582, 1965
14. **Carson FL, Martin JH, Lynn JA:** Formalin fixation for electron microscopy: A re-evaluation. Am J Clin Pathol 59(3):365–373, 1973
15. **McDowell EM, Trump BF:** Histologic fixatives suitable for diagnostic light and electron microscopy. Arch Pathol Lab Med 100:405, 1976
16. **Burns WA, Bretschneider AM, Morrison AB:** Embedding in large plastic blocks—Diagnostic light and potential electron microscopy on the same block. Arch Pathol Lab Med 103:177–179, 1979
17. **Agodoa L, Striker G, Chi EY:** Glycol methacrylate-embedding of renal biopsy specimens for light microscopy. Am J Clin Pathol 64:655–660, 1975
18. **Chi EY, Smuckler EA:** A rapid method for processing liver biopsy specimens for 2 μ sectioning. Arch Pathol Lab Med 100:457–462, 1976
19. **Bennett HS, Wyrick AD, Lee SW, McNeil JH Jr:** Science and art in preparing tissues embedded in plastic for light microscopy with special reference to glycol methacrylate, glass knives and simple stains. Stain Technol 51:71–97, 1976
20. **Zeitoun P, Lehy T:** Utilization of paraffin-embedded material for electron microscopy. Lab Invest 23:52–57, 1970
21. **Takeda M:** Virus identification in cytologic and histologic material by electron microscopy. Acta Cytol (Baltimore) 13:206–209, 1969
22. **Bretschneider AM, Burns WA:** Ultrastructure of silver-stained, paraffin-embedded sections. 37th Annual Proceedings of the Electron Microscopy Society of America, 352–353, 1979

4

BASIC CELL STRUCTURE

BRUCE MACKAY

Much of diagnostic electron microscopy is the recognition and interpretation of departures from normal in the ultrastructure of cells and tissues. Knowledge of the basic fine structure of normal cells is thus a necessary prerequisite. Detailed accounts can be found in texts on histology and cell biology.

UNITS OF MEASUREMENT

Millimeters are hopelessly large units with which to measure structures seen with the electron microscope, and much smaller subdivisions must be used. A micrometer (μm), or micron, is one thousandth of a millimeter, and the diameter of a red blood cell is 8 μm. A nanometer (nm), or millimicron, is one thousandth of a micrometer, and the smallest cytoplasmic secretory granules are in the order of 100 nm. There are ten Angstrom units (A) in a nanometer, and an actin microfilament is approximately 60 A in thickness.

MEMBRANES

Much of the substance of cells seen with the electron microscope consists of membranes. The cytoplasm and nucleus are limited by membranes, and the cytoplasm is subdivided into many functional compartments by elaborate membrane systems. There has been much speculation on the molecular composition of biologic membranes, but a unit membrane appears to be a bimolecular leaflet of lipid, approximately 75 A thick, with interspersed proteins. It presents a trilaminar appearance in high magnification electron micrographs. Membranes are viewed as being in a semifluid state, such that protein molecules can move freely within them and project from the surface to form receptor sites with specific affinities. Some substances diffuse freely through a semipermeable membrane, while others require the intervention of transport molecules and the expenditure of energy. Energy is also required to maintain membranes in a state of biological equilibrium through gradients of ionic diffusion. Many enzymes are located in or on membranes.

The cell membrane, often called the plasmalemma, is not usually in contact with any of the cytoplasmic membrane systems, though in a striated muscle fiber, the inner of the two layers of the sarcolemma is continuous with tubules of the sarcoplasmic reticulum. Carbohydrate may be attached to the outer surface of the cell membrane, and when present in sufficient quantity, can be visualized as a thin surface layer (the glycocalyx), seen well on absorptive cells of the small intestine. A thicker layer of carbohydrate admixed with protein constitutes the basal lamina that clothes the basal aspect of epithelial cells and coats some mesenchymal cells.

NUCLEUS

Nuclei vary greatly in size and shape, and cells such as osteoclasts and foreign body giant cells have many, but their basic structure is the same. In the resting nucleus, the component threads of the chromosomes are mostly dispersed, but foci of activity appear more electron-dense because of increased coiling of the threads. These are termed areas of heterochromatin.

Nuclear RNA is concentrated in the nucleolus. Like the nucleus, the nucleolus varies in size and appearance, but characteristically it appears dense and well circumscribed. It consists of thread-like and amorphous components. Nucleoli are positioned at the nucleolar organizing sites of the chromosomes, where the gene segments for ribosomal RNA are located. They form the subunits of ribosomes, and move to the nuclear membrane to discharge them into the cytoplasm. Cells that are actively engaged in protein synthesis possess large and often multiple nucleoli, which are a common finding in reactive cells and malignant tumors.

The nucleus is surrounded by two unit membranes, and the space between them is the nuclear envelope. It is usually less than 20 nm, but the width varies (Fig. 4-1). The outer layer of the nuclear envelope is continuous with the endoplasmic reticulum and bears ribosomes on its cytoplasmic aspect. After cell division, the nuclear membrane is reconstituted from the reticulum. The two layers of the membrane fuse around small orifices, called the nuclear pores (Fig. 4-1), which can make up more than 20% of the nuclear surface and are presumed to be sites where material can pass from the nucleus to the cytoplasm. Sac-like reduplication of the nuclear membrane produces nuclear blebs, which are seen more frequently in hematopoietic cells than in other cells. More elaborate festooning of the nuclear membrane can extend out into the cytoplasm to form complex systems of tubules and cisternae (Fig. 4-1), often with nuclear pores and attached ribosomes; these annulate lamellae have been seen in a variety of pathologic processes, including tumors, and are not of diagnostic significance.

CYTOPLASM

Organelles and inclusions in the cytoplasm lie within the ground substance, which is an aggregation of molecules and fluid that undergoes frequent sol-gel transformations and varies in electron density with its degree of hydration. Many membrane systems, including those of the mitochondria and endoplasmic reticulum subdivide the cytoplasm into functional compartments.

Mitochondria are the energy-producing organelles. They occur in almost every type of cell and range in number up to more than 1000. Mitochondria contain DNA

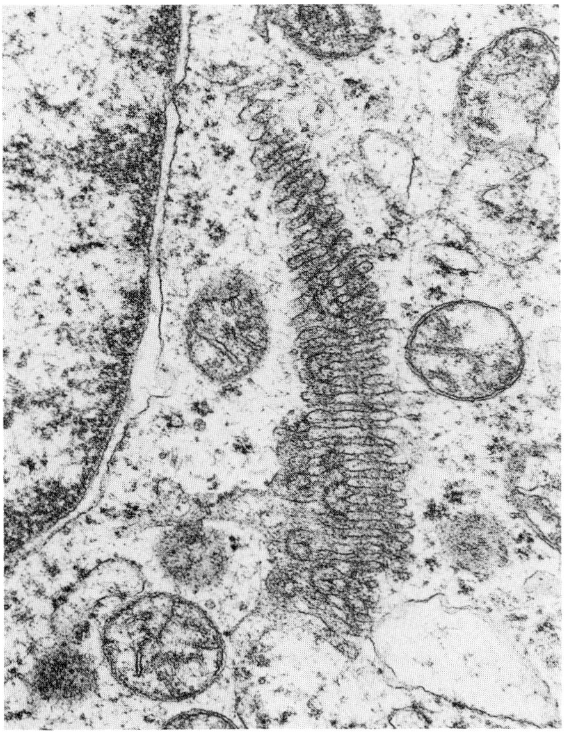

FIG. 4-1. Part of the nuclear envelope is visible. It varies slightly in width, and a nuclear pore can be seen. Annulate lamellae are produced by replication of the nuclear membrane. ×22,000.

and are self-replicating, and this in part accounts for variations in shape that can be seen among them. However, most are oval or elongated. A mitochondrion is composed of two concentric unit membranes, the inner of which forms cristae, a series of shelf-like projections into the interior or matrix. In some mitochondria, particularly those of cells engaged in steroid synthesis, cristae are tubular in shape. The density of the mitochondrial matrix varies. It often contains small electron-dense bodies thought to be sites of cation concentration. Larger densities occur in adrenocortical cells and their tumors (see Chapter 10), and irregular dense areas in the matrix are seen in cells under hypoxic conditions.

Some cells accumulate large numbers of mitochondria, although the reason is not well understood. They are seen in the thyroid gland (Hürthle cells), in the salivary glands (oncocytes), and in disorders of these and other organs (see Chapter 10). Mitochondria are susceptible to degenerative processes, which cause them to bleb and swell. This is a common finding in tissues whose fixation has been delayed or is inadequate.

The endoplasmic reticulum consists of an anastomosing system of sacs and tubules (cisternae), whose interior is effectively segregated from the ground substance of the cytoplasm. Much of the reticulum is studded on its outer aspect by ribosomes, and continuities between this rough or granular endoplasmic reticulum and the outer layer of the nuclear envelope have been mentioned. Many ribosomes also lie free within the cytoplasmic ground substance. Ribosomes are composed of RNA, and their function is to synthesize short chains of amino acids, which form the building blocks of protein molecules. Information on the amino acid sequences of the proteins to be formed by the ribosomes is conveyed from the nucleus by messenger RNA molecules, and using these blueprints the ribosomes organize amino acids from the cytoplasmic pool into matching chains.

A cell that forms proteins which will be passed outside the cell for use in some other location, or which will be retained within membrane-bound sacs in the cell as primary lysosomes, possesses considerable amounts of rough endoplasmic reticulum. Much of the cytoplasm of a plasma cell or an exocrine gland cell (Fig. 4-2) is occupied by flattened, parallel cisternae. Fibroblasts contain randomly oriented cisternae that distend when the cell is actively synthesizing components of the connective tissue matrix. Aggregates of electron-dense material may accumulate within the cisternae of lipid-forming cells (see Chapter 10). A cell that is producing proteins for its own use generally has sparse endoplasmic reticulum, but has many free ribosomes that group within the ground substance as polyribosomes. The transformed lymphocyte is an example.

In cells that synthesize proteins for export, such as those of the pancreatic acini, the peptide fragments manufactured by the ribosomes pass into the lumen of the cisternae where they cluster to form polypeptides. As the protein molecule is fabricated, it migrates to the Golgi complex, which is composed of a stack of flattened, bowl-shaped sacs of smooth-surfaced membranes close to the nucleus. In cells with cytoplasmic polarity, the Golgi complex lies to the apical side of the nucleus. Material accumulates in the Golgi complex, where it is concentrated, some carbohydrate may

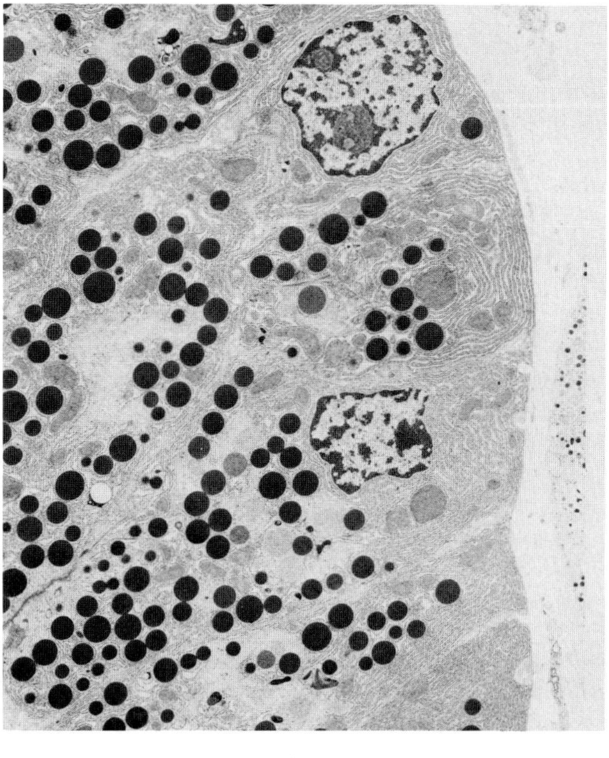

FIG. 4-2. Part of an exocrine acinus from the pancreas. The cells contain plentiful rough endoplasmic reticulum and many large dense secretory granules. ×3550.

be added, and the product is packaged by being invested with a unit membrane. The spherical secretory granules so formed appear as dense-core, membrane-limited bodies (Figs. 4-2 and 4-3). They migrate toward the apical surface of the cell and on receipt of an appropriate stimulus are discharged by a process of exocytosis. In polypeptide-forming endocrine cells, the granules are liberated from the surface closest to the nearest sinusoid. Misplaced exocytosis may be seen in certain pituitary tumors.

Secretory granules vary in size (Fig. 4-3), but they are characteristically spherical, with a uniformly electron-dense core. The range of diameters of granules in a particular type of cell is usually narrow, but their size can be influenced by the secretory activity of the cell. One that is rapidly synthesizing and secreting does not have time to form large granules or to accumulate many granules within its cytoplasm.

Most cells contain at least small quantities of smooth or agranular endoplasmic reticulum in addition to the membranes of the Golgi complex, but in cells engaged in steroid hormone formation, such as the interstitial cells of the testis, smooth reticulum may occupy much of the cytoplasm and the granular endoplasmic reticulum is sparse. Smooth reticulum can also be seen in liver cells, where it is believed to participate in the formation of large lipid molecules and to aid in detoxifying drugs. It is plentiful in muscle cells, where it transmits waves of depolarization from the surface to the myofibrils, and localizes pools of calcium. Distended tubules of the sarcoplasmic reticulum in heart muscle cells provide an indication of the development of cardiotoxicity in patients receiving anthracyclene drugs such as adriamycin (Fig. 4-4).

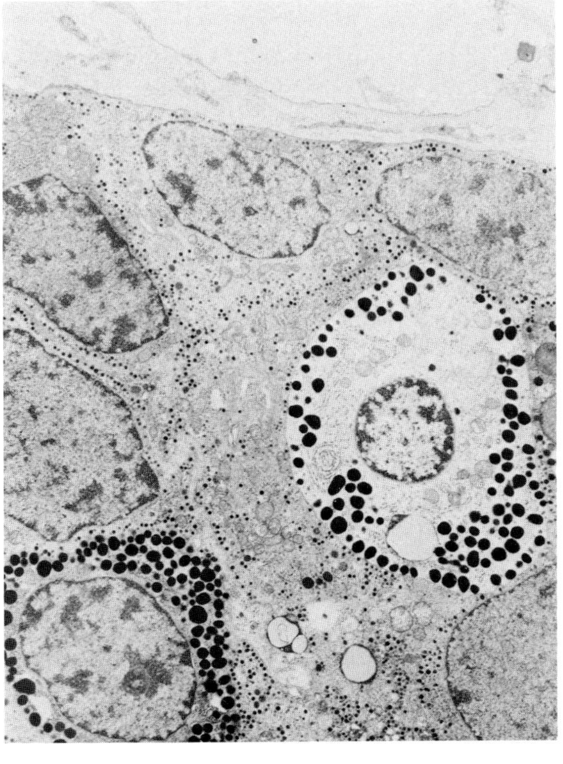

FIG. 4-3. Considerable variation in granule size can be seen among these anterior pituitary cells. ×3450.

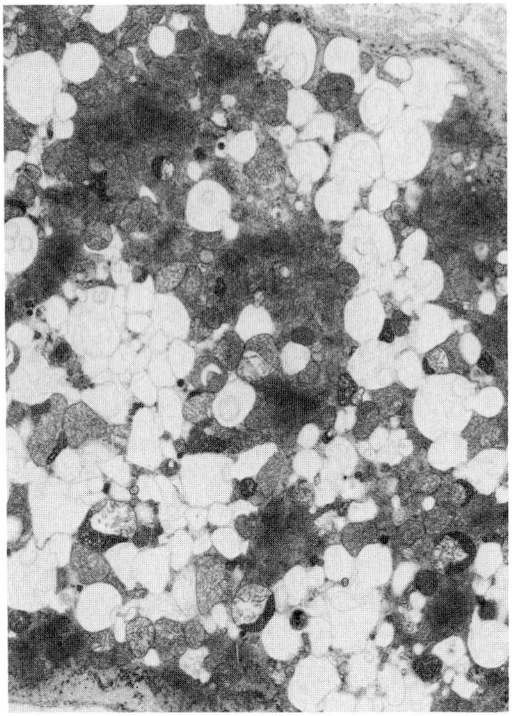

FIG. 4-4. The vacuoles in this heart muscle cell are produced by distension of tubules of the sarcoplasmic reticulum. The patient was receiving adriamycin. ×5000.

Lysosomes are cell organelles concerned with the digestion of unwanted material either brought into the cell by phagocytosis as in neutrophils and macrophages, or effete constituents of the cytoplasm such as worn-out mitochondria. Lysosomes are membrane-bound sacs of lytic enzymes synthesized by the ribosomes and packaged in the Golgi complex. A newly formed primary lysosome is a spherical electron-dense body that appears identical to a secretory granule. When a primary lysosome engulfs phagocytosed material, it becomes a secondary lysosome or phagosome, and is termed an autophagic vacuole when its contents are from within the cell, or a heterophagic vacuole when they have been phagocytosed. Sometimes lysosomes are unable to completely digest their contents, and the residual bodies may persist as dense aggregates in the cytoplasm.

The centrioles are small, paired, cylindrical organelles that have the ability to synthesize microtubules. They perform this function when a cell divides, and the microtubules that constitute the core of a cilium are formed by basal bodies derived from the centrioles. Microtubules are present in most cells as the internal framework or cytoskeleton, but they may not be clearly visualized in thin sections. Tonofilaments resemble actin microfilaments but are twice their diameter. They are seen to good advantage in squamous cells, where they insert on the desmosomes (Fig. 4-5).

Filaments are present in the cytoplasm of many cells. They are characteristically slender microfilaments, roughly 60 A thick, composed of the contractile protein actin, and their presence presumably confers some motility on the cell. Actin myofilaments form the contraction ring that constricts the cytoplasm of a dividing

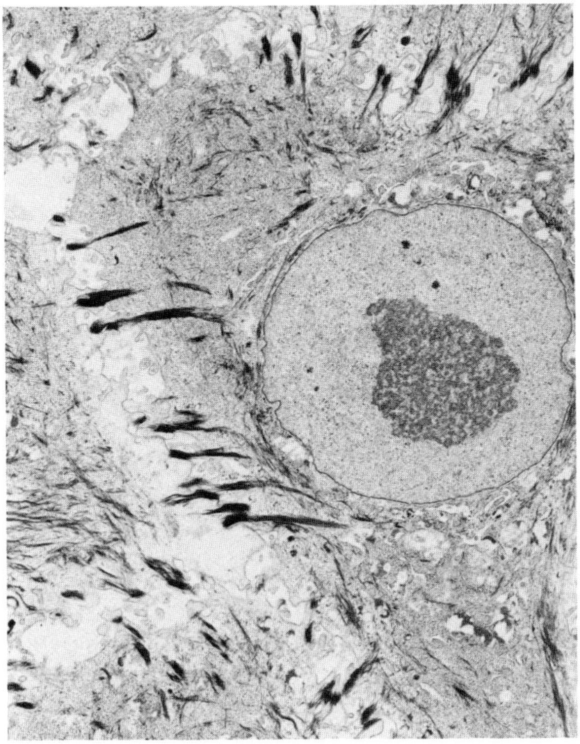

FIG. 4-5. Squamous epithelial cells united by desmosomes with tonofilament bundles. ×4100.

cell, and in the columnar absorptive cells of the small intestine a band of microfilaments occupies the apical cytoplasm and extends up into the microvilli. The myofilaments of muscle cells are more complex admixtures of actin with myosin and other proteins.

Glucose may be stored in cells in the form of glycogen. It is plentiful in some cells, e.g., hepatocytes. Its ultrastructural appearance depends on the mode of preparation of the tissue. It can be seen as dense granules (Fig. 4-6) or electron-lucent zones (see Chapter 10). Lipid droplets are generally spherical, but they vary in size and electron density. Lipid tends to accumulate in hypoxic cells. Mucin appears as clear zones in the cytoplasm (Fig. 4-7), often exhibiting dense centers that can simulate secretory granules at low magnification. As mucin accumulates, it becomes more diffuse. In goblet cells (Fig. 4-7), it displaces the organelles to the base of the cell as it balloons the apical cytoplasm.

CELL SURFACE

Each cell is enclosed by its cell or plasma membrane, which defines the profile of the cell and can be readily visualized with the electron microscope. Some cells have smooth profiles, while others show varying degrees of irregularity that may be designed to increase the cell surface and facilitate passage of material through the cell

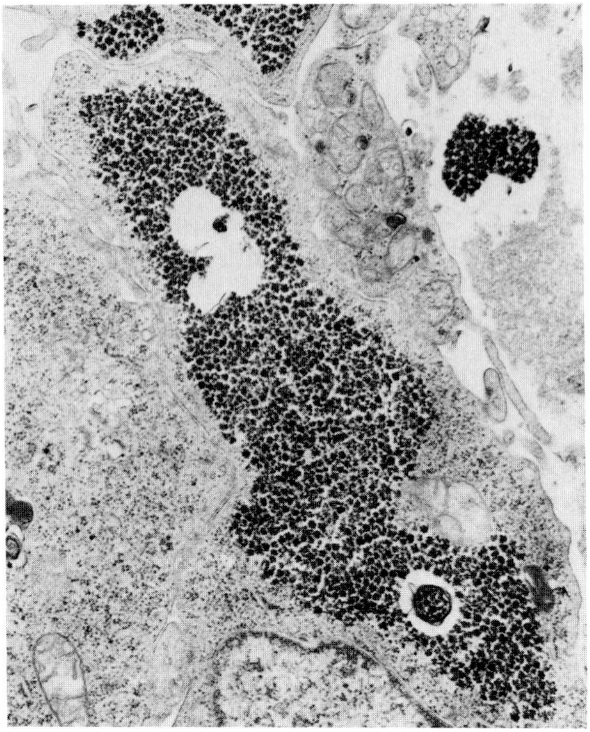

FIG. 4-6. Aggregates of cytoplasmic glycogen appear as electron-dense granules in this preparation. ×7800.

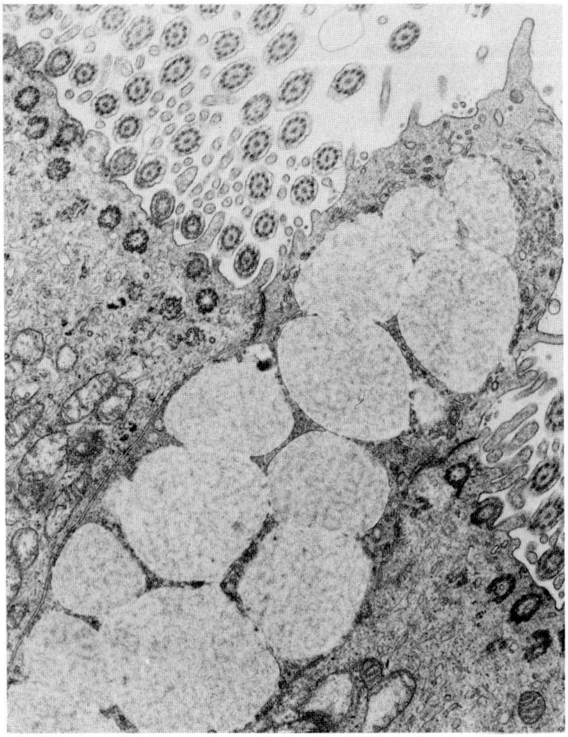

FIG. 4-7. Ciliated cells surround a mucin-filled goblet cell in the respiratory mucosa. ×12,000.

membrane. In the renal tubules, for example, the basal aspect of the cell is thrown into a series of deep folds, and long mitochondria lie alongside the folds, providing energy for the active transport of material across the membrane. Closely apposed cell membranes may interdigitate, as in the epithelial cells of some glands. More organized projections of the cell surface are the microvilli, cilia, and cytoplasmic processes.

Microvilli are short finger-like projections that increase the surface area of a cell. They are readily visible on the absorptive cells of the small intestine, where their regularity simulates a picket fence (Fig. 4-8). The microvilli of the gut have an internal core of microfilaments that extends upward from the terminal web. Long, slender microvilli devoid of microfilament cores are seen on mesothelial cells. Irregular projections of the cell cytoplasm that simulate microvilli are a common finding on cells in some pathologic tissues.

Cilia are longer, thicker, and better organized than microvilli, and are actively motile. Their synchronous contraction sweeps material along the surface of a mucosa, such as the lining of the respiratory passages. A cilium contains an orderly arrangement of microtubules (Fig. 4-6) produced by its basal body. A tapering rootlet of contractile filaments extends downward from the basal body (Fig. 4-9).

A cytoplasmic process is much longer and thicker than a microvillus or cilium. Processes are common in neural cells and their tumors (see Chapter 10). The branching cytoplasmic arms of the glomerular podocyte are particularly elaborate (see Chapter 5, Fig. 5-3).

The basal lamina that separates epithelial cells from the underlying connective

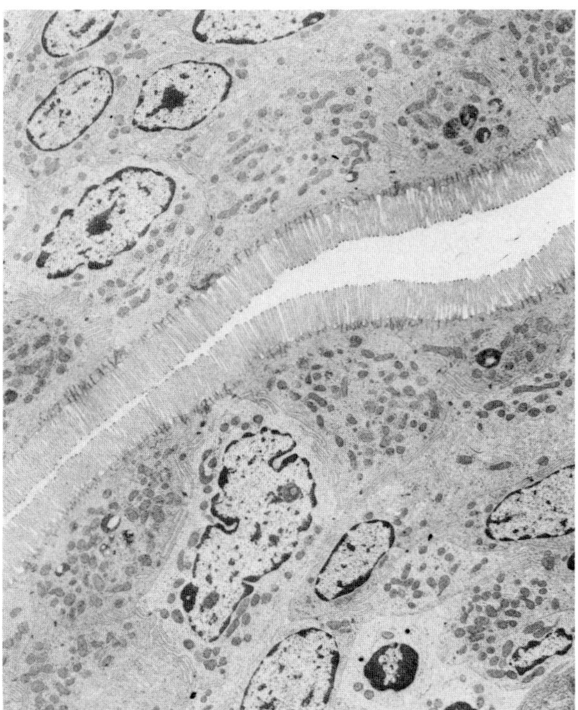

FIG. 4-8. Arrays of regular microvilli on small intestinal epithelial cells. ×3250.

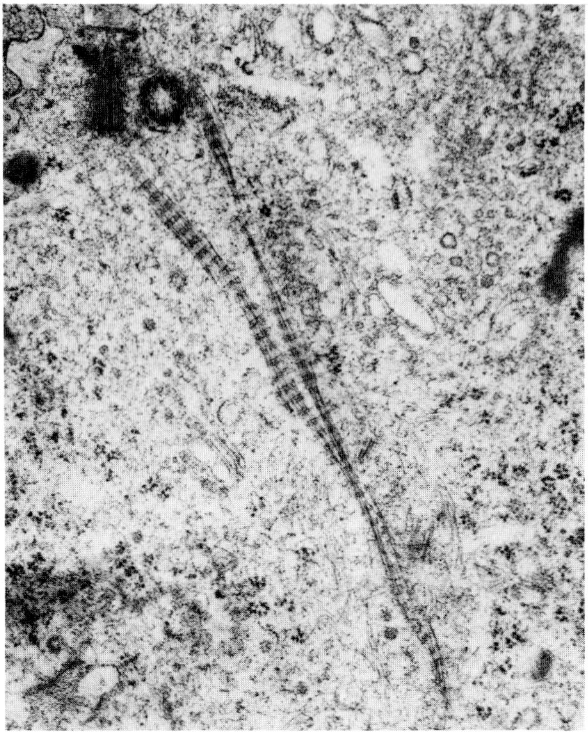

FIG. 4-9. Example of ciliary rootlets. ×25,000.

tissue consists of carbohydrate with some proteins, and is apparently formed by the epithelial cell. A well-defined basal lamina is present between epithelial and endothelial layers of the renal glomerulus, and in the pulmonary alveoli it is interposed between pneumocytes and the endothelial cells of septal capillaries.

Cells are held together in tissues by various factors, including electrical forces and cement substances such as sialic acid. Where greater cohesion is required, specialized cell junctions adapted for the purpose are formed. The desmosome that unites squamous (Fig. 4-5) and other epithelial cells is formed by densities of the apposed cell membranes with an intervening intermediate line, and tonofilaments of the cytoskeleton attach to the inner aspects of the plasmalemmal densities. The cell membranes of adjacent intestinal epithelial cells are united at their luminal borders by tight junctions, by which the membranes fuse to seal off the intercellular space from the lumen. Below the tight junction, the membranes form an attachment (the zonula adherens), into which microfilaments of the terminal web insert. These contact sites, together with an adjacent desmosome, constitute a junctional complex. Modified cell junctions are formed in certain specialized tissues.

STROMA

The collagen of connective tissue is formed by fibroblasts. Precursor molecules (tropocollagen), synthesized within the cytoplasm, aggregate outside the cell to form collagen fibrils with their characteristic 640-A periodicity. Collagen is capable of

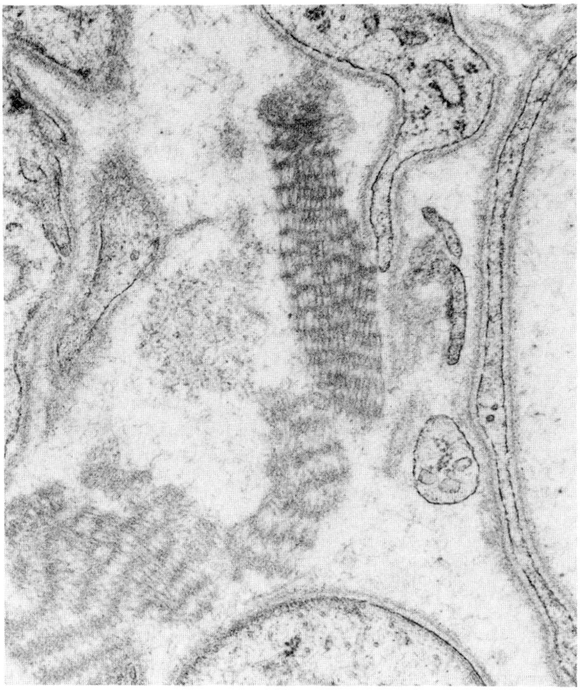

FIG. 4-10. Segment-long-spacing collagen from a schwannoma. ×21,-000.

dissociation under certain environmental conditions, and may recombine in a different morphologic form. An example is the segment-long-spacing collagen that is common in the stroma of some neurogenic neoplasms (Fig. 4-10).

Elastin is a contractile protein with many similarities in its molecular composition to collagen. It appears in electron micrographs as irregular aggregates of homogeneous-appearing material with interspersed denser zones. At high magnification it has a fine filamentous composition without periodicity.

CELL SPECIALIZATION

As cells specialize for specific functions, they undergo modifications in their morphology that can be extensive, as in skeletal muscle cells and neurons. A detailed description of the ultrastructure of these and other specialized cells would be beyond the scope of this book, and reference should be made to a comprehensive histology textbook.

ADDITIONAL READINGS

1. **Lentz TL:** Cell Fine Structure. An Atlas of Drawings of Whole-Cell Structure. Philadelphia, WB Saunders, 1971
2. **Weiss, L, Greep RO:** Histology, 4th ed. New York, McGraw-Hill, 1977

5

THE USE OF ELECTRON MICROSCOPY IN KIDNEY BIOPSY INTERPRETATION

NELSON G. ORDÓÑEZ

Only after the introduction of electron microscopy to the study of the kidney could the interrelationship of the components of the glomerular capillary wall and the existence of the mesangium be established.[1,2] By enabling the pathologist to recognize changes often not seen under the light microscope, electron microscopy has not only contributed significantly to the understanding of the pathogenesis of renal diseases, but also has proven to be of definitive diagnostic value.

The aim of this chapter is not to provide a comprehensive review of all of the diseases that can affect the kidney, but to emphasize those conditions in which electron microscopy provides morphologic information important for diagnosis. It must be understood that electron microscopy has not replaced the light microscope in the evaluation of renal biopsy specimens. On the contrary, a correct interpretation of renal biopsy specimens requires the correlation of light, electron, and immunofluorescence microscopic information with the patient's clinical findings. Since glomerular diseases are the usual cause of symptoms leading to renal biopsy, and the primary practical value of electron microscopy has been the diagnosis of those conditions that affect the glomerulus, this chapter will be devoted almost exclusively to the pathology of the glomerulus.

NORMAL ULTRASTRUCTURE OF THE GLOMERULUS

The glomerulus is a vascular structure composed of capillary loops arranged in lobules and surrounded by an epithelial-lined basement membrane, the Bowman's capsule. The capillary wall consists of a basement membrane lined on the outer aspect, or urinary side, by visceral epithelial cells, and on the inner aspect, or capillary side, by attenuated fenestrated endothelium (Fig. 5-1). The glomerular basement membrane is a trilaminar structure composed of a wide central electron-dense zone, or lamina densa, bordered by two narrow electron-lucent layers, the lamina rara interna and the lamina rara externa. The main function of the glomerulus is filtration; and the basement membrane is the main component of the filtration barrier.[3,4] In normal adult individuals, the basement membrane measures 3400 A in thickness;[5] it is somewhat thinner in children. The basement membrane does not totally encompass the capillary lumen. The lumen is centrally bordered by the mesangium, which

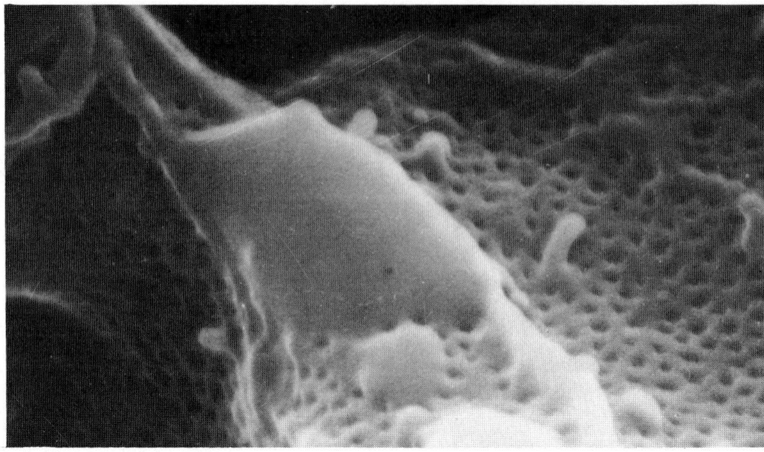

FIG. 5-1. Scanning electron micrograph of a capillary loop demonstrating the fenestrated character of the endothelium. ×20,000.

consists of mesangial cells embedded in a basement membrane-like material, the mesangial matrix (Fig. 5-2). The mesangial cells are phagocytic and contain contractile elements in their cytoplasm that may be involved in the regulation of intraglomerular blood flow.[6–9] Although some ultrastructural differences exist between the endothelial and mesangial cells, these differences usually are not evident in glo-

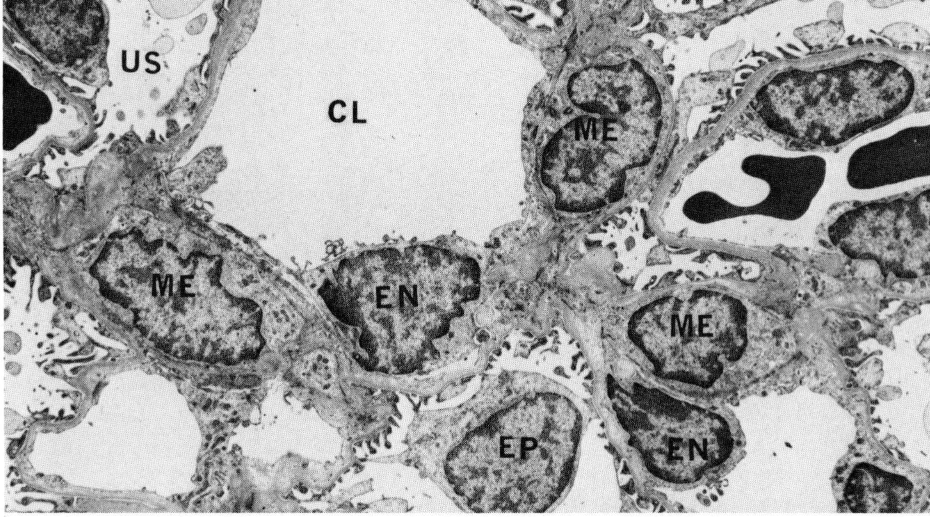

FIG. 5-2. Electron micrograph showing the normal relationship of the different components of the glomerulus: visceral epithelial cell (EP), endothelial cell (EN), mesangium (ME), urinary space (US), and capillary lumen (CL). ×5400.

merular diseases, and differentiation between endothelium and mesangium is usually based on the location of these structures in the glomerulus. Although the mesangial matrix resembles the basement membrane, it differs from the latter by its antigenic characteristics and its resistance to certain toxins, such as habu snake poison which dissolves the mesangial matrix but does not affect the basement membrane.[8,10]

The visceral epithelial cells are involved in basement membrane synthesis and may play an important role in glomerular permeability.[4,11,12] On its urinary side, these cells cover the glomerular basement membrane and attach to the basement membrane by cytoplasmic extensions or foot processes that, under the scanning electron microscope, appear as a complex of interdigitating structures that originate from different epithelial cells (Fig. 5-3). The foot processes are situated approximately 400 A apart, with a connecting slit-pore diaphragm 70 A thick (Fig. 5-4).

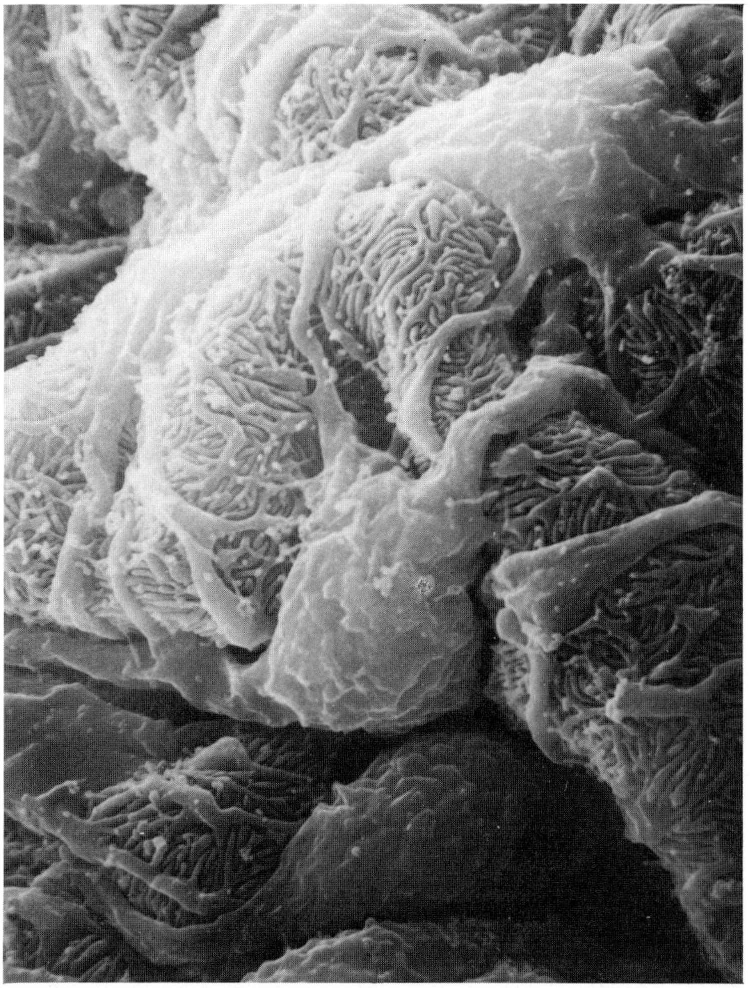

FIG. 5-3. Scanning electron micrograph of a loop covered by epithelial cells with interdigitating foot processes. ×2500.

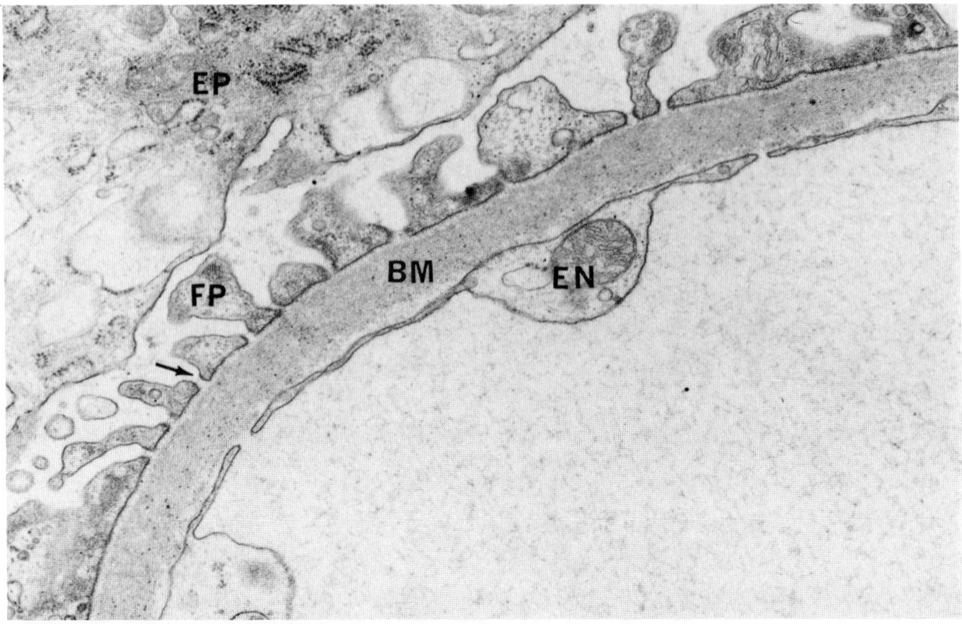

FIG. 5-4. Higher magnification of a normal glomerular loop showing endothelium (EN), basement membrane (BM), and epithelial cell (EP). Slit membrane is evident (arrow) between foot processes (FP). ×19,200.

Bowman's capsule limits the urinary space and is lined by a layer of flattened parietal epithelial cells that become continuous with the epithelium of the proximal convoluted tubule at the glomerular urinary pole, and with the visceral epithelial cells of the glomerular tuft at the vascular pole.

PRIMARY GLOMERULAR DISEASES

Epithelial Cell Disease

Epithelial cell disease accounts for approximately 60 to 80% of cases of idiopathic nephrotic syndrome in childhood and for 20 to 30% of cases in adults.[13–19] Typically, the patients are young children with highly selective proteinuria, which, in the majority of cases, responds to steroids or immunosuppressive therapy.[13,18] There is a clear male predominance of approximately 2.5:1 in children, but no sex difference is seen in adults.[13,14,19] On light microscopy the most prominent changes are seen in the convoluted tubules where large amounts of lipids and protein transport droplets accumulate in the cell cytoplasm. In contrast, the glomeruli are essentially normal. Because of these light microscopic features and the finding of lipoid particles in the urine of these patients, Munk, in 1913, proposed the term *lipoid nephrosis* for this condition.[20] Other common synonyms are *minimal change disease* and *nil lesion,* both referring to the lack of significant morphologic alteration on light microscopy. Only after ultrastructural studies were done on such cases was it established that the glomeruli, rather than the tubules, were the primary site of pathology.

Electron microscopy shows extensive loss of epithelial foot processes, and the epithelial cell cytoplasm appears directly applied to the glomerular basement membrane (Fig. 5-5). Although these changes were originally interpreted as "fusion" of foot processes, scanning electron microscopic studies have demonstrated that the loss of foot processes is due to extensive cell swelling and retraction of the foot processes into the parental cell bodies, and not to actual fusion.[21] In addition to the loss of foot processes, the epithelial cell cytoplasm appears hyperactive, with an increase in cytoplasmic organelles, and frequently cysts and numerous microvilli. Although there is increased electron density of the epithelial cell cytoplasm adjacent to the basement membrane, no immune complex deposits are seen. All the changes are reversible, and biopsy specimens taken during remission cannot be distinguished from normal. Foot process obliteration is not restricted to epithelial cell disease and can be found in other conditions associated with heavy proteinuria or the nephrotic syndrome. The endothelium is usually normal, but focal swelling can be present. The capillary lumina may contain small fibrin thrombi or aggregated platelets.[22] The mesangium usually shows little or no change, but mild increase in mesangial cells and matrix may be found.[23]

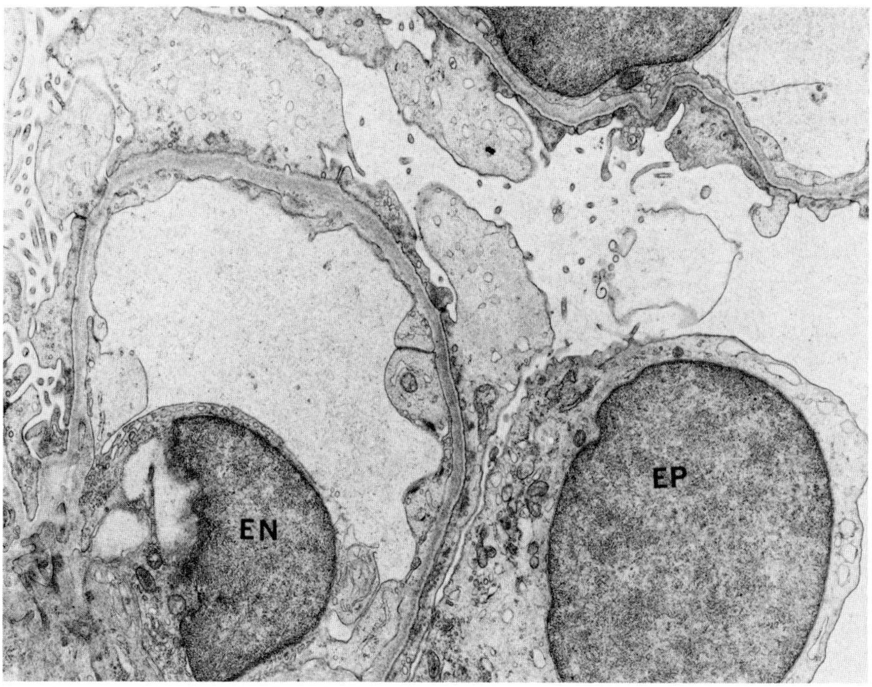

FIG. 5-5. Portion of a glomerulus from a patient with epithelial cell disease. The foot processes are obliterated and numerous microvilli project into the urinary space. Epithelial cell (EP), endothelium (EN). ×9400.

Focal Segmental Glomerulosclerosis

In a 1957 postmortem study of 20 children with the idiopathic nephrotic syndrome, Rich described a progressive form of focal sclerosis that starts as hyaline thickening of the peripheral segments of the tufts in the juxtamedullary glomeruli and later extends to the cortical glomeruli.[24] Subsequent studies demonstrated the association of this lesion with a steroid-resistant nephrotic syndrome, and frequent progression to end-stage renal disease.[25–27] Focal glomerulosclerosis is found in 10% of children and 15% of adults with the nephrotic syndrome.[28] The segmental sclerosis usually involves one or more lobules of the glomerular tuft near the axial region. The segmental sclerosis is frequently associated with localized adhesions to Bowman's capsule. These areas show increased mesangial matrix and mild mesangial hypercellularity; only when the sclerosis is advanced do these areas become hypocellular. The loops in the sclerosed areas are distorted and contain hyaline deposits and, not infrequently, lipid-laden cells. Ultrastructural examination reveals extensive foot process obliteration involving both segmentally sclerosed and nonsclerosed glomeruli. There is villous hyperplasia, microcyst formation, and an increase in lysosomes and rough endoplasmic reticulum in epithelial cells (Fig. 5-6). A variable degree of increase in mesangial matrix is usually present in all glomeruli, and a mild increase in the number of mesangial cells is not an unusual finding; a prominent increase may be associated with rapid progression.[29] The basement membrane is wrinkled, thickened, or folded. The endothelial cells may have cytoplasmic fat vacuoles (Fig. 5-7). Subendothelial, paramesangial, homogeneous deposits, frequently containing myeloid bodies or lipid droplets, are present. Occasionally, there are areas of detachment of the epithelial lining. These spaces can be quite wide and can accumulate multilayered basement membrane-like material and cellular debris.[30] Immunofluorescence shows segmental deposits containing IgM and C3.[28,31] Treatment failure occurs in over 80% of the patients.[28,31,32] Recurrence of the nephrotic syndrome soon after transplantation has been reported on several occasions.[33,34] The question of whether focal glomerulosclerosis is a separate entity or a progression of the minimal change disease has not yet been resolved.

Membranous Nephropathy

Membranous nephropathy is a histopathologic entity characterized by epimembranous and/or intramembranous immune complex deposits and variable basement membrane thickening without mesangial proliferation. Most cases are idiopathic, but an identical glomerular lesion can occur in association with a wide variety of conditions such as drug reactions, infections, systemic lupus erythematosus, and tumors.[35–39] Therefore, the finding of this lesion in a renal biopsy specimen is not an end-point in diagnosis, but rather a stimulus to search for an associated condition. This is particularly important because in some of these conditions the lesion is reversible or may have different prognostic and therapeutic implications.

Idiopathic membranous nephropathy is twice as more common in males and usually occurs in middle-aged adults.[18,38–43] The nephrotic syndrome is the most common form of presentation.[38,40,42,43] During the course of the disease, 90 to 97% of the patients develop the nephrotic syndrome.[38,42,43] Macroscopic hematuria is rare,

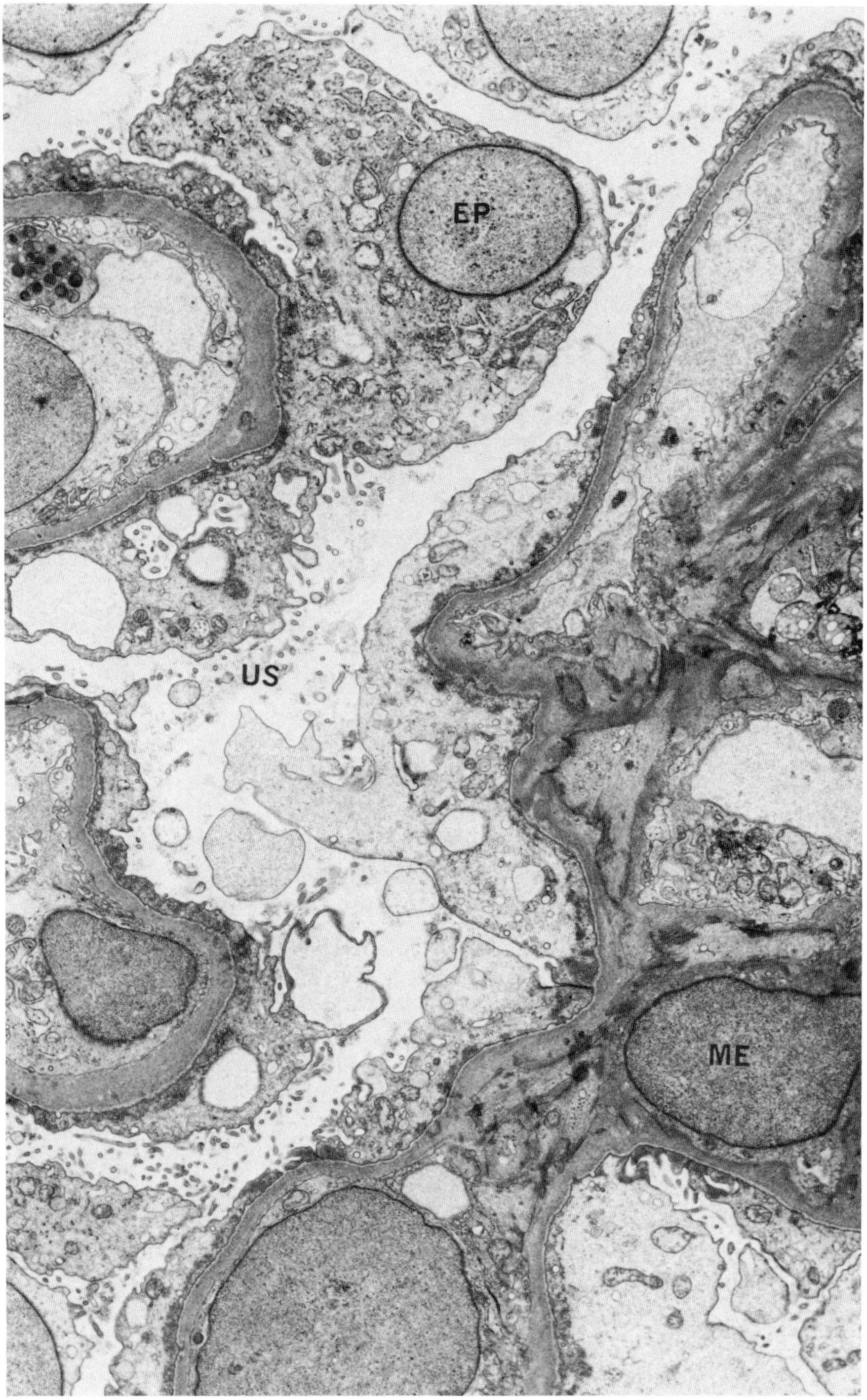

FIG. 5-6. Renal biopsy specimen from a patient with focal glomerulosclerosis. There is foot process obliteration with an increase in epithelial cell organelles, microcyst formation, and microvillous reaction. The basement membrane is folded, and the epithelial cell cytoplasm adjacent to the basement membrane has a focal increase in electron density. Epithelial cell (EP), mesangium (ME), urinary space (US). ×6700.

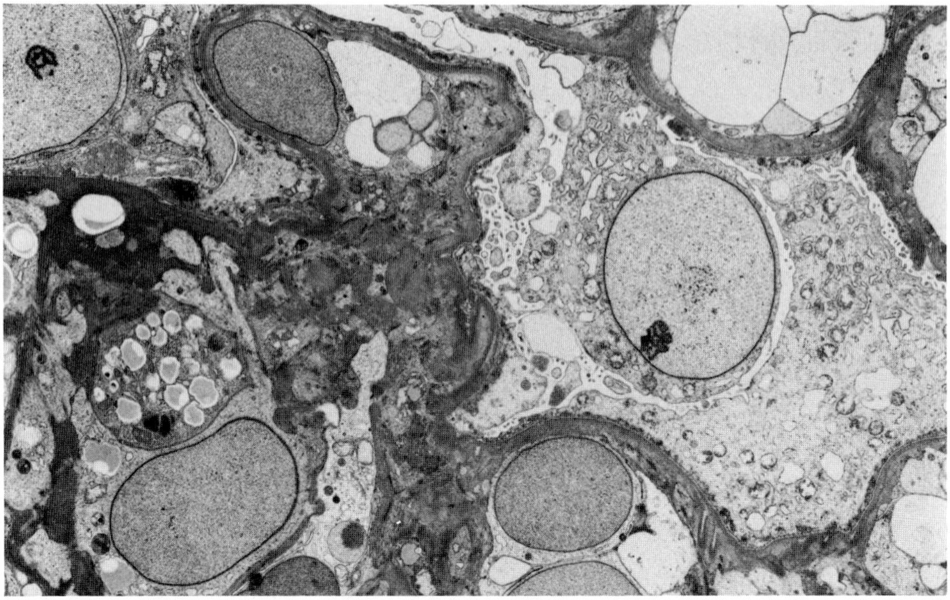

FIG. 5-7. Segmentally sclerosed area showing an increase in mesangial matrix and foot process obliteration. Note also a foam cell with intracytoplasmic lipid droplets. ×3600.

but 25 to 90% of the patients present microscopic hematuria sometime during the disease.[18,38,41,43] Chronic proteinuria with episodes of nephrotic syndrome usually occurs for many years, eventually leading to renal failure. Membranous nephropathy occurs in 30% of the adults[18,40,41] and in 5% of the children with nephrotic syndrome.[44] Spontaneous remission of the disease has been reported to occur in about 25% of the cases,[37,42,43,45] most frequently in children,[46,47] especially if presentation has been with proteinuria without the nephrotic syndrome.[48]

By light microscopy, the glomeruli appear slightly enlarged and normocellular. Based on the morphologic changes present in the capillary wall, four stages that correlate with the clinical evolution of the disease can be recognized.[38] In Stage I, there is no significant change in the thickness of the basement membrane, and the diagnosis of minimal change disease can be made incorrectly if light microscopy alone is utilized. By electron microscopy, sparse deposits involving short segments of the capillary wall are present between the epithelial cell cytoplasm and the lamina densa. The latter appears homogeneous and regular. The deposits can be irregular, dome-shaped, or appear like small humps, with a well-circumscribed line between the lamina densa and the epithelial cell. The foot processes over the deposits are obliterated but may appear normal in areas free of deposits (Fig. 5-8).

In Stage II, the capillary wall is thickened. There are numerous subepithelial deposits, which appear separated by extensions of the basement membrane (Fig. 5-9). They often are confluent and usually involve all the capillary loops. The deposits do not stain with methenamine silver, giving the basement membrane extensions a "spike" appearance. The foot processes are extensively obliterated, and the epithelial

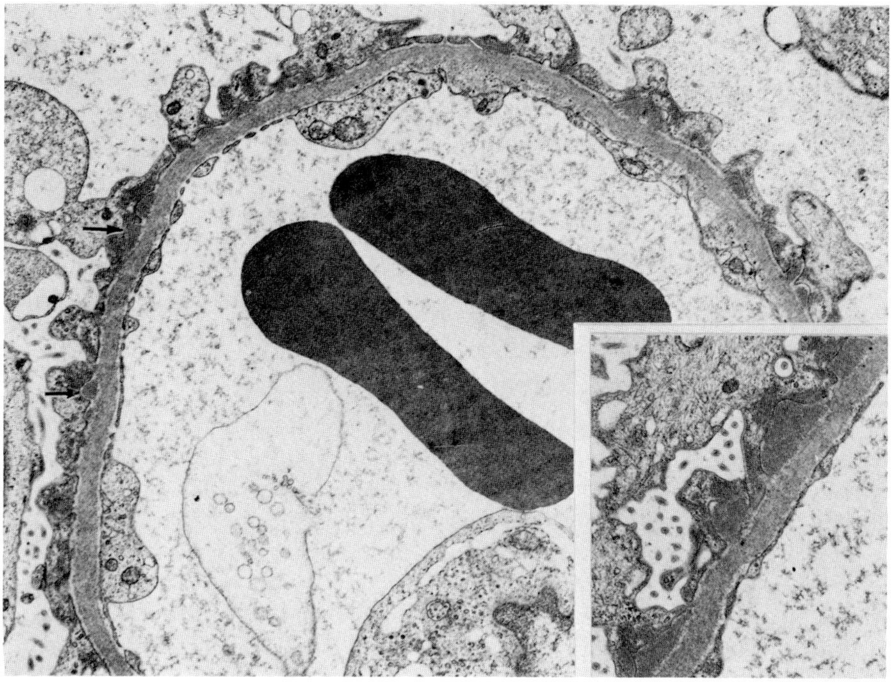

FIG. 5-8. Electron micrograph from a patient with secondary syphilis and membranous nephropathy Stage I. The basement membrane is not thickened. Small sparse subepithelial deposits (arrows) are separated from the basement membrane by a thin clear zone (insert). The foot processes over the deposits are obliterated. ×9400. Insert ×16,000.

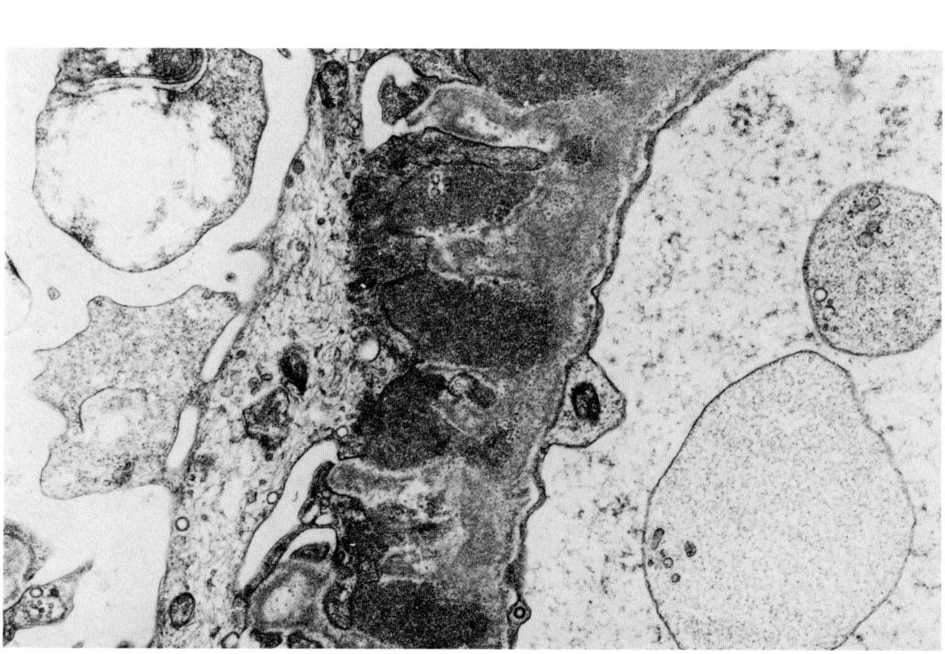

FIG. 5-9. Membranous nephropathy Stage II. Spike formation is present between subepithelial deposits, which lie very close to each other. ×17,500.

cytoplasm shows microvillous hyperplasia, microcyst formation, and well-developed endoplasmic reticulum. The epithelial cytoplasm adjacent to the basement membrane is condensed and may show numerous lysosomes, which probably represent protein transport droplets. The endothelium may appear moderately swollen but otherwise is normal. In most cases, the deposits vary in size and are finely granular, but on rare occasions may appear as parallel fibrotubular structures. The finding of these fibrillar deposits has important diagnostic significance because they often are associated with cryoglobulinemia.[49] Deposits showing a fingerprint-like pattern similar to those found in lupus can be found on rare occasions in idiopathic membranous nephropathy.[50]

In Stage III, the basement membrane is markedly thickened, producing narrowing of the capillary loops. The deposits seem to be incorporated into the basement membrane at this stage. The lamina densa is markedly distorted by the deposits, which are covered by basement membrane-like material. Often this newly formed basement membrane is reticular and less electron-dense than the lamina densa. At this stage, the deposits frequently show varying degrees of resolution with rarefaction. Spherical structures, which have been confused with virus particles, may be found during the stage of dissolution of the deposits. Individual deposits may also contain striated membranous bodies, probably originating from degenerated, trapped cellular components (Fig. 5-10).[51]

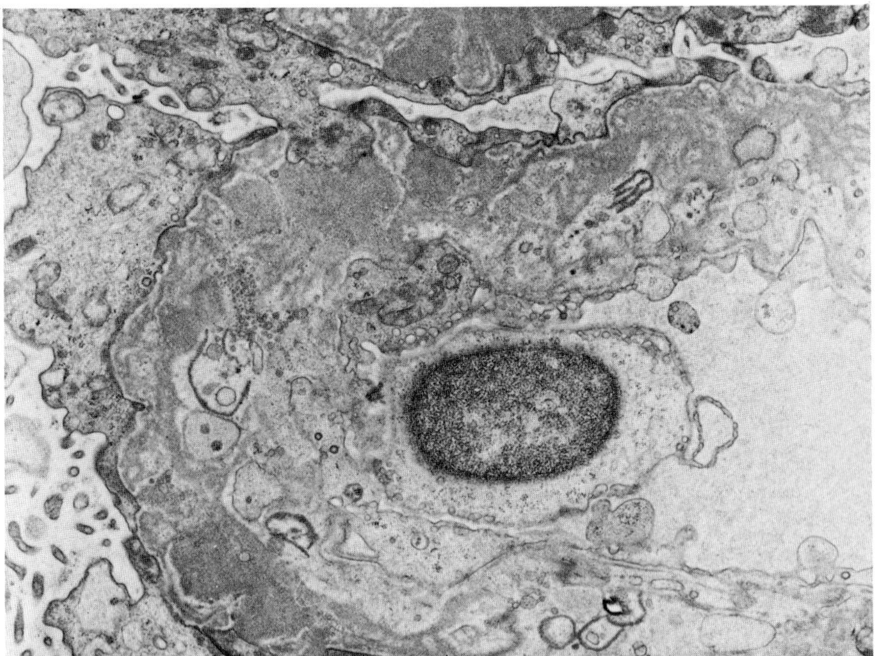

FIG. 5-10. Membranous nephropathy Stage III. The basement membrane is markedly thickened. The deposits are covered by a thin lamina densa and a lamina rara externa layer. Numerous striated membranous structures within partially dissolved deposits are discernible. ×12,500.

During the late stage (Stage IV), the deposits gradually lose their electron density, and the basement membrane becomes vacuolated (Fig. 5-11). In this final stage, the basement membrane is markedly thickened and irregular. Deposits may be absent. The capillary lumina become obliterated and the glomerular tufts solidified. At this stage, the differential diagnosis may be difficult and includes the different types of chronic glomerulonephritis and diabetic glomerulosclerosis. Immunofluorescence microscopy in all stages of membranous nephropathy gives a typical, generalized, peripheral granular pattern of IgG and C3 and sometimes with C1, C4, IgM, and IgA. Reaction for IgA is usually not intense. If the reaction is strong and is associated with the early components of complement ($C1_q$ and C4), the possibility of lupus membranous nephropathy has to be raised. Although the degree of proteinuria does not correlate with the staging of the renal lesions, there is correlation between staging and prognosis. Patients with Stage I or Stage II idiopathic membranous nephropathy more often undergo clinical remission than those with late stages of the disease. On rare occasions, membranous nephropathy has recurred in renal allografts.[52,53] Also, it has been reported de novo in transplanted kidneys.[54]

Diffuse Endocapillary Proliferative Glomerulonephritis

The term *diffuse endocapillary proliferative glomerulonephritis* is used to describe lesions characterized by mesangial and endothelial proliferation. Although this category has, by common usage, become virtually synonymous with acute, poststreptococcal glomerulonephritis, it may also be associated with infection by bacteria

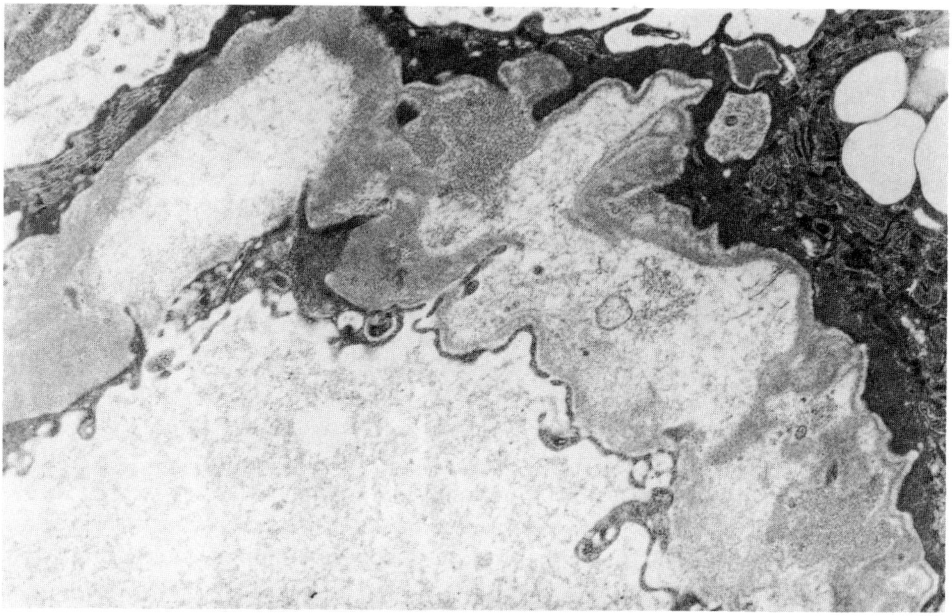

FIG. 5-11. Membranous nephropathy Stage IV. The basement membrane is thickened with large electron-lucent areas and small amounts of residual deposits. ×15,500.

(*Staphylococcus, Meningococcus,* and *Pneumococcus*),[50,55,56] and probably with complicating viral infections (varicella, mumps, and hepatitis-B).[57–59] The disease classically occurs one to two weeks after infection with Group A streptococci of a nephritogenic strain.[60] It usually affects children between the ages of 5 and 15 years, and, less frequently, adults. Clinically, it is manifested by edema, hematuria, proteinuria, hypertension, low serum complement, and impaired renal function. The disease has a good prognosis, especially in children, and only a relatively small percentage of patients progress to chronic renal failure after many years of the acute episode.[61,62] However, in a recent study of a large group of patients, chronic renal failure and hypertension occurred in over half of the patients followed for over a 10-year period.[63] Progression is more common in adults and in those patients who develop crescentic glomerulonephritis. There is evidence that crescentic postinfectious glomerulonephritis has a better prognosis than other forms of crescentic glomerulonephritis, but in the individual case it is difficult to predict the outcome.[64]

Histologically, the glomeruli are enlarged and bloodless and have a tendency to fill Bowman's space. The glomerular intracapillary cellularity is increased due to mesangial proliferation and, to a lesser extent, to an increase in the number and size of endothelial cells. The results are a narrowing of the capillary lumina and accentuation of the lobular glomerular pattern. A variable leukocytic infiltrate can contribute to the capillary obstruction. Neutrophils may be seen in direct contact with the capillary basement membrane, the endothelial cell cytoplasm being disrupted or absent at these points (Figs. 5-12 and 5-13).[65] Besides polymorphonuclear leukocytes, there is an increased number of mononuclear cells with large oval nuclei, which, by light microscopy, cannot be distinguished from endothelial and mesangial cells. By electron microscopy, they have been shown to be monocytes. Occasional eosinophils may also be present. The endothelial cell cytoplasm is swollen and the fenestrae are inconspicuous. Capillary thrombosis and disruption of the basement membrane also can be present.[66] The proliferating mesangial cells are separated from each other by strands of mesangial matrix and, on occasion, may extend focally into the subendothelial space.

One of the most characteristic ultrastructural features is the presence of subepithelial, finely granular, dome-shaped, electron-dense deposits called *humps.*[67] The deposits are separated from the lamina densa by a recognizable lamina rara externa and are surrounded by epithelial cell cytoplasm with a zone of increased density adjacent to the deposits. The foot processes of the epithelium overlying the humps are usually obliterated (Fig. 5-13). In general, there is a correlation between the number of humps and the degree of polymorphonuclear infiltration; however, humps may occur in loops lacking polymorphonuclear leukocytes. Humps with atypical shape and size and varying texture may be seen and have been associated with severe inflammation and delayed resolution.[68] In some loops, the humps may be confluent and may appear separated by overlying epithelial cytoplasm. At other times, they may be separated by extensions of basement membrane simulating the spikes of membranous nephropathy. Intramembranous deposits, sometimes in continuity with the humps, and small subendothelial and mesangial deposits are not uncommon. The humps generally disappear in six to eight weeks, but have been identified, although infrequently, for as long as six months after the onset of disease.[67,69] During the resolving phase, the basement membrane may show focal irregularities with decreased electron density (Fig. 5-14). The capillary lumina become patent, and endo-

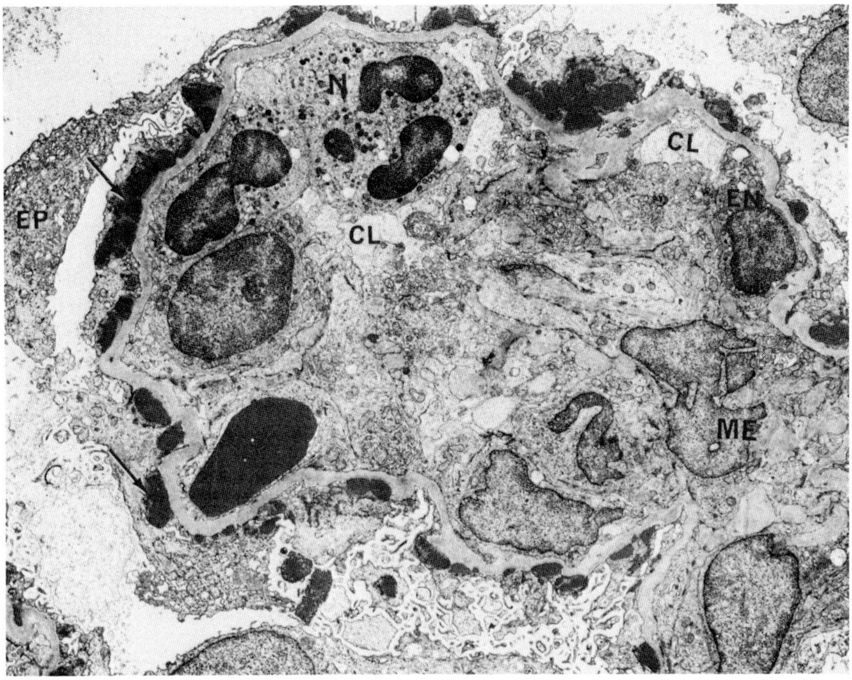

FIG. 5-12. Hypercellular glomerular segment from a patient with acute poststreptococcal glomerulonephritis. The capillary lumen is narrowed because of endocapillary cell proliferation and polymorphonuclear leukocyte infiltration. Numerous subepithelial deposits are present along the epithelial surface of the basement membrane (arrows). Capillary lumen (CL), epithelial cell (EC), endothelial cell (EN), mesangium (ME), neutrophilic leukocyte (N). ×3600.

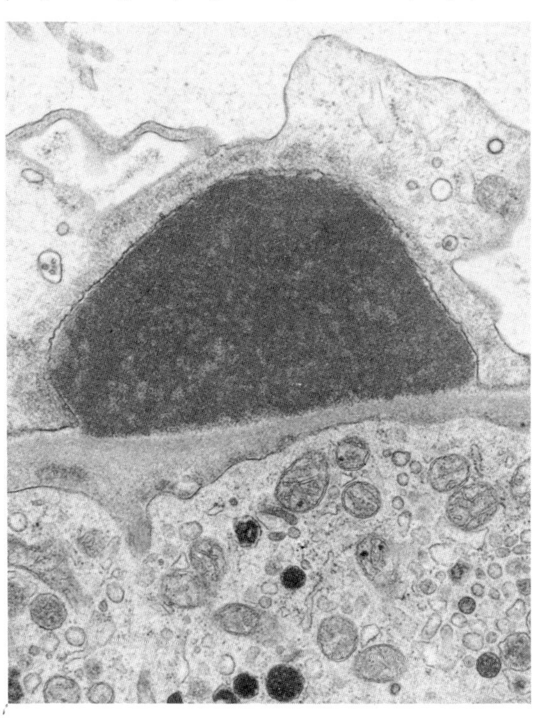

FIG. 5-13. Hump-like deposit in acute poststreptococcal glomerulonephritis. The epithelial cell exhibits loss of foot processes and shows aggregates of electron-dense material within the cytoplasm. The capillary loop is denuded from its endothelium, and the lumen appears obliterated by a polymorphonuclear leukocyte that is in direct contact with the basement membrane. ×15,000.

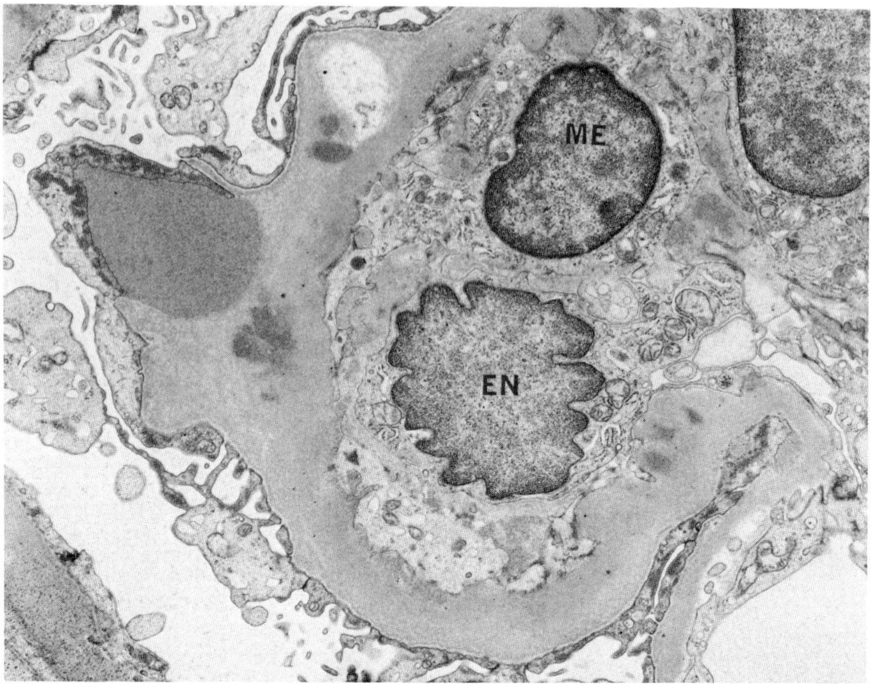

FIG. 5-14. Biopsy specimen from a patient five weeks after an acute episode of poststreptococcal glomerulonephritis showing subepithelial deposit partially surrounded by "spikes" of basement membrane. The electron-lucent area is probably due to mobilized deposits. Small amounts of intramembranous and mesangial deposits are also present. Endothelial cell (EN), mesangium (ME). ×9500.

thelial swelling or polymorphonuclear leukocytic infiltration disappears. Electron-dense deposits may be absent or restricted to the mesangium (Fig. 5-15). The presence of a varying degree of mesangial hypercellularity, together with an increase in mesangial matrix, may continue for years after the acute episode. In some cases, the mesangial prominence is focal and segmental. The immunofluorescence studies during the acute phase of the disease typically reveal granular staining for IgG and C3 along the capillary loop in locations corresponding to the humps. This finding and its analogy with similar deposits in experimental serum nephritis are supportive of an immune complex pathogenesis.[70,71] After a few weeks, the IgG may no longer be demonstrable, and C3 staining gradually diappears from the periphery and becomes restricted to the mesangial areas during the phase of resolution.[50] The absence of IgA and of early reacting complement components ($C1_q$ and C4) permits differentiation of cases of postinfectious glomerulonephritis in the resolving stage from IgA nephropathy and lupus nephritis, respectively.

Mesangiocapillary (Membranoproliferative) Glomerulonephritis

In 1965, a group of children were described as having a chronic form of glomerulonephritis accompanied by a persistent decrease in the serum levels of C3 and in the

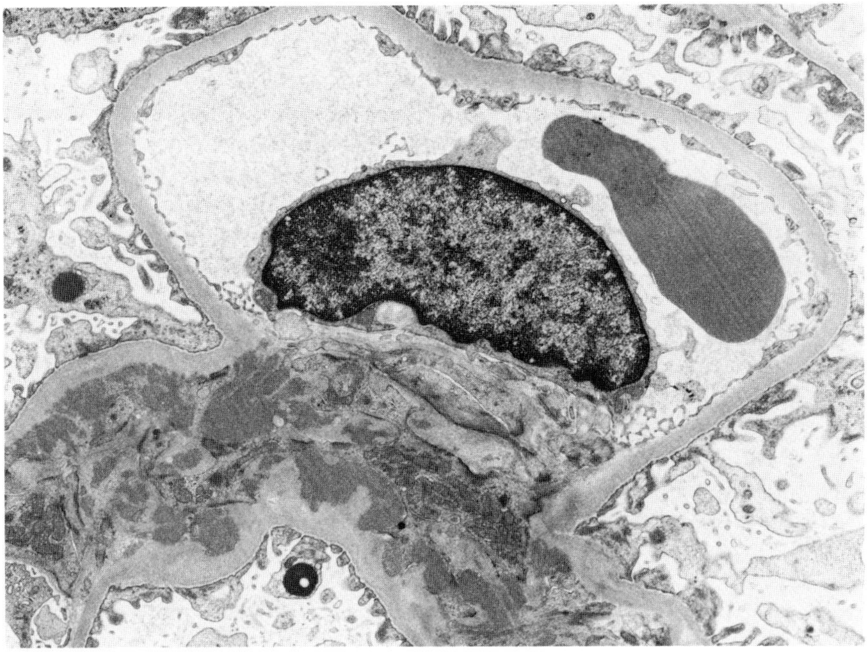

FIG. 5-15. Mesangial deposits in resolving poststreptococcal glomerulonephritis. ×8800.

total hemolytic complement activity.[72,73] The patients with this entity, which was called *persistent hypocomplementemic glomerulonephritis,* had a distinctive histopathologic lesion characterized by a combination of mesangial cell proliferation and capillary wall thickening. Because this lesion has subsequently been found to occur in the absence of hypocomplementemia, the terms *membranoproliferative glomerulonephritis* and *mesangiocapillary glomerulonephritis* (MCGN) have replaced that of *hypocomplementemic glomerulonephritis.* The condition is said to account for about 20% of all cases of chronic glomerulonephritis in pediatric groups and about 5% in adults.[74,75] It also accounts for 5 to 13% of childhood cases and 6% to 16% of adult cases of primary nephrotic syndrome.[76] Heavy proteinuria, usually associated with the nephrotic syndrome, and hematuria are found in approximately three-fourths of the patients when first evaluated. The remaining cases are most commonly detected because of asymptomatic proteinuria and/or microhematuria. The clinical course of the patient is slow, indolent, and marked by episodes of nephritic or nephrotic syndrome, finally progressing to renal failure over a period of 10 or more years.[77] Many patients have persistent depletion of C3, which is often associated with a circulating C3 nephritic factor.[78,79]

Based on the morphology of the glomerular lesion, MCGN has been subdivided into three types.[80] In the first type, the glomeruli are diffusely enlarged, with thickening of the capillary walls and marked mesangial proliferation that produces lobulation of the tuft. The capillary wall thickening is due to interposition of mesangial cells and matrix throughout the subendothelial zone and produces a double-contoured or tram-track appearance in the periphery of the loop, demonstrable with

PAS and silver stains. Electron microscopy reveals that the basement membrane is intact and shows subendothelial dense deposits.[77,81–83] This variety is called MCGN with subendothelial dense deposits, or MCGN Type I (Fig. 5-16).

In the second type, the thickening is due to the presence of slightly argyrophilic PAS-positive electron-dense intramembranous deposits. There is circumferential mesangial interposition, which usually is much less prominent than that observed in MCGN Type I. The dense material, which lacks the granularity usually seen in the immune complex deposits, is centrally located in the glomerular basement membrane in a ribbon-like fashion.[84,85] In some areas, the material may be discontinuous or may show great variation in thickness. Similar deposits are usually present in the mesangium and form homogeneous nodules or, less frequently, subepithelial humps.[86] Focal involvement of Bowman's capsule and tubular basement membrane is common. This variety warrants the name of dense deposit disease, MCGN with intramembranous dense deposits, or MCGN Type II (Fig. 5-17).

In the third type, which has only recently been described, the basement membrane is disrupted by focal, massive, granular deposits that are randomly located within the basement membrane and extend to both subepithelial and subendothelial regions (Fig. 5-18).[87,88] The extent of the disruption of the basement membrane can be better

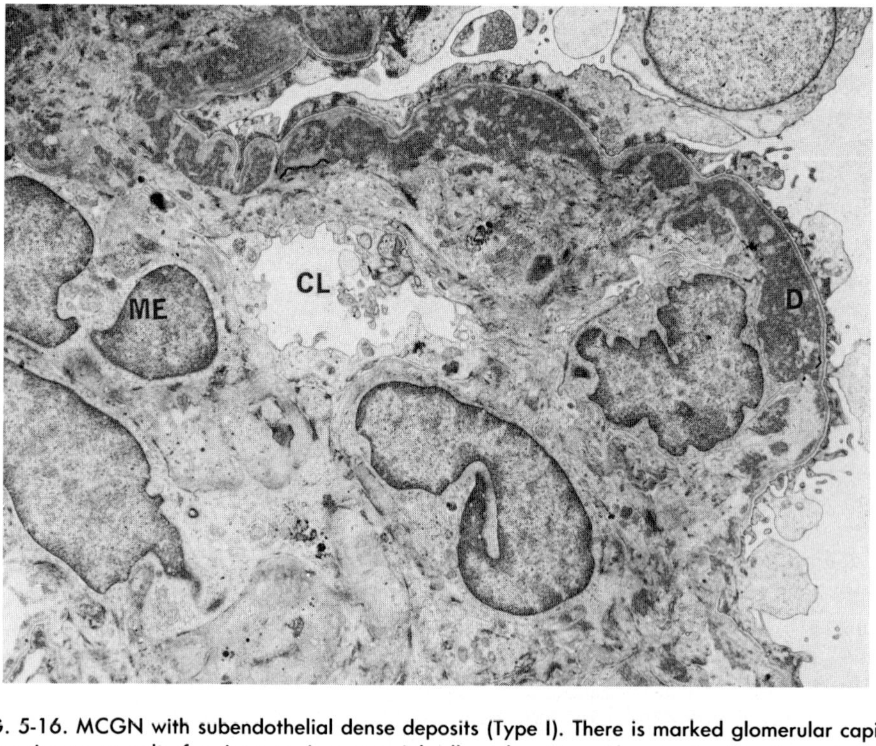

FIG. 5-16. MCGN with subendothelial dense deposits (Type I). There is marked glomerular capillary narrowing as a result of an increase in mesangial cells and matrix, with extension into the periphery of the loop. Capillary lumen (CL), mesangium (ME), subendothelial deposits (D). ×9000.

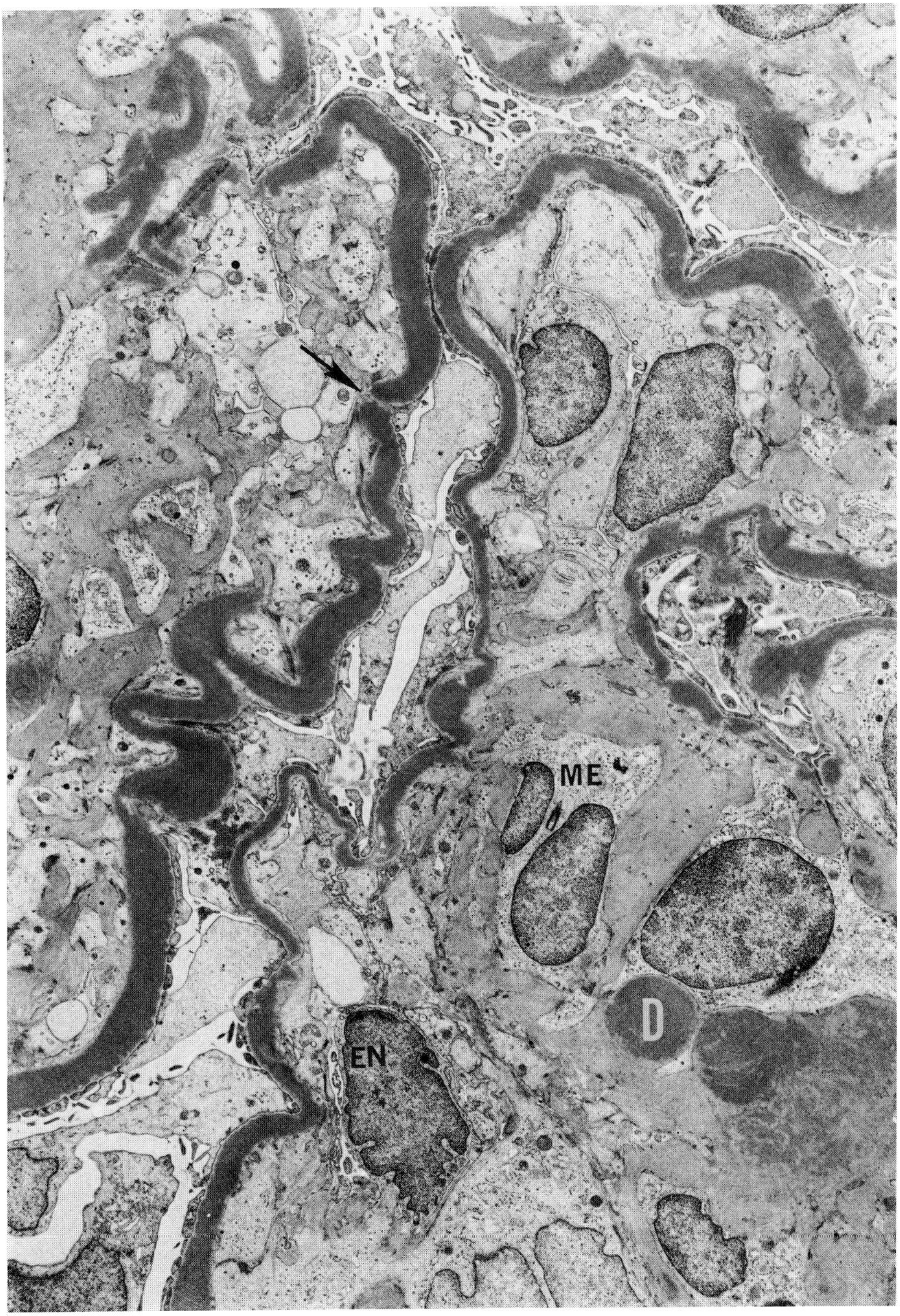

FIG. 5-17. MCGN with intramembranous dense deposits (Type II). Intramembranous deposits of a highly osmiophilic material with focal interruptions (arrow). The capillary lumen is obstructed by an increase in mesangial cells and matrix. Nodular deposits (D) are also present in the mesangium. Endothelial cell (EN), mesangium (ME). ×7900.

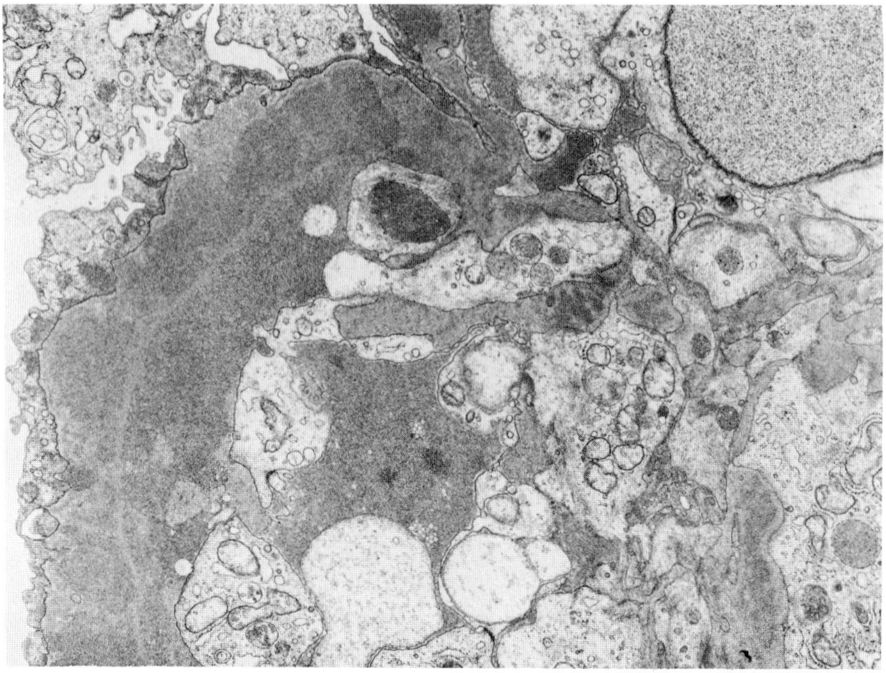

FIG. 5-18. MCGN Type III. There is massive accumulation of intramembranous deposits with extension into the subendothelial and subepithelial areas. The foot processes are obliterated. ×10,500.

appreciated by electron microscopy with silver-impregnated tissue.[87] All varieties of MCGN may show mesangial deposits and prominent epithelial foot process obliteration.

By immunofluorescence, C3 is constantly found in all types of MCGN. However, while Types I and III have a granular, peripheral, and segmental or generalized distribution,[81,89] in dense deposit disease there is linear, continuous staining along the periphery of the glomerular lobule, which is typically associated with bright, granular, mesangial deposits.[50,86] Early complement components, IgG and IgM, are often found in Types I and III.[89] In contrast, immunoglobulins, C1, and C4 are almost always absent in Type II.[85,90–92]

Therapy in general has not been successful, and recurrence after transplantation often occurs, especially in dense deposit disease.[34,93] Recently, it has become evident that although MCGN with subendothelial dense deposits and probably Type III MCGN are a heterogeneous group of conditions caused by immune complex deposits, dense deposit disease is a distinct clinicopathologic entity characterized by a unique dense transformation of the basement membrane.[50,83]

Focal Glomerulonephritis

Focal glomerulonephritis is a morphologic term used to indicate that some, but not all, glomeruli have inflammatory changes that usually involve a portion of the tuft (segmental). Thus, focal glomerulonephritis is a histopathologic diagnosis and does not have a distinct clinical counterpart. It may occur as a disease limited to the kidney or as part of a systemic process. Some cases, like IgA nephropathy, may present focal, segmental mesangial proliferation, while others, such as periarteritis nodosa, Wegener's granulomatosis, and anaphylactoid purpura, often show capillary thrombosis with necrosis, with or without epithelial crescents (focal necrotizing glomerulonephritis). Recent advances in recognition of patterns of lesions by light, immunofluorescence, and electron microscopy have allowed a morphologic differentiation of many of these entities. If the findings do not fit the diagnostic criteria of any of the specific diseases listed in Table 5-1, then the morphologic diagnosis rendered must be idiopathic focal glomerulonephritis.

Diffuse Mesangioproliferative (Mesangiopathic) Glomerulonephritis

Diffuse proliferation of the mesangial cells and matrix without significant involvement of capillary walls or lumina occurs in a variety of renal diseases, such as IgA nephropathy, anaphylactoid purpura, systemic lupus erythematosus, and during the resolving stage of postinfectious glomerulonephritis. The differential diagnosis of these conditions is usually achieved by the correlation of light, electron, and immunofluorescent microscopic findings with the patient's clinical data. Sincc most of these entities are described in other parts of the chapter, only IgA nephropathy will be discussed here.

IgA (Berger's) Nephropathy. In 1969, Berger described a disease entity that is characterized by diffuse, focal, segmental mesangial proliferation by light microscopy.[94] By immunofluorescence microscopy, diffuse mesangial deposits of IgA often associated with IgG and C3, and less frequently IgM, are present.[95–97] By electron microscopy, electron-dense deposits in the mesangium and less frequently in the capillary loops, are found (Fig. 5-19). The disease can occur at any age, but is most common in children and young adults.[95,96] The clinical presentation is usually one of repeated episodes of microscopic or gross hematuria, often in association with nonspecific respiratory tract infections. Mild proteinuria on occasion may reach nephrotic proportions.[95] Although earlier reports emphasized the benign course of the disease, recent studies with long-term follow-up showed that in a significant proportion of patients, especially adults, the disease progressed to chronic glomerulonephritis and renal failure.[95,96,98] Approximately half of the patients who receive transplants present with recurrence of the disease in their donor kidneys.[96,99,100]

The etiology of IgA nephropathy is unknown. The association with upper respiratory infections suggests an infectious etiology. Immunofluorescence studies support an immune complex pathogenesis. The absence of $C1_q$ and C4 suggests that complement is activated via the alternate pathway. The finding of cutaneous vascular deposition of IgA in patients with Berger's nephropathy suggests a relationship between anaphylactoid purpura and IgA nephropathy. Although this relationship is obscure, some authors have proposed that the latter condition may represent a monosymptomatic form of anaphylactoid purpura.[101]

TABLE 5-1
Differential Diagnosis of Focal Glomerulonephritis

Diagnosis	*Light Microscopy*	*Immunofluorescence*	*Electron Microscopy*
Antiglomerular basement membrane GN	FNGN*	Linear peripheral IgG, linear or irregular C3	
IgA (Berger's) nephropathy	FMPGN† or FNGN	Mesangial IgA with or without IgG, C3	Mesangial deposits
Anaphylactoid (Henoch-Schönlein) purpura	FMPGN or FNGN	Mesangial IgA with or without IgG, C3	Mesangial deposits
Systemic lupus erythematosus	FMPGN or FNGN, hematoxylin bodies, "wire loops"	Mesangial or mesangial and peripheral IgG with varying $C1_q$ C4, C3, IgA, IgM	Subendothelial, mesangial, peritubular, and fingerprint-like deposits
Resolving postinfectious GN	FMPGN	Mesangial, ± peripheral granular C3	Mesangial, ± capillary wall deposits
Subacute bacterial endocarditis	FNGN	Peripheral granular C3 with or without IgG, IgM	Intracapillary deposits, with occasional "humps"
Polyarteritis nodosa	FNGN		
Wegener's granulomatosis	FNGN		
Hereditary GN	No significant changes or FMPGN		Splitting of the glomerular basement membrane

* FNGN = Focal necrotizing glomerulonephritis.
† FMPGN = Focal mesangial proliferative glomerulonephritis.

Crescentic Glomerulonephritis

Crescentic glomerulonephritis is a morphologic term used in association with a clinical syndrome characterized by rapidly deteriorating renal function, often resulting in uremia and death within a few weeks or months (rapidly progressive glomerulonephritis). The formation of epithelial crescents may occur in a number of different diseases, but in general the diagnosis of crescentic glomerulonephritis is used to designate those cases in which at least 60% of the glomeruli are involved by crescents.

In early stages, the crescents are composed of large cells that occupy a fraction of the entire Bowman's space. Most of the cells have clear cytoplasm containing a small number of organelles; others are smaller and darker with abundant rough endoplasmic reticulum. The cells can be attached to each other by cell junctions, or can be separated by electron-dense, proteinaceous material or by large amounts of fibrin. Mitotic figures sometimes can be seen, as well as multinucleated giant cells, which, in

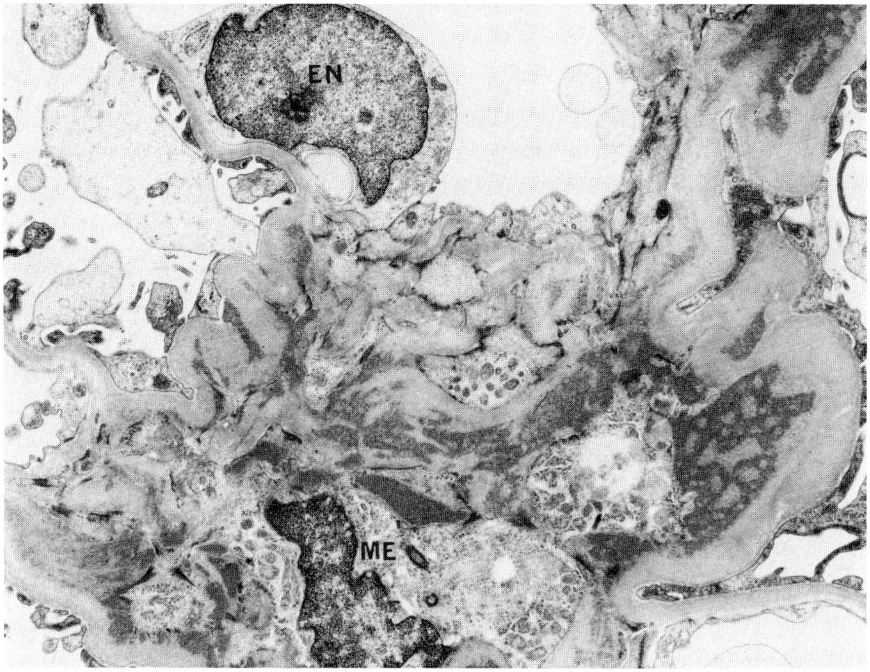

FIG. 5-19. Mesangial deposits in IgA nephropathy. Endothelial cell (EN), mesangium (ME). ×9000.

some cases, can be prominent. The finding of these cells has no diagnostic value, although on rare occasions, granulomatous crescentic glomerulonephritis has been reported in association with polyarteritis.[102] The origin of the crescentic cells is controversial. Tissue cultures have suggested that the cells of the crescents are derived from macrophages,[103] but labeling[104] and ultrastructural studies do not support that origin and indicate that these cells most likely originate from the parietal epithelial cells of Bowman's capsule.[105–107] As the disease progresses, the extracapillary proliferating cells are transformed into fibroepithelial crescents. On light microscopy, the cells are elongated and are surrounded by argyrophilic, PAS-positive material that, by electron microscopy, has a basement membrane-like appearance.[106] The capillary tuft is usually collapsed and may show cell proliferation. Ruptures and defects of the basement membrane are usually seen (Fig. 5-20).[108] It has been suggested that these breaks are probably produced by the lytic actions of complement and leukocytic enzymes, or perhaps by the action of antibasement membrane antibodies.[109] The escape of plasma components through the gaps and the deposit of fibrin into Bowman's space could be the mechanism responsible for the proliferation of the epithelial cells.

The prognosis of patients with crescent glomerulonephritis depends on the percentage of glomeruli involved by crescents, or the underlying cause of the glomerular disease,[63, 110] or both. Therefore, the prognosis and diseases can be differentiated using light, electron, and immunofluorescent microscopic criteria.[64] Diffuse endoca-

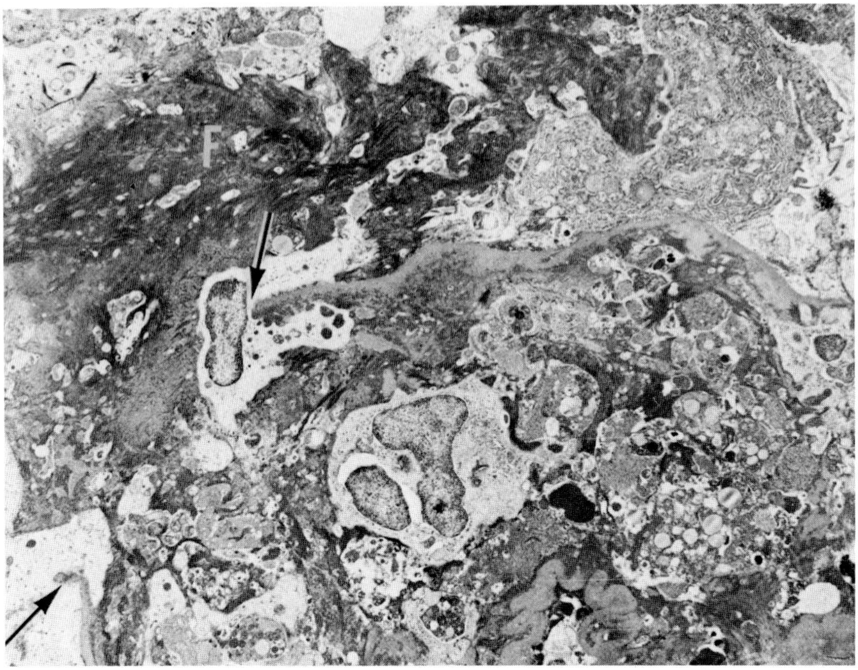

FIG. 5-20. Portions of a glomerulus from a patient with crescentic glomerulonephritis showing rupture of the basement membrane (arrows) as well as intracapillary and extracapillary fibrin deposition. Fibrin (F). ×5700.

pillary proliferation associated with humps containing C3, with or without IgG, is characteristic of postinfectious glomerulonephritis, while peripheral mesangial interposition associated with subendothelial deposits or intramembranous electron-dense deposits is indicative of Types I and II mesangiocapillary glomerulonephritis. Bright mesangial deposits of IgA are usually present in both IgA nephropathy and anaphylactoid purpura. Patients with systemic lupus erythematosus can be recognized by the presence of hematoxylin bodies and of large amounts of subendothelial (wire loop) and extraglomerular deposits, usually containing various types of immunoglobulins, C3, $C1_q$, and C4. Crescentic glomerulonephritis is an uncommon complication of membranous nephropathy. The presence of subepithelial electron-dense deposits will permit recognition of the nature of the lesion. Immunofluorescence studies in these cases have shown both granular and linear peripheral patterns of staining with IgG. The demonstration of antibasement membrane antibodies in some of these patients suggests that they may be involved in the pathogenesis of the crescentic glomerulonephritis of some cases of idiopathic membranous nephropathy.[50,111]

In contrast to the above-described entities, which often present a diffuse distribution of the glomerular lesions, some systemic diseases, such as Goodpasture's syndrome, polyarteritis nodosa, and Wegener's granulomatosis often are manifested as focal necrotizing crescentic glomerulonephritis with minimal endocapillary proliferation. There are no diagnostic ultrastructural features of the glomerular lesion of Goodpasture's syndrome. A diffuse increase in the density of the glomerular basement membrane, associated with electron-lucent widening of the lamina rara interna, is suggestive but nonspecific.[50] The diagnosis in these cases depends on the finding of a linear peripheral pattern on immunofluorescence and the demonstration of circulating antibasement membrane antibodies.[64]

Although crescentic glomerulonephritis may occur in both classic and microscopic types of polyarteritis nodosa, it is more common in the latter condition than in the former. Electron microscopy in these cases fails to demonstrate immune complex deposits, and the immunofluorescence is usually negative. The diagnosis in these cases requires the demonstration of a necrotizing arteritis in the arcuate or larger blood vessels in the classic type, or the finding of arteritis affecting interlobular arteries and arterioles in the microscopic form. The glomerular and vascular lesions of Wegener's granulomatosis are identical to those of microscopic polyarteritis. Although subepithelial electron-dense deposits and granular reaction for IgG have occasionally been described,[112] these are not common features.[50,113] Studies of large numbers of patients will be required before a definitive ultrastructural and immunofluorescent diagnostic pattern can be outlined.

The crescentic glomerulonephritis of thrombotic microangiopathies, including the hemolytic-uremic syndrome and thrombotic thrombocytopenic purpura, appears related to extensive fibrin deposition in glomerular capillaries and blood vessels. The characteristic ultrastructural feature in these cases is a narrowing of the capillary lumina that occurs because of irregular subendothelial deposition of amorphous material of less electron density than the basement membrane.[113] On occasion, patients with malignant nephrosclerosis may present numerous crescents in addition to the fibrinoid necrosis of arterioles and onionskin-like intimal hyperplasia of small arteries. The recognition of the vascular lesion and the absence of deposits or positive immunofluorescence for immunoglobulins allow us to distinguish these cases from other forms of crescentic glomerulonephritis.

If after using the morphologic and clinical criteria outlines, a case of crescentic glomerulonephritis cannot be classified as any specific entity, it should be designated as idiopathic crescentic glomerulonephritis.[64] Fortunately, the number of these unclassified cases has been reduced since immunofluorescence and electron microscopy became part of the routine evaluation of renal biopsy specimens.

GLOMERULONEPHRITIS OF SYSTEMIC DISEASES

Lupus Nephritis

It has become clear in recent years that the course of patients with systemic lupus erythematosus (SLE) depends on the type and severity of renal involvement. Also, it is widely accepted that in the majority of cases the histologic features present in the biopsy specimen serve as a good indicator of the natural history and response to treatment. Based on the distribution of the glomerular proliferation by light microscopy and the localization of immune deposits by immunofluorescence and electron microscopy, the renal lesions in SLE have been categorized as follows: (1) minimal or no glomerular lesion, (2) mesangial proliferative glomerulonephritis, (3) focal and segmental proliferative glomerulonephritis, (4) diffuse proliferative glomerulonephritis, and (5) membranous nephropathy.[114,115]

The biopsy specimens of patients with minimal glomerular lesions are characterized by normal glomeruli by light microscopy, and by mild, nonspecific ultrastructural changes. Deposits are absent. Approximately 15% of the biopsy specimens of lupus patients with renal involvement show mesangial glomerulonephritis.[50] In these cases, there is proliferation of mesangial cells and the mesangial matrix without associated glomerular necrosis. Electron-dense deposits are located almost exclusively in mesangial sites.

Patients with active proliferative glomerulonephritis, whether it is focal or diffuse, frequently have certain typical lesions. These include: (1) the presence of wire loops, an eosinophilic thickening of the capillary wall caused by subendothelial accumulation of immune complexes (Fig. 5-21); (2) hyaline thrombi, representing massive deposits in the capillary lumen (Fig. 5-22); and (3) hematoxylin bodies, lilac or violaceous amorphous structures that are the only histologic feature pathognomonic of SLE. Unfortunately, hematoxylin bodies are only found in about 8% of renal biopsy specimens in lupus.[116] Although they occur most often in the glomeruli, they can also be present in the proximal tubular epithelium and renal interstitium, as well as in other organs, such as the endocardium, lymph nodes, spleen, skin, pancreas, and

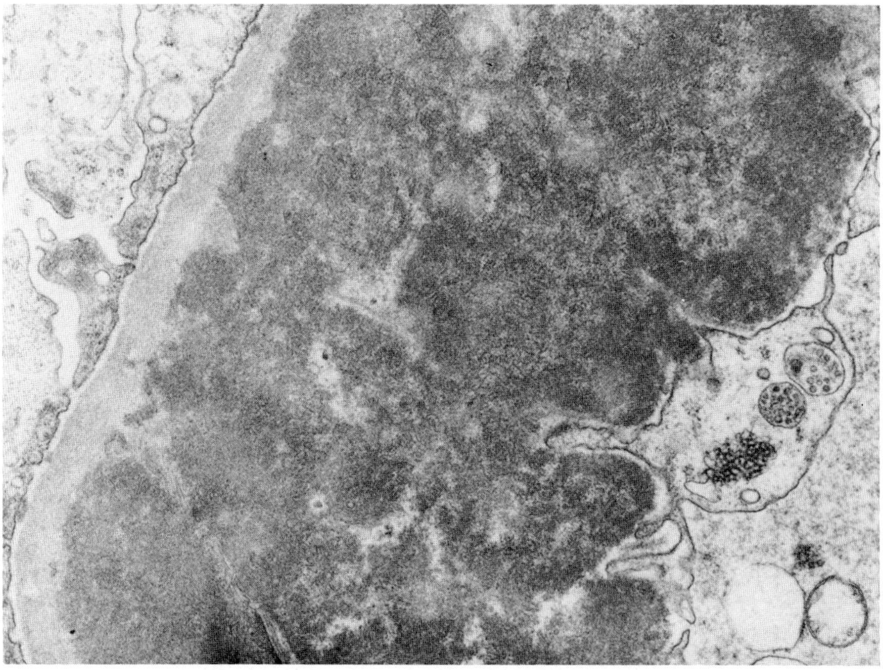

FIG. 5-21. Glomerular capillary loop with abundant subendothelial electron-dense deposits, which correspond to the "wire-loop" lesion seen by light microscopy. Note the aggregate of microtubular structures within endothelial cells. ×18,500.

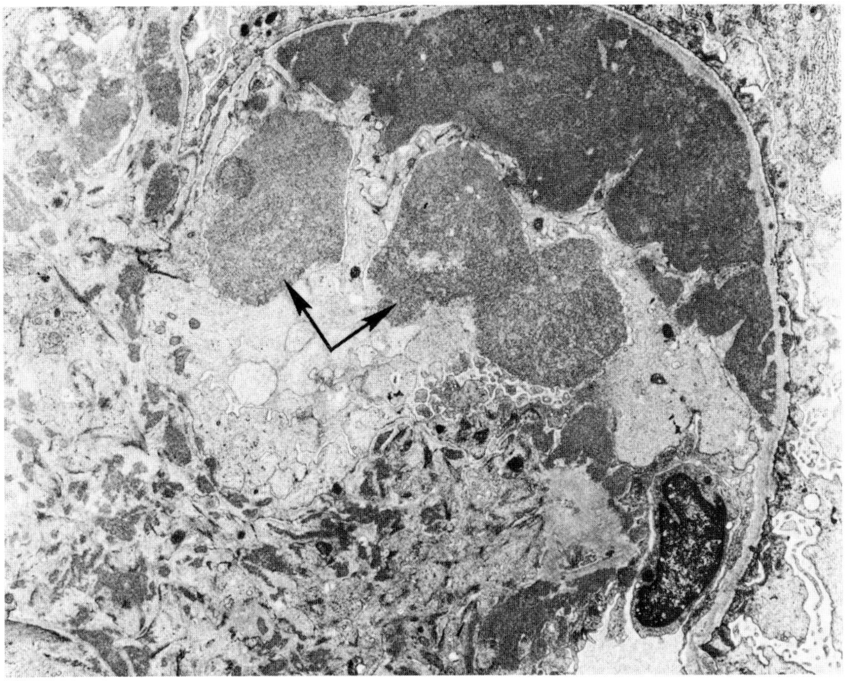

FIG. 5-22. Portion of a glomerulus from a patient with active lupus nephritis. The capillary lumen is obliterated by complexes that correspond to the hyaline thrombi seen by light microscopy (arrows). Large amount of subendothelial deposits (right) and mesangial deposits (left) are also present. ×8000.

blood vessels.[117,118] Ultrastructural studies demonstrate that they are composed of nuclear and cytoplasmic elements that are probably derived from leukocytes that have been ingested, with or without immune complexes, by the mesangial cells, which are known to have phagocytic activity. These observations indicate that the hematoxylin bodies are probably an in vivo counterpart of the LE phenomenon.

In focal and segmental lupus glomerulonephritis, the histologic activity is confined to portions of some glomeruli. The lesion consists of a proliferation of endothelial and mesangial cells, usually at the periphery of the glomerular tuft and often associated with capsular adhesions, infiltration of polymorphonuclear leukocytes, and focal necrosis. There are large amounts of mesangial deposits, often accompanied by subendothelial, subepithelial, and intramembranous deposits. This form of renal involvement occurs in about 20% of patients with SLE.[50] Patients with minimal glomerular lesions, mesangial lupus glomerulonephritis, and focal lupus glomerulonephritis usually have microscopic hematuria and mild proteinuria. The nephrotic syndrome occurs only rarely. In the past, it was generally accepted that these lesions seldom evolved into a diffuse form, and that progression to renal insufficiency was unusual.[119] However, due to the recently accepted practice of performing biopsies on lupus patients without clinical renal involvement, and as more serial studies have appeared, it has become evident that patients with mesangial or focal lupus glomerulonephritis often progress to a more severe, diffuse proliferative form.[120,121] Various studies have shown that this progression may occur in about 30% of the cases.[115] The average five-year survival rate of patients with focal proliferative lupus glomerulonephritis is 60 to 87%.[50]

In diffuse lupus glomerulonephritis, all glomeruli are involved. There is accentuation of the lobular pattern. Widespread mesangial and endothelial cell proliferation, adhesions between Bowman's capsule and glomerular capillary tufts, wire loop lesions, and hyaline thrombi are present. In the most active cases, these changes are pronounced and may be associated with extensive necrosis, epithelial cell crescents, hematoxylin bodies, and vasculitis. In addition, inflammation commonly extends to the tubules and interstitium. The most characteristic electron microscopic finding is the presence of subendothelial electron-dense deposits, usually associated with mesangial and sometimes subepithelial and intramembranous electron-dense deposits. There are several reports indicating a strong correlation between the number and location of the deposits and the progression of the disease.[122–124] In these studies, patients with massive deposits, especially when located in the subendothelial area, have a poorer prognosis than those with few deposits that are mainly confined to areas other than the subendothelial zone. Some patients with diffuse lupus glomerulonephritis may show changes identical to those present in cases of idiopathic mesangiocapillary glomerulonephritis.[50] Diffuse lupus glomerulonephritis has been demonstrated in approximately half of the biopsy specimens from lupus patients with renal involvement.[125–128] This type of lesion progresses to renal failure within two years of onset in over 50% of patients. The morbidity caused by this type of renal disease is increased by the common occurrence of nephrotic syndrome, which is usually unresponsive to steroid therapy. The average five-year survival for patients with diffuse glomerulonephritis is 38 to 83%.[50] In general, the diffuse proliferative glomerulonephritis is uncommon in patients with mild or no proteinuria and minimal urinary sediment abnormalities; however, there are reports of patients with active histologic lesions who have a normal urinary sediment and normal renal function tests at the time of the biopsy.[127]

The glomerular lesion in membranous lupus nephropathy is, in most cases, histologically indistinguishable on light microscopy from the lesion of the idiopathic form. The glomeruli often appear normal or show only a slight increase in mesangial matrix. The capillary walls are thickened due to immune complex deposition in the subepithelial space. In contrast to the idiopathic form of membranous nephropathy, in which the deposits are typically limited to subepithelial or intramembranous locations, small amounts of mesangial deposits may exist in lupus membranous nephropathy (Fig. 5-23). In addition, the location of the deposits in relation to the glomerular basement membrane in lupus nephropathy is more variable and often changes from loop to loop. Membranous glomerular lesion is the least common manifestation of glomerular involvement and has been found in only 5 to 10% of biopsy specimens in patients with SLE.[50,128] Patients with this lesion often develop proteinuria and the nephrotic syndrome at some point in the course of their disease. Although renal failure ultimately occurs, it usually develops gradually. The average five-year survival of these patients is 80 to 100%.[50]

Electron-dense deposits may also be found around tubules, tubular capillaries, and in the walls of larger blood vessels (Fig. 5-24). Tubular deposits may induce a

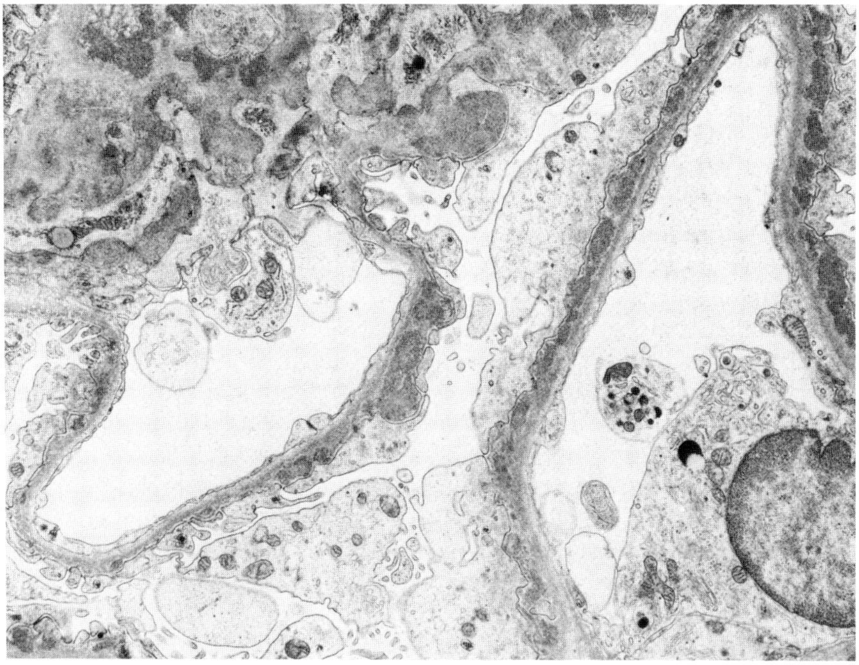

FIG. 5-23. Membranous lupus nephropathy. Note the presence of subepithelial and intramembranous deposits along the loop and the relatively smaller amounts in the mesangial matrix (left upper corner). ×7200.

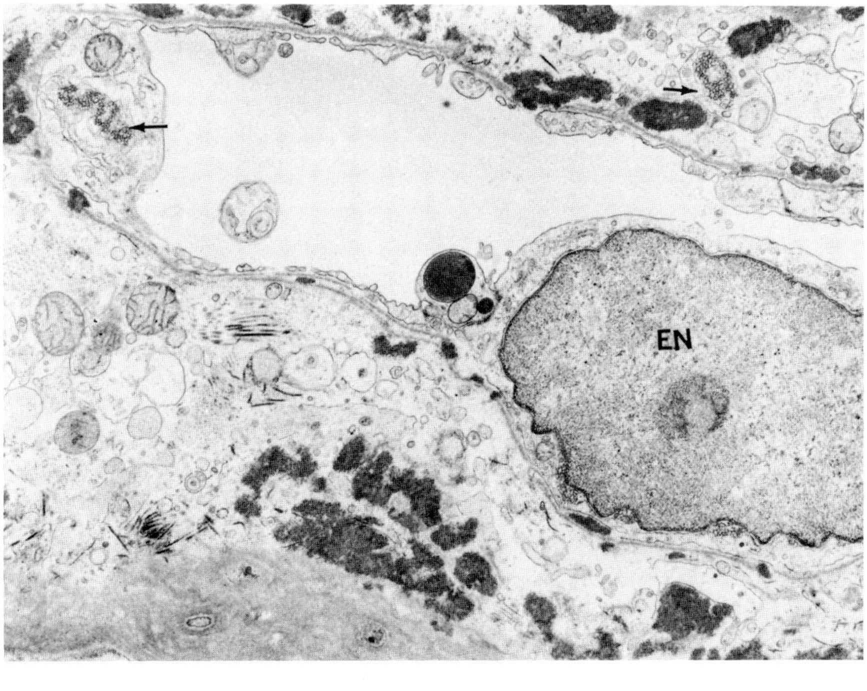

FIG. 5-24. Deposits around interstitial capillary and tubular basement membrane (left lower corner). Note the presence of microtubules within cell cytoplasm (arrows). Endothelial cell (EN). ×9500.

focal interstitial nephritis. Regardless of their location within the kidney, these deposits may occasionally have a paracrystalline appearance forming fingerprint-like structures that are probably secondary to the presence of phospholipids (Fig. 5-25).[129] These are not specific for lupus and have been produced experimentally by repeated immunization of mice with egg albumin.[130] The identification of extraglomerular electron-dense deposits or deposits with fingerprint-like appearance is always suggestive of SLE. Clusters of interwoven microtubular structures have been reported in the endothelial cell cytoplasm in glomeruli and interstitial capillaries in most biopsy specimens from patients with SLE (Fig. 5-21 and 5-24).[50,131] They also have been found in extrarenal locations (skin, skeletal muscle, lymph nodes, and circulating lymphocytes),[132,133] as well as in one-fourth of the kidney biopsy material from patients without any clinical or laboratory evidence of SLE.[130] Although these microtubular structures have a morphologically superficial resemblance to the internal protein core of paramyxoid virus, ultracytochemical and tissue culture studies have failed to demonstrate a viral origin for them.[134] The present belief is that they probably represent a nonspecific cell response to injury.

Anaphylactoid (Schönlein-Henoch) Purpura

Patients suffering from anaphylactoid (Schönlein-Henoch) purpura are usually children or young adults with skin, articular, and visceral manifestations produced by a systemic vasculitis of the leukocytoclastic type.[135,136] Renal manifestations occur in

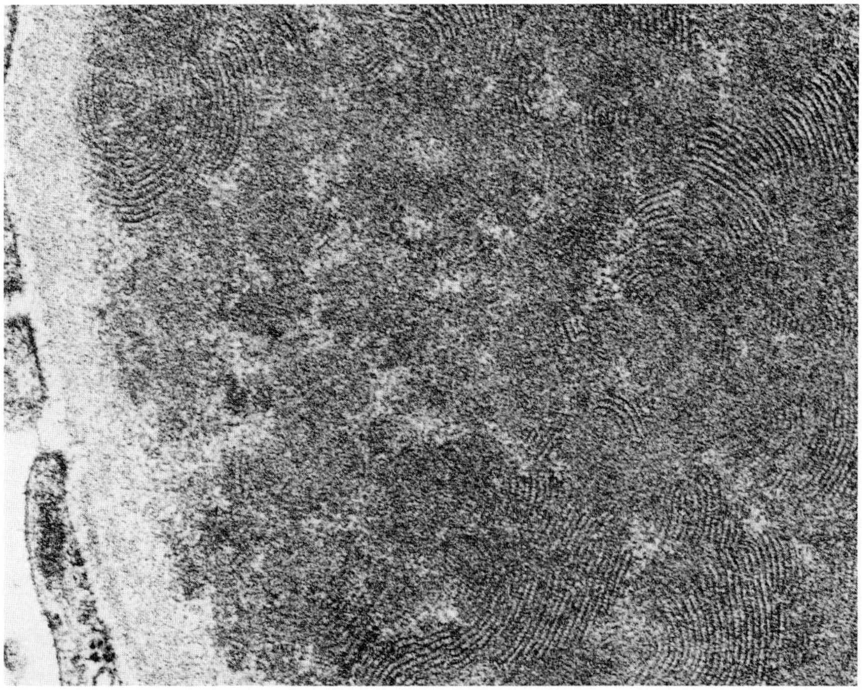

FIG. 5-25. Subendothelial deposits of fingerprint-like pattern. ×49,000.

20 to 70% of the cases[136] and range from mild hematuria and proteinuria with no reduction of renal function to severe renal insufficiency and nephrotic syndrome. Approximately 50 to 20% of the patients progress to end-stage disease.[80,137] In general, there is a good correlation between the severity of the lesion and clinical manifestations. Patients who have the nephrotic syndrome usually progress to chronic renal failure.[135,136] By light microscopy, the most characteristic changes are mesangial proliferation, with or without epithelial crescents. Focal and segmental patterns of involvement are not unusual. Occasional cases may present a pattern of mesangiocapillary glomerulonephritis. Electron microscopy most frequently reveals mesangial deposits that may extend to the subendothelial area.[138,139] In the mesangium, the deposits are located within the matrix or beneath the basement membrane. Subepithelial deposits can also be seen.[140] When in this location, the deposits may resemble those seen in membranous nephropathy,[136] or they may appear similar to the humps of postinfectious glomerulonephritis.[139] The most distinctive diagnostic feature of the disease is the demonstration of IgA by immunofluorescence. As in Berger's nephropathy, the deposits can also contain IgG, IgM, C3 and properdin. The differential diagnosis between these two conditions can be made on clinical grounds only.

Diabetic Glomerulosclerosis

Approximately 10% of all patients with diabetes mellitus die in chronic renal failure.[141] Proteinuria, usually of the nonselective type, is the most common and consistent abnormality. The nephrotic syndrome can occur in 6 to 40% of the cases.[142,143] Hypertension is common and usually complicates advanced renal failure. A moderate degree of microscopic hematuria is not a rare finding. The majority of patients who eventually develop diabetic glomerulopathy have had the disease for a duration of 10 to 20 years. Thus, the incidence of diabetic glomerulopathy could be expected to be much higher in juvenile-onset diabetes than in adult-onset diabetes. Diabetic microangiopathy throughout the body is the most distinctive morphologic change in the diabetic. This angiopathy is characterized by an increased amount of vascular basement membrane material. The most striking lesions in diabetic nephropathy are found in the glomeruli and blood vessels. This includes diffuse glomerulosclerosis, nodular glomerulosclerosis, and the so-called insudative lesions. Included in this latter group are fibrin caps, capsular drops, and arteriolar hyalinosis. Diabetic glomerulosclerosis is the general term for these lesions, and although they are considered to be an expression of the microangiopathy, it is uncertain at the present time whether these various lesions represent variants of one pathologic process or arise independently.

Diffuse glomerulosclerosis is the most common lesion observed in diabetic nephropathy.[144] There is a diffuse increase in the mesangial matrix with thickening of the capillary walls. Glomerular basement membrane thickening is the most important finding in diabetic nephropathy and is often present in both diffuse and nodular of glomerulosclerosis. Although minor degrees of basement membrane thickening are not specific for diabetes, measurement of the basement membrane width is the most satisfactory quantitative method of assessing early diabetic glomerular lesions.[145] In advanced stages, the basement membrane may reach many times its normal thickness, often with accentuation of the normal fibrillar structure (Fig. 5-26).[146]

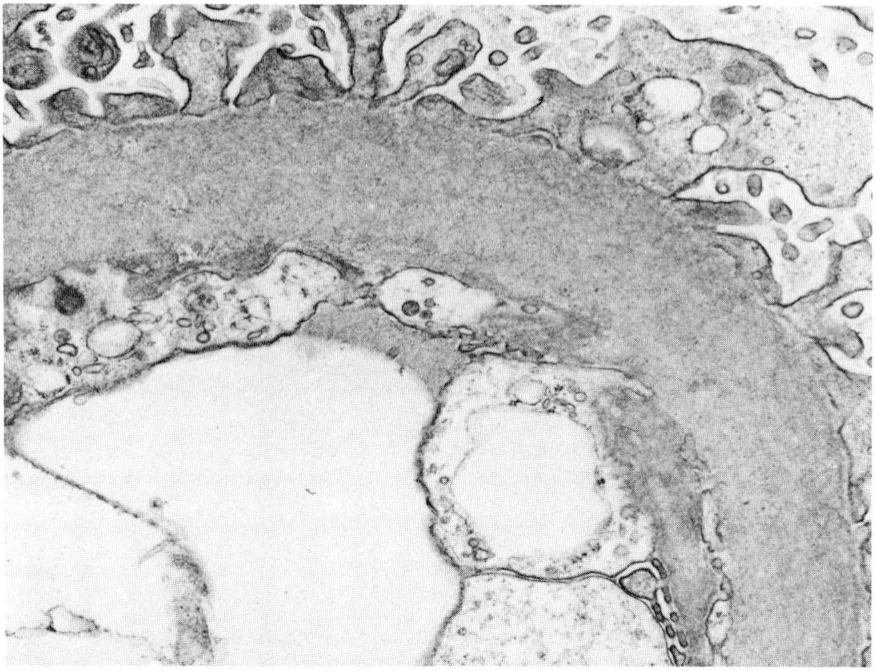

FIG. 5-26. Severely thickened glomerular basement membrane in diffuse diabetic glomerulosclerosis. ×18,000.

The basement membrane and mesangial changes by electron microscopy occur before light microscopic alterations can be diagnosed. Whether or not the mesangial lesion precedes abnormalities in the peripheral basement membrane remains controversial. However, a recent morphometric electron microscopic study done on renal biopsy specimens from short-term juvenile diabetics indicates that the earliest morphologic abnormality in diabetic glomerulopathy is widening of the peripheral basement membrane with a simultaneous increase in the amount of basement membrane-like material in the mesangial regions.[147] Mesangial cell proliferation can be seen in early diffuse diabetic glomerulosclerosis.[148] These cases should be differentiated from diffuse proliferative glomerulonephritis. Several studies have indicated that the clinical parameters of renal function are better correlated with diffuse glomerulosclerosis than with the nodular lesion.[143,144,149]

Nodular glomerulosclerosis is considered to be the characteristic lesion of diabetic glomerulosclerosis. Histochemically, the nodules react like basement membranes and mesangial matrix. Ultrastructurally, they are composed of greatly increased mesangial matrix containing lipid droplets, variable electron-dense granules, calcium deposits, and collagen fibers (Fig. 5-27). The capillary loops surrounding the nodules may have reduction of the lumina due to expansion of the mesangium, or may be aneurysmally dilated. Sometimes the capillary loops may have peripheral mesangial extension. Both diffuse and nodular diabetic glomerulosclerosis can be found to-

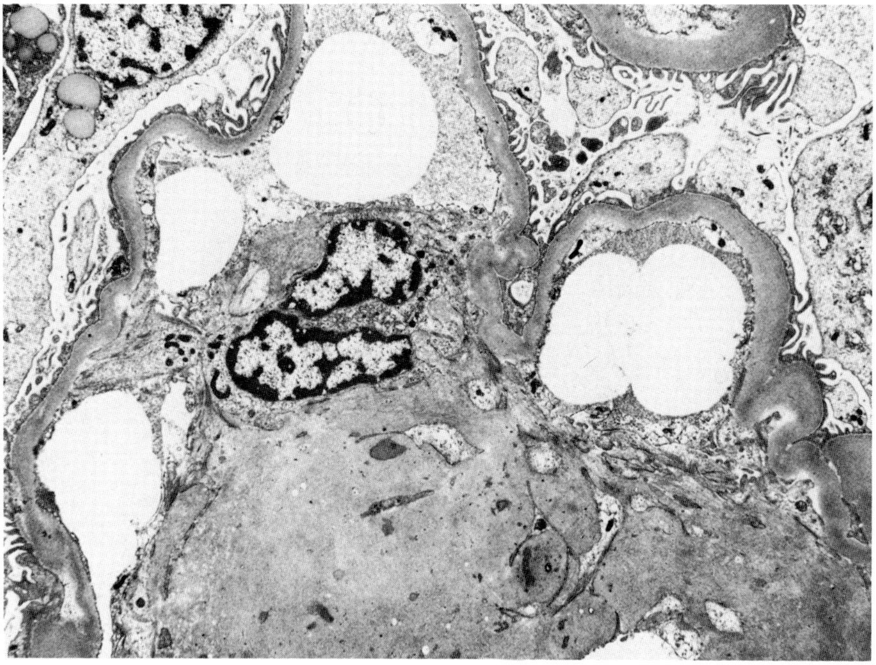

FIG. 5-27. Nodular diabetic glomerulosclerosis (Kimmelsteil-Wilson). The massive enlargement of the mesangium is due to an increase in matrix. Note pronounced thickening of glomerular basement membrane. ×5400.

gether, not only in the same patient, but also in the same glomerulus. Although nodular glomerulosclerosis is considered almost pathognomonic of diabetic nephropathy, identical light microscopic lesions can be found in cases of multiple myeloma and kappa light chain nephropathy.[150–152] However, in this latter disorder, the nodules appear to be composed of fibrillar electron-dense material that stains with antiserum against kappa light chain and lacks the basement membrane-like material characteristic of diabetic nodular glomerulosclerosis. Diffuse or nodular deposits of amyloid can sometimes be confused with diabetic glomerulosclerosis by light microscopy, but the ultrastructural identification of amyloid fibrils establishes the diagnosis.

The insudative lesions are the least specific of the glomerular changes in diabetes. Ultrastructurally, they appear as masses of finely granular, electron-dense material, often containing lipid droplets (Fig. 5-28). They usually are located at the periphery of the loop in a subendothelial location (fibrin caps), within the basement membrane of Bowman's capsule (capsular drops), or sometimes in the mesangium or glomerular basement membrane. In blood vessels, they are more extensive in the subintimal and media layers, but can involve the adventitia.[153] Histochemical and immunofluorescence studies indicate that this insudative material represents infiltration of plasma constituents including protein, lipids, and mucopolysaccharides.[154, 155]

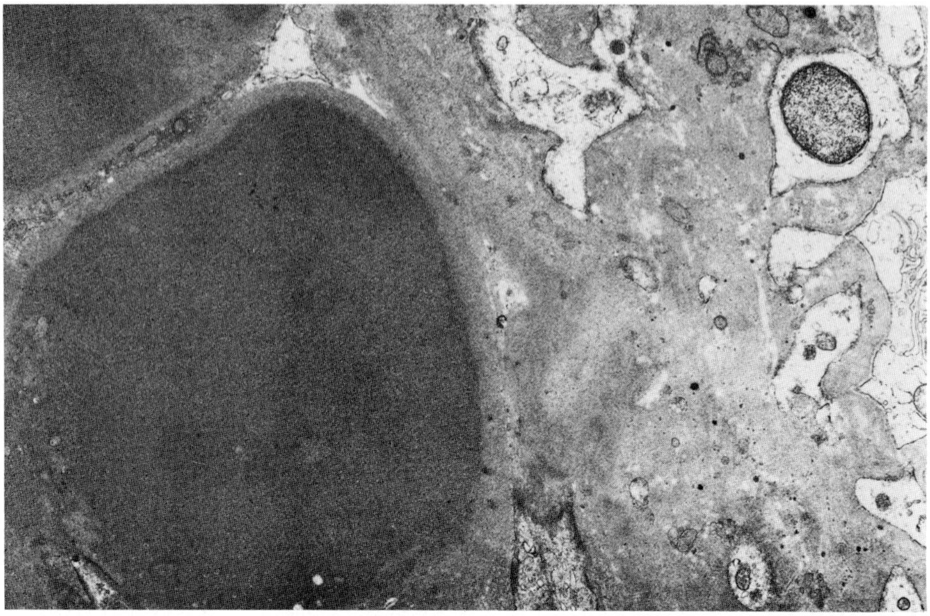

FIG. 5-28. Insudative lesion ("fibrin caps") in diabetic glomerulosclerosis. Note marked increase of mesangial matrix. ×5800.

Amyloidosis

Amyloidosis is a term used to designate the group of conditions characterized by extracellular deposition of fibrillar proteins that have a β-pleated sheet configuration on x-ray diffraction analysis.[156] Recent studies indicate that amyloid is not a material having a constant chemical composition but is instead a group of proteins sharing common physical characteristics.[157] The deposits can be distributed throughout the body, involve many organs and tissues (systemic amyloidosis), or may be limited to a single site (localized amyloidosis). Due to its insolubility and relative resistance to proteolytic digestion under physiologic conditions, amyloid interferes with physiologic functions, produces gradual destruction of vital organs, and can lead to death.

Systemic amyloidosis with renal involvement can be secondary to a variety of chronic inflammatory diseases, such as osteomyelitis, bronchiectasis, tuberculosis, leprosy, rheumatoid arthritis, Crohn's disease, and ulcerative colitis; it can also be secondary to certain malignancies such as Hodgkin's disease, malignant lymphoma, and bronchogenic and renal carcinoma. It may also appear as an idiopathic primary entity, occurring either sporadically or in heredofamilial form.[156,158] Renal involvement occurs in over two-thirds of the patients with secondary amyloidosis, in 10 to 15% of cases with multiple myeloma, and in 18 to 28% of primary amyloidosis.[159-162] Nonselective proteinuria, with or without the nephrotic syndrome, is the most common manifestation of renal involvement by amyloidosis. The most significant deposition of amyloid in the kidneys occurs in glomeruli, but it can also be seen around tubules, in the interstitium, and in blood vessels. Although several histochemical techniques can be used to detect amyloid (crystal violet, thioflovin T or S, and Congo red), none is absolutely specific, and all require a critical amount of amyloid before a positive reaction can be produced.

In early cases of renal involvement, when the conventional histochemical preparations fail to detect the deposits of amyloid, electron microscopy is the only method available to diagnose amyloidosis. Ultrastructurally, the amyloid fibrils in tissue sections are arranged randomly and appear as rigid, nonbranching rods measuring 80 to 100 A in diameter and 300 to over 10,000 A in length. Sometimes a beaded structure with a periodicity of 55 A can be seen in the fibrils. On cross section, the fibrils appear to have hollow cores. Initially, the amyloid deposition occurs in the mesangium between mesangial cells, between mesangial and endothelial cells, and between mesangial cells and the basement membrane. From the mesangium, amyloid deposition extends into the subendothelial zone and throughout the basement membrane (Figs. 5-29 and 5-30).[163] A spicular appearance can sometimes be produced and can, on light microscopy, resemble membranous nephropathy.[164] The epithelial foot processes adjacent to the deposits of amyloid are usually obliterated. In a few reported cases, some investigators found fibrils with the ultrastructural characteristics of amyloid in the urinary sediment from patients with renal amyloidosis, indicating the value of this procedure for diagnostic purposes.[165] However, more recent studies have demonstrated fibrils ultrastructurally indistinguishable from amyloid which are in the urine of patients with nephrotic syndrome due to amyloidosis.[166]

Hemolytic Uremic Syndrome

Hemolytic uremic syndrome (HUS) is a term used to describe a condition characterized by microangiopathic hemolytic anemia, thrombocytopenia, and renal failure.[167,168] Although originally described in children, it can occur in adults. It is more common before the age of four, and both sexes are affected equally. In adults, it usually occurs during the postpartum period or in association with exogenous estrogen exposure, particularly oral contraceptive preparations.[169–171] Characteristically, the renal and systemic manifestations of the disease appear simultaneously or a few days after a febrile episode of diarrhea and vomiting, or, less frequently, after an upper respiratory tract infection. The renal involvement is manifested by an abrupt onset of oliguria accompanied by proteinuria, hematuria, and hypertension. Coombs negative hemolytic anemia, thrombocytopenia, and bleeding diathesis usually occur at this stage of the disease. Hepatosplenomegaly, jaundice, and central nervous system involvement occur in about half of the cases.

Biopsy specimens taken early in the course of the disease show fibrinoid necrosis, intimal and subintimal fibrin deposits with red cell insudation, thrombosis, and endothelial cell proliferation in the small arteries and arterioles. A characteristic change is aneurysmal dilatation, accompanied by proliferation, of the ascending arteriole (glomeruloid body).[50] The glomeruli show acute ischemic changes, and some of them may appear infarcted. There is mesangial endothelial swelling with mild, irregular cell proliferation that is sometimes accompanied by crescent formation. The capillary walls related to the areas of infarction may show aneurysmal capillary dilatation.[171]

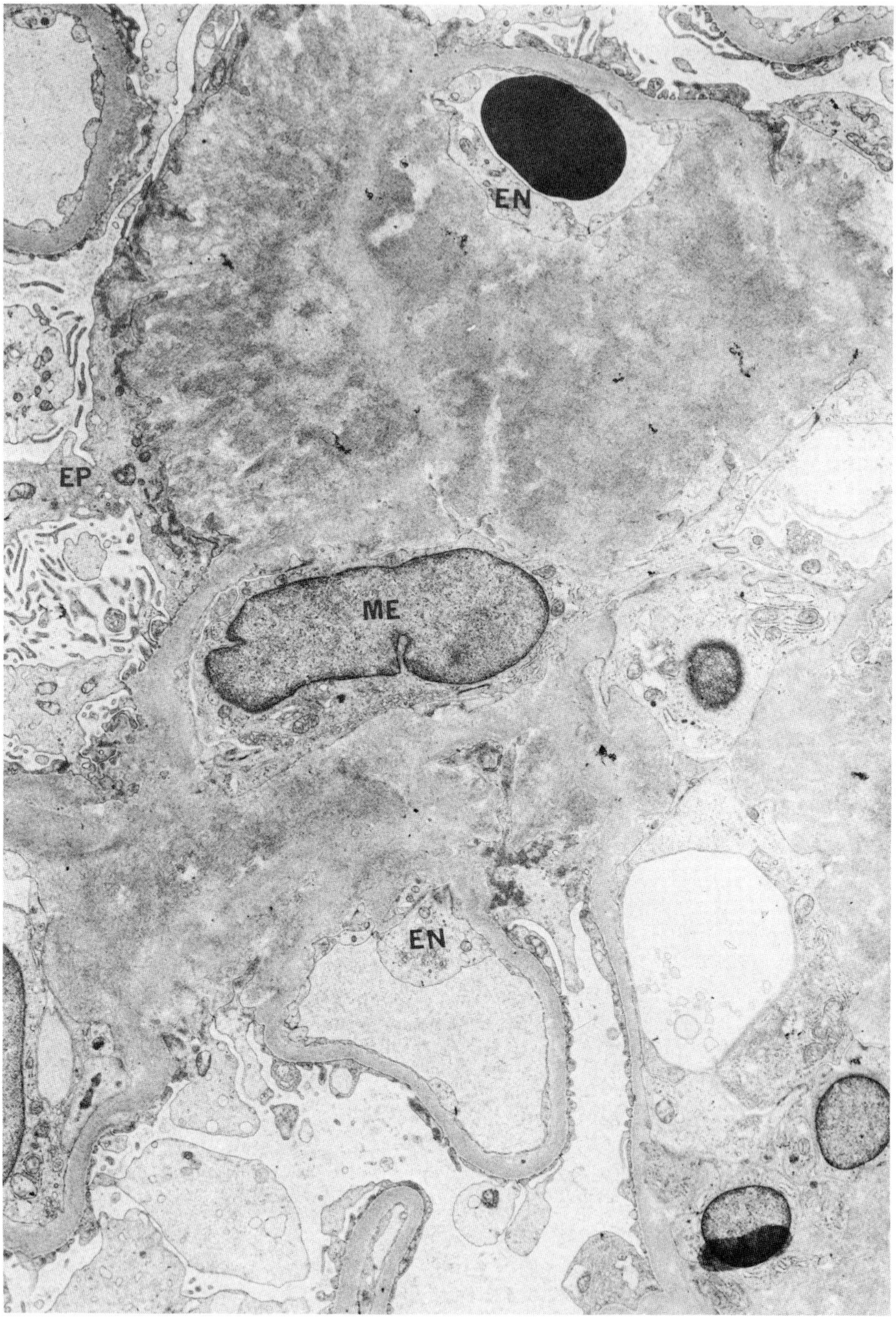

FIG. 5-29. Massive mesangial amyloid deposit with extension into the subepithelial and subendothelial zones. Epithelial cell (EP), endothelium (EN), mesangium (ME). ×11,000.

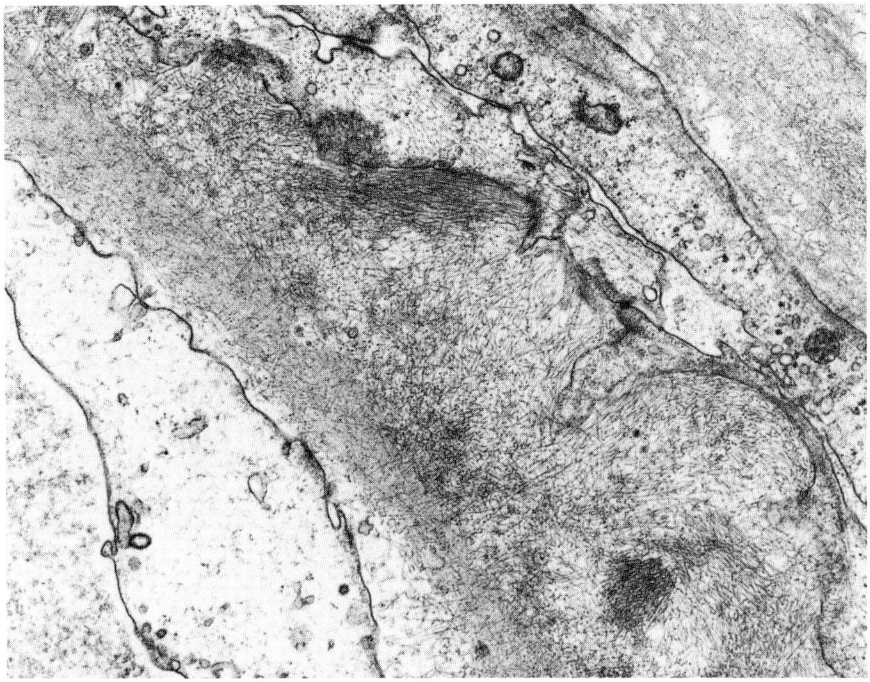

FIG. 5-30. Subepithelial and intramembranous amyloid fibrils. ×19,000.

The most characteristic feature of the disease is found on electron microscopy. It consists of narrowing of the capillary lumen due to a prominent widening of the subendothelial space, which becomes filled with pale, finely particulated or fibrillar material and which, on immunofluorescence microscopy, stains positively for fibrinogen.[50, 173, 174] An irregular, thin layer of basement membrane-like material can be seen between the endothelial cytoplasm and the electron-lucent area (Fig. 5-31). Due to its argyrophilic characteristics, this material can produce a double contour on light microscopy, thus simulating mesangiocapillary glomerulonephritis. The capillary lumina may be occluded by fibrin and platelet aggregates. In these thrombosed areas, the endothelium is often disrupted. The mesangial cells are swollen and hypertrophic and show numerous phagolysosomes.[172] The epithelial cell cytoplasm is usually hyperactive and may show extensive foot process obliteration in the areas of collapse (Fig. 5-32). In later stages, there is a prominent, mucinous, onionskin-like, obliterating endarteritis involving interlobular arteries and arterioles. The vascular lesions are very similar to those seen in progressive systemic sclerosis and malignant hypertension. Because widespread glomerular subendothelial widening can also be present in these conditions, the differential diagnosis between progressive systemic sclerosis, malignant hypertension, and hemolytic uremic syndrome may be extremely difficult and often is dependent upon careful analysis of clinical data. Neither the pathogenesis nor the etiology of hemolytic uremic syndrome is known. Infectious agents, hereditary factors, and such drugs as ergotamine compounds and

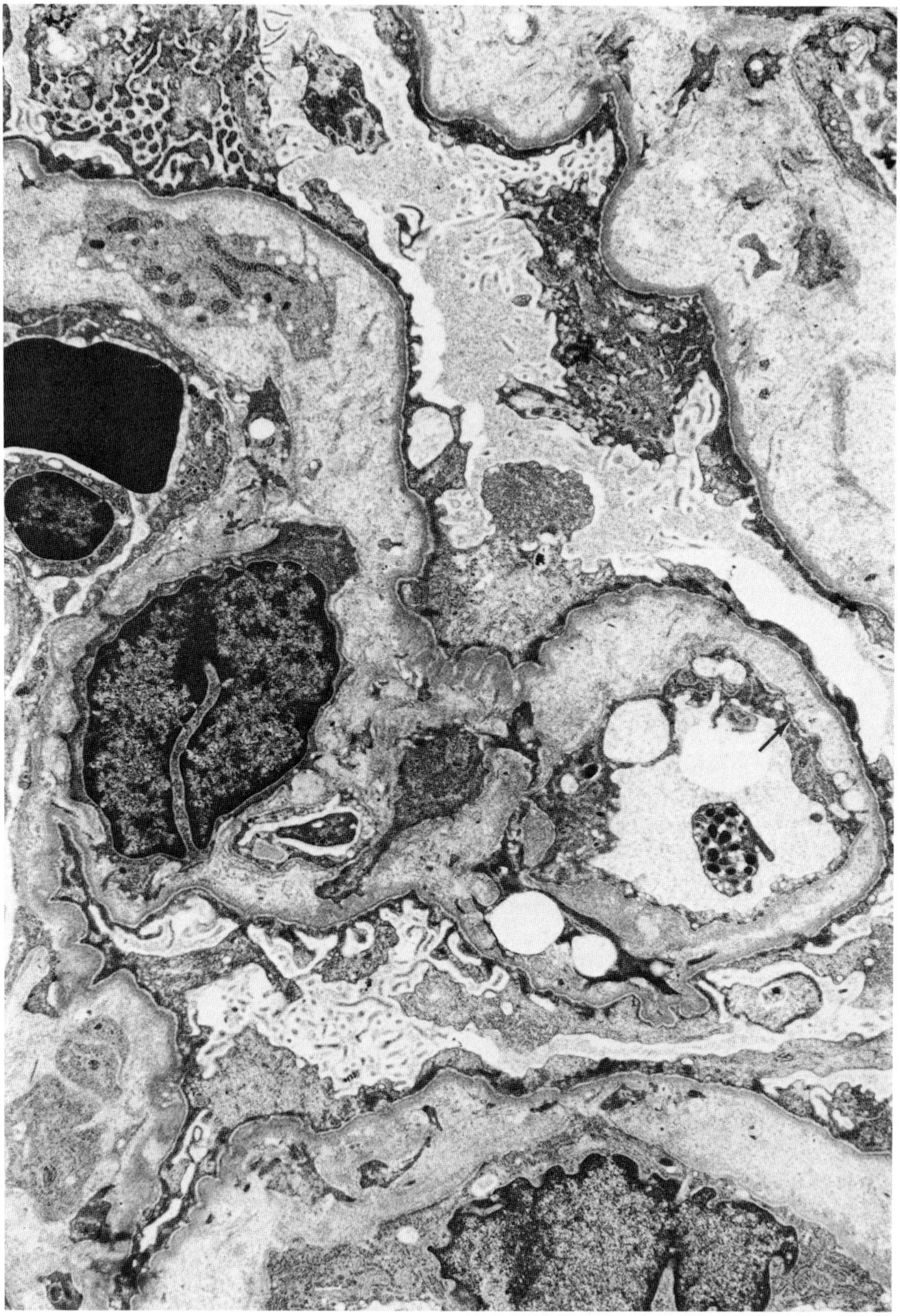

FIG. 5-31. Portion of a glomerulus from a patient with HUS. The lamina rara interna is markedly thickened and contains pale granular and focally fibrillar material, as well as fragments of cell cytoplasm and newly formed, thin lamina densa layer (arrow). The lumen is narrowed, and there is a moderately severe obliteration of foot processes. ×8200.

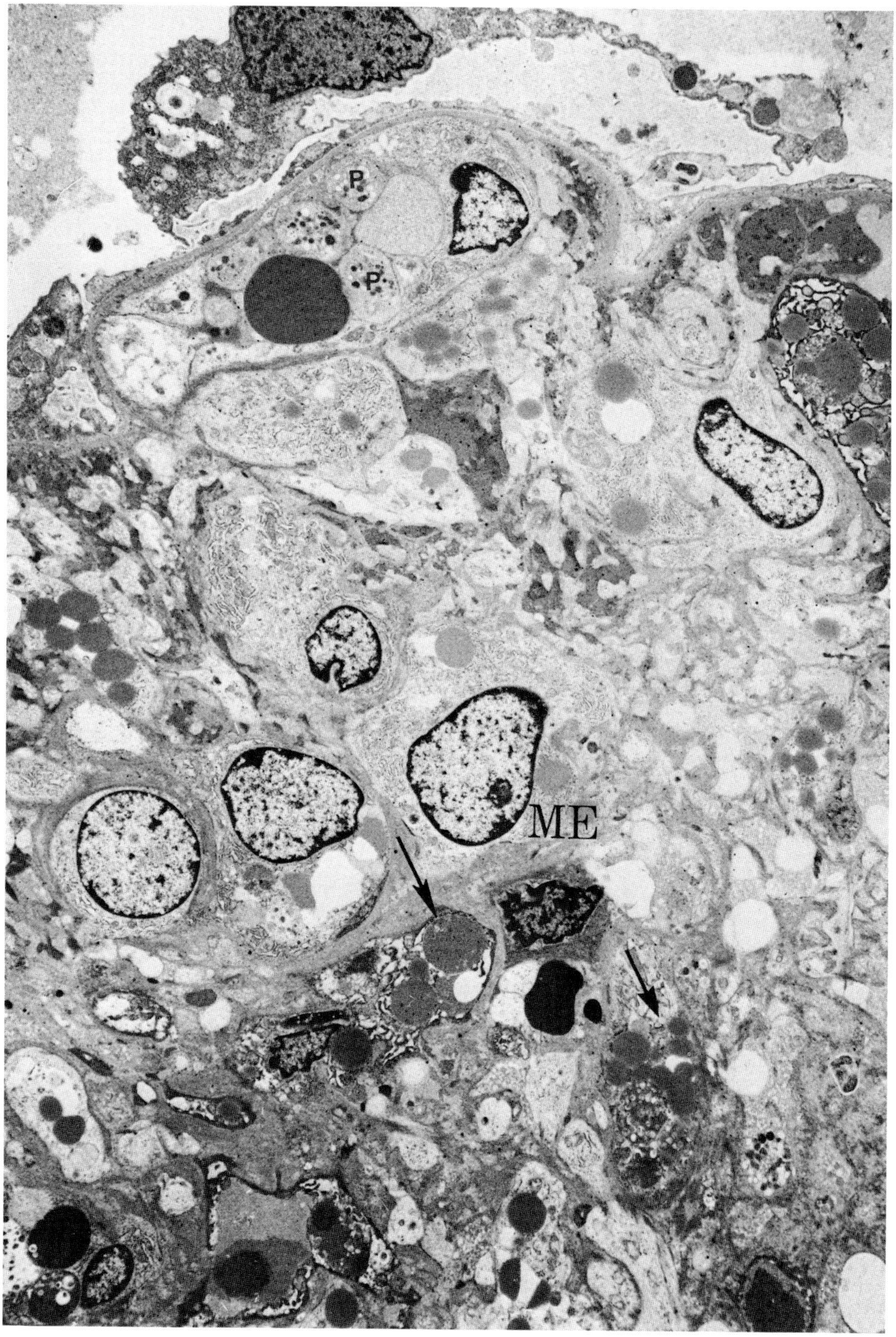

FIG. 5-32. Hemolytic uremic syndrome. The capillary lumen contains several platelets (P). The mesangium is prominent and edematous, and shows numerous phagolysosomes (arrows). The foot processes are obliterated. Endothelial cell (EN), mesangium (ME). ×3300.

oral contraceptives have been implicated. The vascular and glomerular changes, and certain similarities to the Shwartzman reaction, suggest that intravascular coagulation is the underlying process.[175] However, the factors that may precipitate the intravascular coagulation remain unknown.

Preeclamptic Nephropathy

The primary changes in preeclamptic nephropathy are confined to the glomeruli, which are characteristically enlarged and swollen with little or no hypercellularity. The capillary loops have decreased vascularity and appear bloodless because of cytoplasmic changes in the intracapillary cells. Visceral epithelial cells may be prominent as a result of cytoplasmic swelling and an increased number of protein transport droplets. At times, the glomerular swelling results in herniation of the glomerular tuft into the neck of the proximal tubule ("pouting").[176] The reactive endothelial and mesangial cells contain a variety of cytoplasmic vacuoles, and the swollen cytoplasm narrows or obliterates the capillary lumen and may cause some widening of the lobular stalk. The most prominent endothelial cell vacuoles are clear and are lined by a single membrane. Although the early description of this change resulted in the labeling of this lesion with the term *endotheliosis,* it is now clear that the mesangial cells are usually more affected.[50] The mesangial change is more variable and consists of a wide variety of lysosomal structures, some of which have a myeloid figure appearance. Others appear as bland lipids with variable amounts of residual density, and some are completely clear (Fig. 5-33).[177] When the mesangial cell hypertrophy is

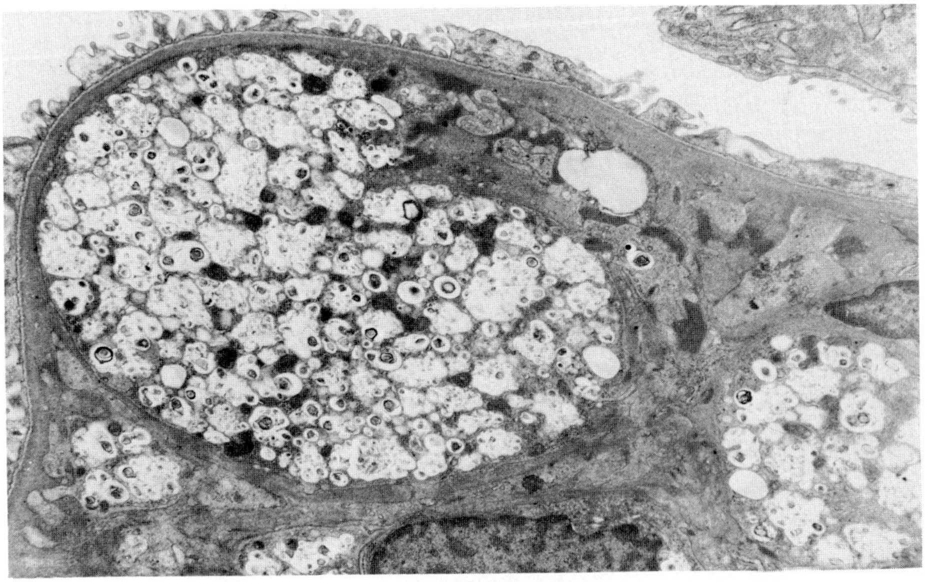

FIG. 5-33. Preeclamptic nephropathy. The capillary lumen is obliterated by swollen endothelial cells containing myeloid bodies. Similar lysosomal changes are also present in mesangial cells. Note the deposition of granular electron-dense material beneath the basement membrane. ×8000.

severe, mesangial interposition between the endothelium and the basement membrane may occur. A variable amount of amorphous extracellular electron-dense material separates the swollen cells at points of severe narrowing of the capillary lumen. These mesangial and subendothelial deposits usually have immunofluorescence characteristics of fibrinogen. Distinct strands of fibrin are seen in the most severe cases. On occasion, translucent subendothelial zones containing loose fibrillar material have occurred. After a significant postpartum interval, large cholesterol clefts may be present in addition to the lipid-containing vacuoles.[50] The renal vasculature, specifically the afferent arteriole, is not sclerotic, and the endothelial cells remain normal. The contrast between the swollen, vacuolated glomerular endothelial cell and the adjacent normal endothelium of the arteriole is striking. Patients who have preeclamptic nephropathy superimposed on essential hypertension have hypertensive arterial and arteriolar changes in addition to the glomerular lesions described. The prognosis of patients with preeclampsia is quite good. Glomerular capillary changes disappear within a few weeks after delivery, and the blood pressure usually returns to normal within a one-month period.[50]

Cryoglobulinemia

Cryoglobulins are complexes of immunoglobulins composed of one or more different classes of immunoglobulins that precipitate at low temperatures (4 C) and become soluble again when the temperature is elevated (30 C). The immunoglobulins may be monoclonal (comprising one class of immunoglobulins such as IgG, IgM, IgA, or Bence Jones protein), or they may be mixed, with IgG-IgM being the most common combination. Cryoglobulins are found in the serum of patients with various disease states, such as lymphoproliferative disorders, chronic infection, and autoimmune conditions. However, in approximately 30% of the cases, no associated disorder is found (essential or idiopathic cryoglobulinemia).[178]

A distinctive clinical syndrome characterized by weakness, purpura, arthralgias, hepatosplenomegaly, lymphadenopathy, and glomerulonephritis has been observed in patients with idiopathic mixed cryoglobulinemia. Renal biopsy findings in mixed IgG-IgM essential cryoglobulinemia have included diffuse proliferative glomerulonephritis (often with a mesangiocapillary pattern), focal and segmental glomerulonephritis, and, less frequently, membranous nephropathy.[49,179] Epithelial crescents and intracapillary thrombi and wire loops, which react for immunoglobulins, C3, and early complement components, may produce a picture analogous to lupus glomerulonephritis. Vasculitis and interstitial nephritis may also be present. Electron microscopy often demonstrates mesangial, subendothelial, intramembranous, and, less frequently, subepithelial deposits. In approximately half of the cases, the glomerular deposits and cryoprecipitate have a highly characteristic, organized, fibrillar structure (Fig. 5-34).[49] At the present time, it is thought that IgG-IgM complexes fix, complement, and initiate tissue injury at the site of deposition.

Multiple Myeloma

Involvement of the kidney is common in multiple myeloma. The classic lesion consists of protein casts accompanied by giant cell reaction to the casts in the distal and

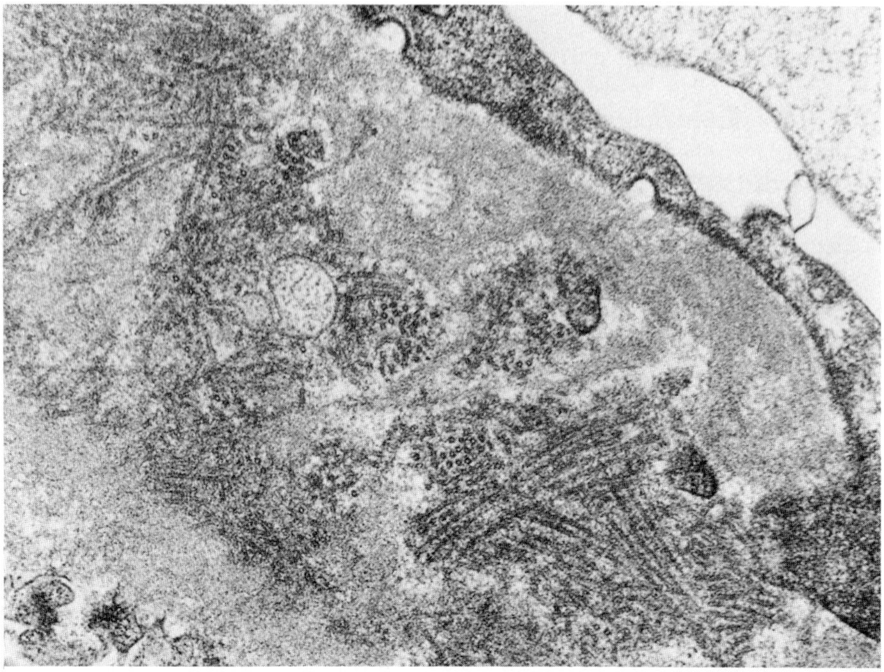

FIG. 5-34. Glomerular basement membrane from a patient with membranous nephropathy and mixed IgG-IgM cryoglobulinemia showing microtubular fibrillar deposits. ×54,000.

collecting tubules. Obstruction produced by these casts is considered to be the main cause of renal impairment. Glomerular lesions and interstitial infiltrates of plasma cells may occur and also contribute to the disturbed renal function. Although in most cases the glomeruli are normal by light microscopy or show nonspecific changes by electron microscopy, some patients may show a variable degree of mesangial widening simulating diffuse or nodular diabetic glomerulosclerosis.[150,151] In some of these cases, electron microscopy may reveal deposits of amyloid or, less frequently, kappa light chain.[152,180] In the latter condition, the deposits appear as a continuous, electron-dense, fine-grained material localized along the inner aspect of the basement membrane and in the mesangium. Similar deposits can be seen in the outer aspect of the tubular basement membrane, in blood vessels, and in the interstitium (Fig. 5-35). Although these deposits may be confused with dense deposit disease, the rather regular distribution of deposits in the latter condition, without significant involvement of the lamina densa, and the typical electron density of the deposits, permit its recognition. Deposition of kappa light chains can occur in other organs. The glomerular (endothelium) is frequently disrupted, and it is not rare to find platelets or fibrin thrombi in the capillary lumina. The foot processes may show extensive obliteration. The mesangial cells may be reactive and, on occasion, may extend into the subendothelial space. Crystalloid structures may sometimes be found in tubular casts and, less frequently, in glomerular epithelial and endothelial cells.[50,180,181] Immunofluorescence studies demonstrate that the structures are usually composed of lambda light chains.

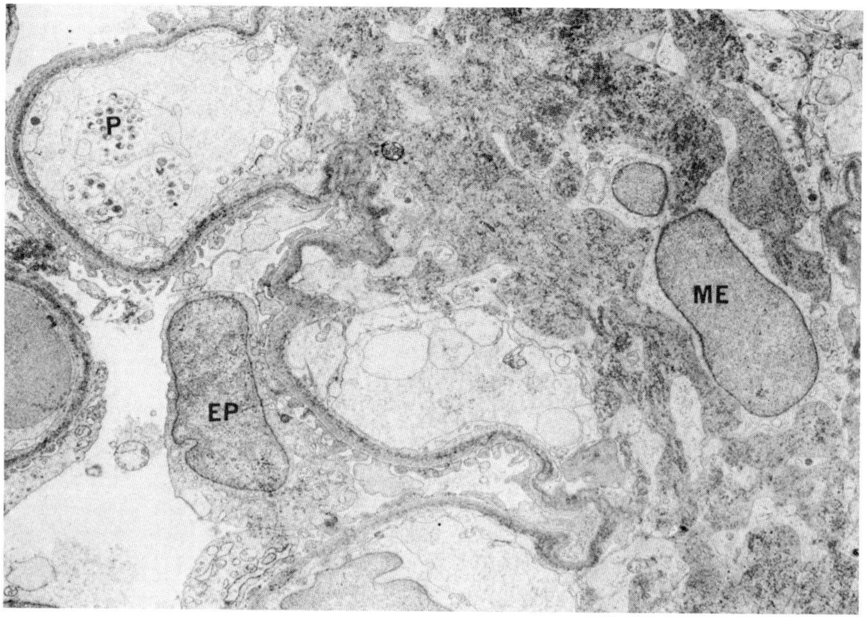

FIG. 5-35. Subendothelial and mesangial deposition of electron-dense material in kappa light chain nephropathy. Epithelial cell (EP), mesangium (ME), platelet (P). ×6000.

Hepatic Glomerulosclerosis

It has become generally accepted that patients with hepatic cirrhosis may develop a rather characteristic pattern of glomerulosclerosis or mild glomerulonephritis. Recent prospective studies have demonstrated a high prevalence of glomerulonephritis in patients with chronic liver disease, which is most often related to alcohol abuse.[182] Clinically severe hepatic glomerulosclerosis may be manifested by proteinuria and renal insufficiency. Glomerular changes, however, may be present in up to three-quarters of cirrhotic patients and need not be manifested by urinary abnormalities.[183]

Examination by light microscopy reveals diffuse or segmental mesangial proliferation, widening of the mesangium, and focal thickening of the capillary wall. The most characteristic findings are the presence of subendothelial and mesangial electron-dense granular deposits containing IgA and particulate structures usually located in the mesangial matrix and subendothelial layer of the basement membrane. These particles lie in the center of the clear zones, which imparts a moth-eaten appearance to the affected areas.[64] In addition, one may see very osmiophilic granules and peculiar circular laminated structures, probably containing calcium (Fig. 5-36).

Hereditary and Congenital Nephropathies

Glomerular lesions associated with hereditary nephropathies have historically been grouped with the focal, nonspecific forms of glomerulonephritis. The prognostic sig-

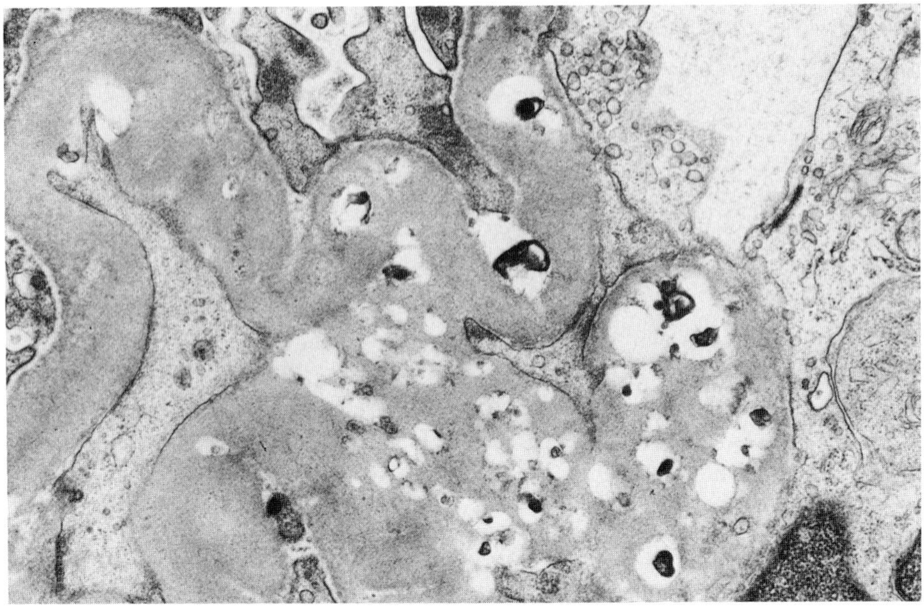

FIG. 5-36. Portion of a glomerulus from a patient with hepatic glomerulosclerosis showing numerous clear zones with irregular dense particles. ×9500.

nificance of differentiating the early patterns of glomerular change was not recognized, and both benign and progressive lesions were grouped as a single pathologic entity. It is now apparent that glomerular pathologic lesions associated with specific hereditary nephropathies are more numerous and varied than previously diagnosed.

Hereditary Nephritis. Hereditary nephritis most frequently occurs with a clinical syndrome characterized by a family history of nephropathy, often associated with perceptive deafness (Alport's syndrome). The disorder can also be associated with other conditions, such as polyneuropathy, hyperprolinemia, ichthyosis, platelet disorders, bone abnormalities, and ophthalmologic defects.[184] Most studies indicate that the syndrome appears to be transmitted as an autosomal dominant with variable penetrance and expression[185–187] and partial sex linkage.[187] Although males and females are affected with equal frequency, the prognosis differs markedly. In men, the disease is usually progressive, leading to renal failure and hypertension, and death occurs in the third or fourth decade; in women, it usually pursues a much more benign and not necessarily fatal course.[187] Although the onset is usually in childhood, there are a few cases in which the disease is manifested either during the neonatal period or later in life. The most common clinical presentation is an episode of gross hematuria frequently following a nonspecific upper respiratory infection, or perhaps the discovery of microscopic hematuria, proteinuria, or deafness. Proteinuria is usually mild, but in the advanced disease may become sufficiently severe to cause the nephrotic syndrome.[186, 188]

In young children, the renal biopsy specimen usually shows no significant changes on light microscopy. The glomeruli may appear normal or show mild mesangial hypercellularity with minor thickening of the capillary walls. By electron microscopy, the lamina densa appears to be split into multiple, interwoven lamellae about 300 to 1000 A wide, enclosing electron-lucent spaces containing dark particles 300 to 500 A in diameter (Fig. 5-37).[189, 190] The nature of these particles is uncertain, but they possibly represent trapped lipid droplets.[181] Sometimes thin basement membrane segments may be found immediately adjacent to, or in between, split areas. The lamina densa in these attenuated areas may be regular or may have varying degrees of rarefaction and reticulation. The subepithelial side of the basement membrane is often irregular, but the endothelial aspect is generally smooth. Similar lesions can be present in the tubular basement membrane and in the basement membrane of Bowman's capsule; however, in these locations they do not have the same diagnostic value.[190] Although the pathogenesis of hereditary nephritis remains obscure, there is strong evidence that the disease primarily affects the glomerular and probably other basement membrane.[184]

Benign Familial Recurrent Hematuria. Benign familial hematuria is a syndrome characterized by recurring episodes of macroscopic or microscopic hematuria without progression to renal failure of other associated abnormalities, and an autosomally dominant pattern of inheritance.[184] Both sexes are affected with similar fre-

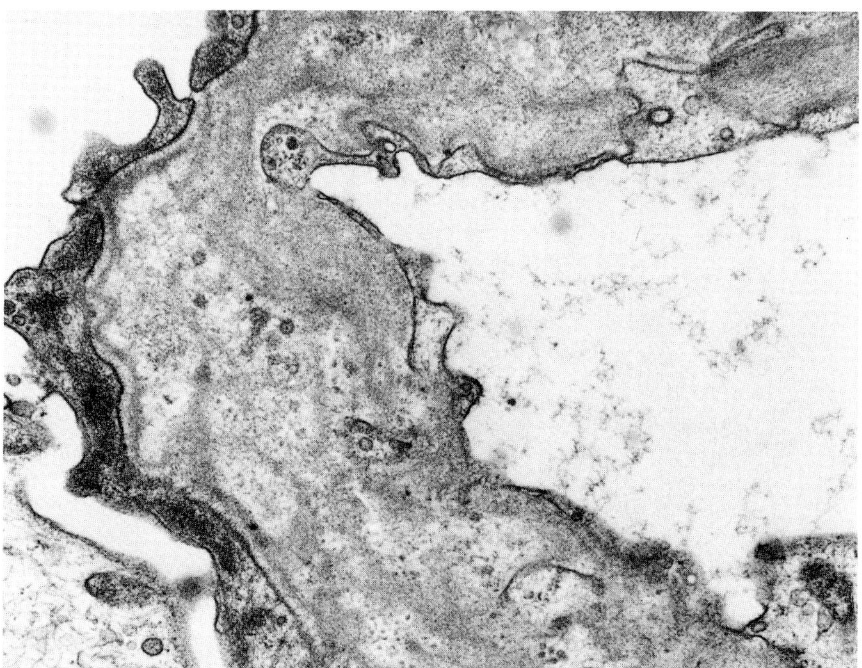

FIG. 5-37. Hereditary nephritis. The lamina densa is irregularly thickened, reticulated, and lamellated. ×27,200.

quency and clinical intensity. The disease can be manifested at any age, although it usually occurs during childhood. Light microscopic examination of renal biopsy specimens from patients with benign familial hematuria generally shows no abnormalities, although some specimens may have minor, nonspecific histologic changes, such as mild mesangial enlargement. Electron microscopic examination shows marked thinning of the glomerular basement membrane with occasional ruptures, which are presumed to be the source of the hematuria (Fig. 5-38).[184, 191] In general, the involved loops have uniform attenuation, and the lamina densa has regular contours. However, in small focal areas, the basement membrane may have mild irregularities and rarefaction. Sometimes the differential diagnosis between benign familial hematuria and early hereditary nephritis can be clinically and histologically difficult. Under the light microscope, both entities can appear identical. In these cases, only electron microscopy can establish the diagnosis. Although focal attenuation of the basement membrane can be seen in hereditary nephritis, it does not have the extension and severity present in cases of benign familial recurrent hematuria.

Congenital Nephrotic Syndrome. The nephrotic syndrome is uncommon in the first year of life. The designation of congenital nephrotic syndrome includes a heterogenous group of diseases and is reserved for those patients who present clinical symptoms within the first weeks of life. Two distinct types have been recognized: infantile microcystic disease (Finnish type) and diffuse mesangial sclerosis. Both forms probably have autosomal recessive inheritance, do not respond to steroid or

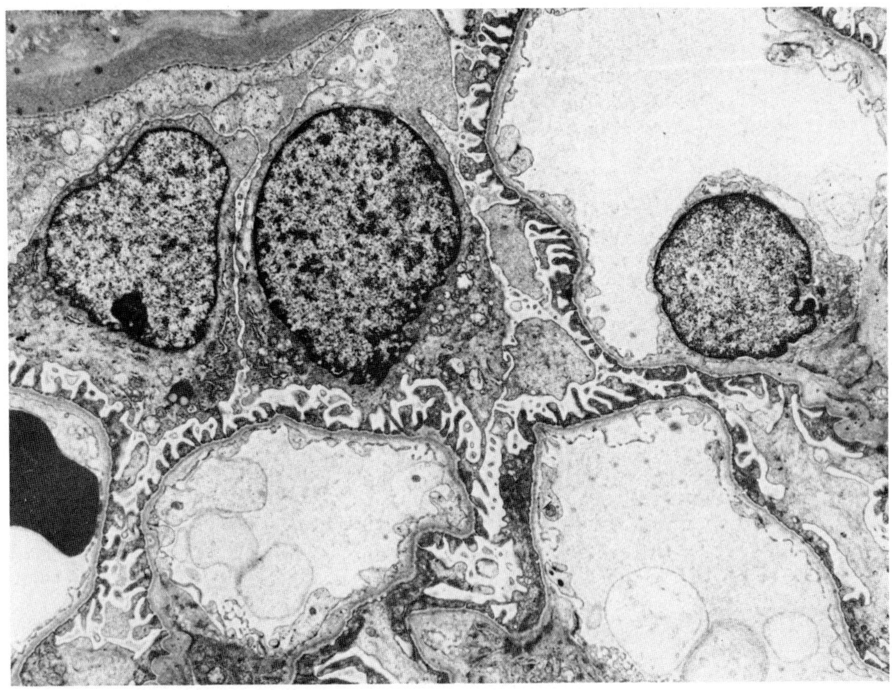

FIG. 5-38. Portion of a glomerulus from a patient with benign familial recurrent hematuria showing thinning of the basement membrane. ×7200.

immunosuppressive therapy, and consequently have a poor prognosis. Renal biopsy is essential for differentiating these forms of congenital nephrotic syndrome from other causes of primary nephrotic syndrome occurring during the neonatal period, such as membranous nephropathy (associated with congenital syphilis or mercury intoxication), congenital toxoplasmosis, mesangial proliferative glomerulonephritis, cytomegalic inclusion disease, epithelial cell disease, and focal segmental glomerulosclerosis.

The congenital nephrotic syndrome of the Finnish type is rare and represents less than 1.5% of the cases of primary nephrotic syndrome in childhood.[13] Both sexes are equally affected. Although about half of the reported patients are from Finland, the disease has been observed in other ethnic groups.[192] The patients are characterized at birth by large placentas, proteinuria, and high susceptibility to infections.[193] Premature birth, mild abnormalities in face and limbs, and poor somatic development are common. Usually the nephrotic syndrome is manifested during the first three months of life. Histologically, the most striking feature is ectasia of the proximal tubules, with flattening of the tubular epithelium. The glomeruli show variable degrees of mesangial proliferation, sclerosis, and distention of Bowman's capsule. Also, there may be an excessive number of immature glomeruli. Electron microscopy demonstrates similar changes to those present in epithelial cell disease. There is extensive obliteration of the foot processes with villous hyperplasia and an increase of mesangial matrix. Immunofluorescence is negative. Patients with diffuse mesangial sclerosis usually develop the nephrotic syndrome after the fourth month of age.[192,194] Renal biopsy specimens show extensive diffuse glomerular sclerosis without cell proliferation. Electron microscopic studies of this condition have not been reported.

Hereditary Onycho-osteodysplasia. Onycho-osteodysplasia, or nail-patella syndrome, is an uncommon, autosomally dominant, hereditary disorder characterized by dysplasia of the fingernails, absence or hypoplasia of the patellae, subluxation of the radial head, iliac horns, and other less common skeletal and ocular abnormalities.[194] About 30% of the patients show renal involvement.[195] Although in early reports the kidney lesion was considered to be relatively benign with symptomatic proteinuria, nephrotic syndrome can occur, and some patients may develop renal failure.[196] Light microscopy of the kidney shows nonspecific changes, such as focal thickening of the glomerular capillary walls and a variable degree of glomerular sclerosis. The characteristic ultrastructural change is the presence of collagen-like fibers within the basement membrane (Fig. 5-39).[197,198] Based on this observation, it has been postulated that the renal involvement could be due to an enzymatic defect in collagen metabolism, with abnormal fibers trapped in the glomerular basement membrane.[198] Other electron microscopic findings include thickening of the basement membrane and focal areas of electron lucency, which give the basement membrane a moth-eaten appearance. Epithelial foot process obliteration can also be present.

Fabry's Disease. Angiokeratoma corporis diffusum universale, or Fabry's disease, is an X-linked disorder of glycosphingolipid metabolism characterized by accumulation of the ceramide trihexoside, which results from a deficiency of ceramide trihexosidase, an α-galactosidase.[199] The skin lesions consist of innumerable small angiokeratomas scattered over the body. The systemic manifestations are a result of the

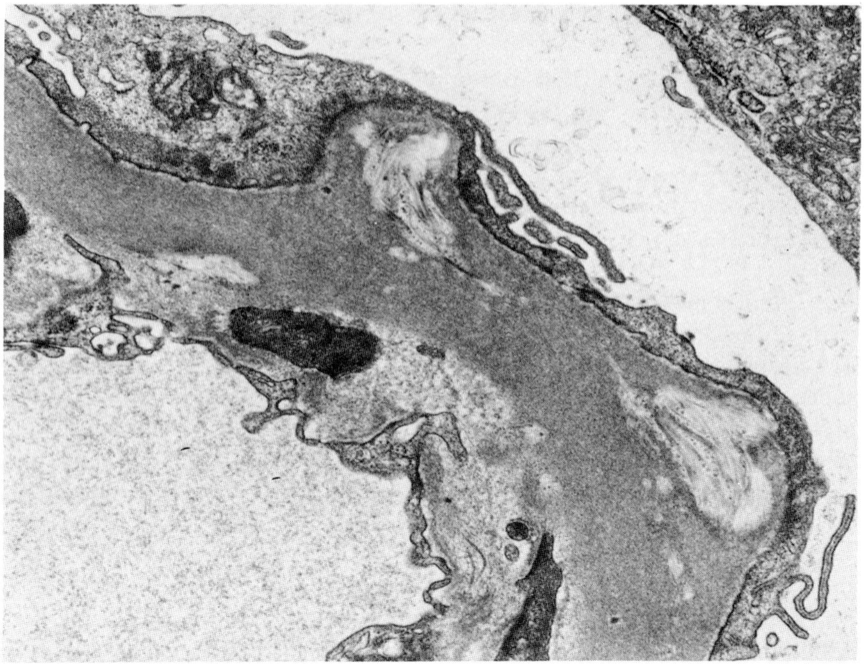

FIG. 5-39. Nail-patella syndrome. The basement membrane is irregularly thickened and contains collagen-like fibers. ×18,000.

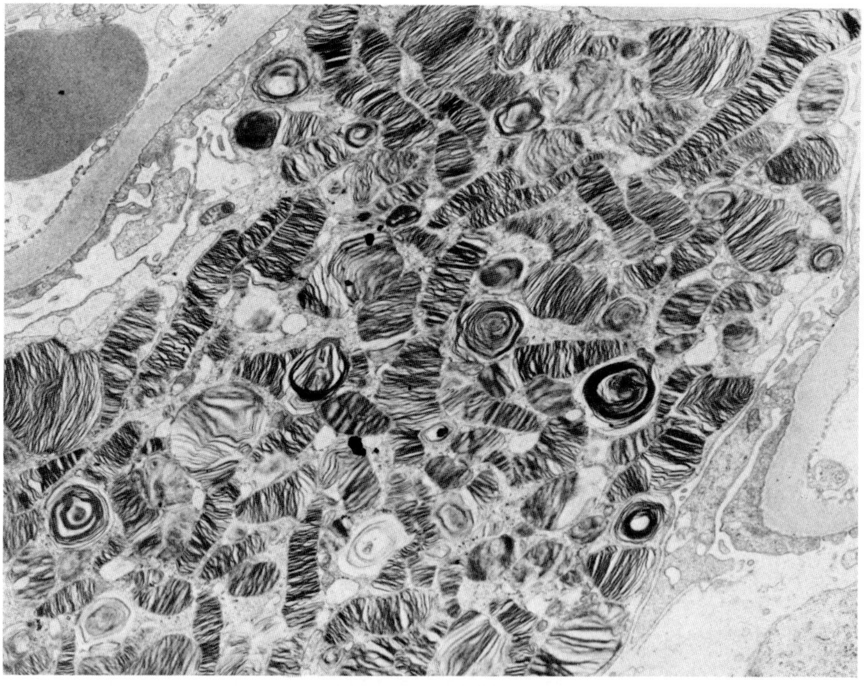

FIG. 5-40. Fabry's disease. The visceral epithelial cell contains numerous lysosomes with characteristic lamellae or myelin-like configuration. ×9000.

deposition of glycosphingolipids in the smooth muscle of the blood vessels, heart, and kidney. Death occurs around the fifth decade from either renal or cardiac failure. On light microscopy, the visceral epithelial cells are enlarged and vacuolated, giving a honeycomb appearance to the glomeruli. Similar changes can also be observed in the parietal epithelium of Bowman's capsule, endothelial cells, and mesangial cells. Ultrastructurally, there are numerous intracytoplasmic laminated inclusion bodies surrounded by a single unit membrane and measuring up to 5 μm in diameter. They are either round with a concentric myelin-like structure, or ovoid with parallel arrangement of layers (Fig. 5-40). Similar inclusions are present in the endothelial and smooth muscle cells of the blood vessels and in the epithelial tubular cells, especially those of the distal convoluted tubule and loops of Henle.[199,200] Although the ultrastructural pattern is almost diagnostic of the disease, similar inclusions have been seen in other conditions,[201] therefore, biochemical studies are required to confirm the diagnosis.

ACKNOWLEDGMENT

The author gratefully acknowledges Dr. Benjamin Spargo, who kindly provided some of the material used to illustrate this chapter.

REFERENCES

1. **Pease DC, Baker RF:** Electron microscopy of the kidney. Am J Anat 87:349, 1950
2. **Yamada E:** The fine structure of the renal glomerulus of the mouse. J Biophys Biochem Cytol 1:551, 1955
3. **Farquhar MG:** The primary filtration barrier: Basement membrane or epithelial slits? Kidney Int 8:197, 1975
4. **Karnovsky MJ:** The structural bases for glomerular filtration. In Churg J, Spargo BH, Mostofi FK, Abell MR (eds): Kidney Disease: Present Status. Baltimore, Williams & Wilkins, 1979, p 1
5. **Østerby R:** Quantitative electron microscopy of the glomerular basement membrane: A methodological study. Lab Invest 25:15, 1971
6. **Leiper JM, Thompson D, McDonald MK:** Uptake and transport of imposil by the glomerular mesangium in the mouse. Lab Invest 37:526, 1977
7. **Becker CG:** Demonstration of actomyosin in mesangial cells of the renal glomerulus. Am J Pathol 66:97, 1972
8. **Scheinman JI, Fish AJ, Matas AJ, Michael AF:** The immunohistopathology of glomerular antigens: II. Glomerular basement membrane, actomyosin and fibroblast surface antigens in normal, diseased and transplanted human kidneys. Am J Pathol 90:71, 1978
9. **Bernik MB:** Contractile activity of human glomeruli in culture. Nephron 6:1, 1969
10. **Suzuki J, Churg J, Grishman E, et al.:** The mesangium of the renal glomerulus. Am J Pathol 43:555, 1963
11. **Walker F:** The origin, turnover and removal of glomerular basement membrane. J Pathol 110:233, 1973
12. **Seiler MW, Rennke HG, Venkatachalam M, Cotran RS:** Pathogenesis of polycation-induced alterations ("fusion") of glomerular epithelium. Lab Invest 36:48, 1977
13. **Habib R, Kleinknecht C:** Primary nephrotic syndrome of childhood: Classification and clinicopathologic study of 406 cases. Pathol Annu 6:417, 1977

14. **Nephrotic syndrome in children:** Prediction of histopathology from clinical and laboratory characteristics at the time of diagnosis: A report of the international study of kidney disease in children. Kidney Int 13:159, 1978
15. **White RHR, Glasgow EF, Mills RJ:** Clinicopathological study of the nephrotic syndrome in childhood. Lancet 1:1353, 1970
16. **Jao W, Pollak VE, Norris SH, et al.:** Lipoid nephrosis: An approach to the clinicopathologic analysis of dismemberment of idiopathic nephrotic syndrome with minimal glomerular changes. Medicine (Baltimore) 53:445, 1973
17. **Seymour AE, Spargo BH, Penska R:** Contributions of renal biopsy studies to the understanding of disease. Am J Pathol 65:549, 1971
18. **Hayslett JP, Kashgarian M, Bensch KG, et al.:** Clinicopathological correlations in the idiopathic nephrotic syndrome due to primary renal disease. Medicine (Baltimore) 52:93, 1973
19. **Cameron JS, Turner DR, Ogg CS, Brown CB:** The nephrotic syndrome in adults with "minimal change" glomerular lesions. Q J Med 43:461, 1974
20. **Munk F:** Klinische diagnostik der degenerativen nierenerkrau kungen. Z Klin Med 78:1, 1913
21. **Arakawa M:** A scanning electron microscopic study of the human glomerulus. Am J Pathol 64:457, 1971
22. **Duffy JL, Cinque T, Grishman E, Churg J:** Intraglomerular fibrin, platelet aggregation, and subendothelial deposits in lipoid nephrosis. J Clin Invest 49:251, 1970
23. **Ludwigen E, Sorensen FH, Olsen S:** A quantitative study of glomeruli in idiopathic nephrosis with minimal or no glomerular lesions. Acta Pathol Microbiol Scand [A] 85:911, 1977
24. **Rich AR:** A hitherto undescribed vulnerability of the juxtamedullary glomeruli in lipoid nephrosis. Bull Johns Hopkins Hosp 100:173, 1957
25. **Hayslett JP, Krassner LS, Bensch KG, et al.:** Progression of "lipoid nephrosis" to renal insufficiency. N Engl J Med 281:181, 1969
26. **Churg J, Habib R, White RHR:** Pathology of the nephrotic syndrome in children: A report of the international study of kidney disease in children. Lancet 1:1299, 1970
27. **Cameron JS, Turner DR, Ogg CS, et al.:** The long term prognosis of patients with focal glomerulosclerosis. Clin Nephrol 10:213, 1978
28. **Newman WJ, Tisher CC, McCoy RC, et al.:** Focal glomerulosclerosis: Contrasting clinical patterns in children and adults. Medicine (Baltimore) 55:67, 1976
29. **Schoeneman MJ, Bennett BP, Griefer I:** The natural history of focal segmental glomerulosclerosis with and without mesangial hypercellularity in children. Clin Nephrol 9:45, 1978
30. **Churg J, Grishman E:** Electron microscopy of glomerulonephritis. Curr Top Pathol 61:107, 1976
31. **Velosa JA, Donadio JV, Holley KE:** Focal sclerosing glomerulonephropathy: A clinicopathologic study. Mayo Clin Proc 50:121, 1975
32. **Lim VS, Sibley R, Spargo BH:** Adult lipoid nephrosis: Clinicopathologic correlations. Ann Intern Med 81:314, 1974
33. **Hoyer JR, Vernier RL, Najarian JS, et al.:** Recurrence of idiopathic nephrotic syndrome after renal transplantation. Lancet 2:343, 1972
34. **Cameron JS, Turner DR:** Recurrent glomerulonephritis in allografted kidneys. Clin Nephrol 7:47, 1977
35. **O'Regan S, Smith M, Drummond KN:** Antigens in human immune complex nephritis. Clin Nephrol 6:417, 1976
36. **Eagen JW, Lewis EJ:** Glomerulopathies of neoplasia. Kidney Int 11:297, 1977
37. **Row PG, Cameron JS, Turner DR, et al.:** Membranous nephropathy: Long-term follow-up and association with neoplasia. Q J Med 44:207, 1975

38. **Ehrenreich T, Churg J:** Pathology of membranous nephropathy. Pathol Annu 3:145, 1968
39. **Rosen S:** Membranous glomerulonephritis: Current status. Hum Pathol 2:209, 1971
40. **Pollak VE, Rosen S, Pirani CL, et al.:** Natural history of lipoid nephrosis and membranous glomerulonephritis. Ann Intern Med 69:1171, 1968
41. **Forland M, Spargo BH:** Clinicopathological correlations of idiopathic nephrotic syndrome with membranous nephropathy. Nephron 6:498, 1969
42. **Franklin WA, Jennings RB, Earle DP:** Membranous glomerulonephritis: Long-term serial observations of clinical course and morphology. Kidney Int 4:36, 1973
43. **Gluck MC, Gallo G, Lowenstein J, Baldwin DS:** Membranous glomerulonephritis: Evolution of clinical and pathologic features. Ann Intern Med 78:1, 1973
44. **Trainin EB, Biochis H, Spitzer A, Greifer I:** Idiopathic membranous nephropathy: Clinical course in children. NY State J Med 76:357, 1976
45. **Noel LH, Zanetti M, Droz DS, Barbanel C:** Long-term prognosis of idiopathic membranous glomerulonephritis: A study of 116 untreated patients. Am J Med 66:82, 1979
46. **Habib R, Kleinknecht C, Gluber MC:** Extramembranous glomerulonephritis in children: Report of 50 cases. J Pediatr 82:754, 1973
47. **Obling H, Greifer I, Bennett BP, et al.:** Idiopathic membranous nephropathy in children. Kidney Int 3:381, 1973
48. **Pierides AM, Malasit P, Morley AR, et al.:** Idiopathic membranous nephropathy. Q J Med 46:163, 1977
49. **Feiner H, Gallo, G:** Ultrastructure in glomerulonëphritis associated with cryoglobulinemia: A report of six cases and review of the literature. Am J Pathol 88:145, 1977
50. **Spargo BH, Seymour AE, Ordóñez NG:** Renal Biopsy Pathology With Diagnostic and Therapeutic Implications. New York, Wiley, 1980
51. **Bariety J, Callard R:** Striated membranous structures in renal glomerular tufts: An electron microscopic study of 340 human renal biopsies. Lab Invest 32:636, 1975
52. **Crosson JT, Wathen RL, Raij L, et al.:** Recurrence of idiopathic membranous nephropathy in renal allografts. Arch Intern Med 135:1101, 1975
53. **Rubin RJ, Pinn VW, Barnes BA, Herrington JT:** Recurrent idiopathic membranous nephropathy. Transplantation 24:4, 1977
54. **Olsen S, Bohmann S-O, Petersen VP:** Ultrastructure of the glomerular basement membrane in long-term renal allografts with transplant glomerular disease. Lab Invest 30:176, 1974
55. **Sato M, Nakazora H, Ofuji T:** The pathogenetic roles of *Staphylococcus aureus* in primary human glomerulonephritis. Clin Nephrol 11:190, 1979
56. **Hyman LR, Jenis EH, Hill GS, et al.:** Alternate C3 pathway activation in pneumococcal glomerulonephritis. Am J Med 58:810, 1975
57. **Yucĕglu AM, Berkovich S, Minkowitz S:** Acute glomerulonephritis as a complication of varicella. JAMA 202:113, 1967
58. **Monteiro GE, Lillecrap CA:** Case of mumps nephritis. Br Med J 4:721, 1967
59. **Brzoke WJ, Krawczyninski K, Nazarewicz T, Morzycka M:** Glomerulonephritis associated with hepatitis-B surface antigen immune complexes in children. Lancet 2:477, 1974
60. **Peter G, Smith AL:** Group A streptococcal infections of the skin and pharynx (pt 1). N Engl J Med 297:311, 1977
61. **Spargo BH:** Pathologic lesions producing the nephrotic syndrome. In Churg J, Spargo BH, Mostofi FK, Abell MR (eds): Kidney Disease: Present Status. Baltimore, Williams & Wilkins, 1979, p 56
62. **Spargo BH, Dodge WF, Travis LB:** The relationship between the clinical course and pathologic features of poststreptococcal glomerulonephritis. In Contributions to Nephrology. Basel, Karger, 1976, vol 2, p 130

63. **Baldwin DS:** Poststreptococcal glomerulonephritis: A progressive disease? Am J Med 62:1, 1977
64. **Spargo BH, Ordóñez NG, Ringus JC:** The differential diagnosis of crescentic glomerulonephritis: The pathology of specific lesions with prognostic implications. Hum Pathol 8:187, 1977
65. **Morita T, Wenzl JE, Kimmelstiel P:** The relationship of neutrophilic and eosinophilic leukocytes to the glomerular capillary basement membrane in acute proliferative glomerulonephritis. Lab Invest 25:445, 1971
66. **Lewy JE, Salinas-Madrigal L, Herdson PB, et al.:** Clinicopathologic correlations in acute poststreptococcal glomerulonephritis: A correlation between renal function, morphologic damage, and clinical course in 46 children with acute poststreptococcal glomerulonephritis. Medicine (Baltimore) 50:453, 1971
67. **Dodge WF, Spargo BH, Bass JA, Travis LB:** The relationship between the clinical and pathological features of poststreptococcal glomerulonephritis: A study of the early natural history. Medicine (Baltimore) 47:227, 1968
68. **Hinglais N, Garcia-Torres R, Kleinknecht C:** Long-term prognosis in acute glomerulonephritis. Am J Med 56:52, 1974
69. **West CD, Ruley EJ, Forristal J, Davis NC:** Mechanisms of hypocomplementemia in glomerulonephritis. Kidney Int 3:116, 1973
70. **Germuth FG Jr, Rodriguez E:** Immunopathology of the Renal Glomerulus: Immune Complex Disease and Antibasement Membrane Disease. Boston, Little Brown, 1973, p 15
71. **Cochrane CG, Koffler D:** Immune complex disease in experimental animals and man. Adv Immunol 16:185, 1973
72. **West CD, McAdams AJ, McConville JM, et al.:** Hypocomplementemic and normocomplementemic persistent (chronic) glomerulonephritis: clinical and pathological characteristics. J Pediatr 67:1089, 1965
73. **Gotoff SP, Fellers FX, Vawter GF, et al.:** The beta 1C globulin in childhood nephrotic syndrome: Laboratory diagnosis of progressive glomerulonephritis. N Engl J Med 273:524, 1965
74. **Habib R:** Glomerular nephropathies in systemic disease. In Habib R, Mathieu H, Broyer M (eds): Pediatric Nephrology. Philadelphia, Saunders, 1974, p 215
75. **Berthoux FC, Carpenter CB, Freyria AM, et al.:** Human glomerulonephritis and the C3 nephritic factor (C_3NeF). Clin Nephrol 5:93, 1976
76. **Jenis EH, Lowenthal DT:** Kidney Biopsy Interpretation. Philadelphia, Davis, 1977, p 51
77. **Habib R, Kleinknecht C, Gubler MC, Levy M:** Idiopathic membranoproliferative glomerulonephritis in children: Report of 105 cases. Clin Nephrol 1:194, 1973
78. **Cameron JS, Glasgow EF, Ogg CS, White RHR:** Membranoproliferative glomerulonephritis and persistent hypocomplementemia. Br Med J 4:7, 1970
79. **Cameron JS:** Clinicopathologic correlations in glomerular disease. In Churg J, Spargo BH, Mostofi FK, Abell MR (eds): Kidney Disease: Present Status. Baltimore, Williams & Wilkins, 1979, p 76
80. **West CD, McAdams AJ:** The chronic glomerulonephritides of childhood. J Pediatr 93:1, 167, 1978
81. **Herdman RC, Pickering RC, Michael AF, et al.:** Chronic glomerulonephritis associated with low serum complement activity (chronic hypocomplementemic glomerulonephritis). Medicine (Baltimore) 49:207, 1970
82. **Bohle A, Gartner HV, Fishbach J, et al.:** The morphological and clinical features of membranoproliferative glomerulonephritis in adults. Virchows Arch [Pathol Anat] 363:213, 1974
83. **Jones DB:** Membranoproliferative glomerulonephritis: One or many diseases? Arch Pathol Lab Med 101:457, 1977

84. **Galle P, Mahieu P:** Electron dense alteration of the kidney basement membranes: A renal lesion specific of a systemic disease. Am J Med 58:749, 1975
85. **Lamb V, Tisher CC, McCoy RC, Robinson RR:** Membranoproliferative glomerulonephritis with dense intramembranous alterations: A clinicopathologic study. Lab Invest 36:607, 1977
86. **Habib R, Gubler MC, Loirat C, et al.:** Dense deposit disease: A variant of membranoproliferative glomerulonephritis. Kidney Int 7:204, 1975
87. **Anders D, Agniola B, Sippel M:** Basement membrane changes in membranoproliferative glomerulonephritis: II. Characterization of a third type by silver impregnation of ultrathin sections. Virchows Arch [Pathol Anat] 376:1, 1977
88. **Strife CF, McEnery PT, McAdams AJ, West CD:** Membranoproliferative glomerulonephritis with disruption of the glomerular basement membrane. Clin Nephrol 7:65, 1977
89. **Nagi AH:** Histologic, ultrastructural, and immunofluorescence glomerulonephritis. J Pathol 106:151, 1972
90. **Davis AE, Schneebergen EE, Grupe WE, McCluskey RT:** Membranoproliferative glomerulonephritis (MPGN type I) and dense deposit disease (DDD) in children. Clin Nephrol 9:184, 1978
91. **Vargas R, Thompson KJ, Wilson D, et al.:** Mesangiocapillary glomerulonephritis with dense "deposits" in the basement membrane of the kidney. Clin Nephrol 5:73, 1976
92. **Kim Y, Vernier RL, Fish AJ, Michael AF:** Immunofluorescence studies of dense deposit disease: The presence of railroad tracks and mesangial rings. Lab Invest 40:474, 1979
93. **Curtis JJ, Wyatt RJ, Bhathena D, et al.:** Renal transplantation in patients with type I and type II membranoproliferative glomerulonephritis: Serial complement and nephritic factor measurements and the problem of recurrence of disease. Am J Med 66:216, 1979
94. **Berger J:** IgA glomerular deposits in renal diseases. Transplant Proc 1:939, 1969
95. **Clarkson AR, Seymour AE, Thompson AJ, et al.:** IgA nephropathy: A syndrome of uniform morphology, diverse chemical features and uncertain prognosis. Clin Nephrol 8:459, 1977
96. **Droz D:** Natural history of primary glomerulonephritis with mesangial deposits of IgA. Contrib Nephrol 2:150, 1976
97. **McCoy RC, Abramowsky CR, Tisher CC:** IgA nephropathy. Am J Pathol 76:123, 1974
98. **Van der Peet J, Arisz L, Brentjens JRH, et al.:** The clinical course of IgA nephropathy in adults. Clin Nephrol 8:335, 1977
99. **Berger J, Taneva H, Nabarra B, Barbanel C:** Recurrence of mesangial deposition of IgA after renal transplantation. Kidney Int 7:232, 1975
100. **Mathew TG, Mathews EF, Hobbs JB, Kincaid-Smith P:** Glomerular lesions after renal transplantation. Am J Med 59:177, 1975
101. **Baart de la Faille-Kuyper EH, Kater L, Kuijten RH, et al.:** Occurrence of vascular IgA deposits in clinically normal skin of patients with renal disease. Kidney Int 9:424, 1976
102. **Buchanan N, Berkowitz F, Gold C:** Granulomatous glomerulonephritis and fulminant polyarteritis nodosa in a child. S Afr Med J 50:1057, 1976
103. **Thompson NM, Holdsworth SR, Glasgow EF:** The macrophage in the development of experimental crescentic glomerulonephritis: Studies using tissue culture and electron microscopy. Am J Pathol 94:223, 1979
104. **Cattell V, Jamieson SW:** The origin of glomerular crescents in experimental nephrotoxic serum nephritis in the rat. Lab Invest 39:584, 1978
105. **Morita T, Suzuki Y, Churg J:** Structure and development of the glomerular crescents. Am J Pathol 72:349, 1973
106. **Bohmann S-O, Olsen S, Petersen VP:** Glomerular ultrastructure in extracapillary glomerulonephritis. Acta Pathol Microbiol Scand [A] 82(suppl 249):29, 1974

107. **Rosen S:** Crescentic glomerulonephritis: Occurrence, mechanisms, and prognosis. Pathol Annu 10:37, 1975
108. **Stejskal J, Pirani CL, Okada M, et al.:** Discontinuities (gaps) of the glomerular capillary wall and basement membrane in renal diseases. Lab Invest 28:149, 1978
109. **Davis CA, McEnery PT, Maby S, et al.:** Observations on the evolution of idiopathic rapidly progressive glomerulonephritis. Clin Nephrol 9:91, 1978
110. **Whitworth JA, Morel-Maroger L, Mignon F, Richet G:** The significance of extracapillary proliferation: Clinicopathologic review of 60 patients. Nephron 16:1, 1976
111. **Moorthy AV, Zimmerman SW, Burkholder PM:** Association of crescentic glomerulonephritis with membranous glomerulopathy. Clin Nephrol 6:319, 1976
112. **Horn RG, Fauci AS, Rosenthal AS, Wolff SM:** Renal biopsy pathology in Wegener's granulomatosis. Am J Pathol 74:423, 1974
113. **Howell SB, Epstein WV:** Circulating immune complexes in Wegener's granulomatosis. Am J Med 60:259, 1976
114. **Appel GB, Silva FG, Pirani CL et al.:** Renal involvement in systemic lupus erythematosus (SLE). A study of 56 patients emphasising histologic classification. Medicine (Baltimore) 57:371, 1978
115. **Pirani CL, Silva FG:** The kidney in systemic lupus erythematosis and other collagen diseases: Recent progress. In Churg J, Spargo BH, Mostofi FK, Abell MR (eds): Kidney Disease: Present Status. Baltimore, Williams & Wilkins, 1979, p 98
116. **Cohen AH, Zamboni L:** Ultrastructural appearance and morphogenesis of renal hematoxylin bodies. Am J Pathol 89:105, 1977
117. **Pollak AD:** Some observations of the pathology of systemic lupus erythematosus. J Mt Sinai Hospital (NY) 26:224, 1959
118. **Shünrah W:** Morphologie und lokalisation der hämmatoxylin körperchen bei lupusnephritis. Virchows Arch [Pathol Anat] 259:331, 1973
119. **Muehrche RC, Kark RM, Pirani CL, Pollak VE:** Lupus nephritis: A clinical and pathological study based on renal biopsies. Medicine (Baltimore) 36:1, 1957
120. **Ginzler EM, Micastri AD, Chen CK, et al.:** Progression of mesangial and focal to diffuse lupus nephritis. N Engl J Med 291:693, 1974
121. **Mahajan SK, Ordóñez NG, Spargo BH, Katz AI:** Changing histopathology patterns in lupus nephropathy. Clin Nephrol 10:1, 1978
122. **Comerford FR, Cohen AS:** The nephropathy of systemic lupus erythematosus: An assessment by clinical, light and electron microscopy criteria. Medicine (Baltimore) 46:425, 1967
123. **Grishman E, Porush JG, Lee SL, Churg J:** Renal biopsies in lupus nephritis: Correlation of electron microscopic findings with clinical course. Nephron 10:25, 1973
124. **Dujovne I, Pollak VE, Pirani CL, Dillard MG:** The distribution and character of glomerular deposits in systemic lupus erythematosus. Kidney Int 2:33, 1972
125. **Pollak VE, Pirani CL:** Renal histologic findings in systemic lupus erythematosus. Mayo Clin Proc 44:630, 1969
126. **Baldwin DS, Gluck MC, Lowenstein J, Gallo G:** Lupus nephritis: Clinical course as related to morphologic forms and their transitions. Am J Med 62:12, 1977
127. **Mahajan SK, Ordóñez NG, Feitelson PJ, et al.:** Lupus nephropathy without clinical renal involvement. Medicine (Baltimore) 56:493, 1977
128. **Donadio JV, Burgess JH, Holley KE:** Membranous nephropathy: A clinicopathologic study. Medicine (Baltimore) 5:527, 1977
129. **Wood C:** Crystalline phospholipid deposits in renal glomeruli in glomerulonephritis. Am J Pathol 66:59A, 1972
130. **Okumura K:** Induction of a disease resembling systemic lupus erythematosus in 57BL/GJ mice by prolonged immunization with egg albumin. Acta Pathol Jpn 23:695, 1973
131. **Bariéty D, Richer DD, Appay MD, et al.:** Frequency of endothelial "virus-like" parti-

cles: An electron microscopy study of 376 human renal biopsies. J Clin Pathol 26:21, 1973

132. **Imamura M, Block SR, Mellors RC:** Electron microscopic study of distinctive structures in peripheral blood lymphocytes obtained from twins with lupus erythematosus. Am J Pathol 81:561, 1975
133. **Garancis JC, Komorowski RA, Bernhard GC, Straumfjord JV:** Significance of cytoplasmic microtubules in lupus nephritis. Am J Pathol 64:1, 1971
134. **Schaff Z, Barry DW, Grimley PM:** Cytochemistry of tubuloreticular structures in lymphocytes from patients with systemic lupus erythematosus and in cultures of human lymphoid cells: Comparison to a paramyxovirus. Lab Invest 29:577, 1973
135. **Meadow SR, Glasgow EF, White RHR, et al.:** Henoch-Schönlein nephritis. Q J Med 41:241, 1972
136. **Levy M, Broyer M, Arsan A, et al.:** Anaphylactoid purpura nephritis in childhood: Natural history and immunopathology. Adv Nephrol 6:183, 1976
137. **Meadow SR:** The Prognosis of Henoch-Schönlein nephritis. Clin Nephrol 9:87, 1978
138. **Sinniah R, Feng PH, Chem BTM:** Henoch-Schönlein syndrome: A clinical and morphological study of renal biopsies. Clin Nephrol 9:219, 1978
139. **Heaton JM, Turner DR, Cameron JS:** Localization of glomerular "deposits" in Henoch-Schönlein nephritis. Histopathology 1:93, 1977
140. **Brun C, Bryld C, Fenger L, Jørgensen JF:** Glomerular lesions in adults with Schönlein-Henoch syndrome. Acta Pathol Microbiol Scand [A] 79:569, 1971
141. **Berkman J, Rifkin HL:** Newer aspects of diabetic microangiopathy. Annu Rev Med 17:83, 1966
142. **Henderson LL, Spargue RB, Wagner HP:** Intercapillary nephrosclerosis. Am J Med 3:131, 1947
143. **Hatch FE, Watt MF, Kramer NC, et al.:** Diabetic glomerulosclerosis: A long-term follow-up based on renal biopsies. Am J Med 31:216, 1961
144. **Watkins PJ, Blainey JD, Brewer DB, et al.:** The natural history of diabetic renal disease: A follow-up study of a series of renal biopsies. Q J Med 41:437, 1972
145. **Østerby R:** Morphometric studies of the peripheral basement membrane in early juvenile diabetes: I. Development of initial basement membrane thickening. Diabetologia 8:84, 1972
146. **Dachs S, Churg J, Mautner W, Grishman E:** Diabetic nephropathy. Am J Pathol 44:155, 1965
147. **Østerby R:** A quantitative electron microscopic study of mesangial regions in glomeruli from patients with short-term juvenile diabetes mellitus. Lab Invest 29:99, 1977
148. **Iidaka K, McCoy J, Kimmelstiel P:** The glomerular mesangium: A quantitative analysis. Lab Invest 19:573, 1968
149. **Gellman DD, Pirani CL, Soothill JF, Muehrche RC, Kark RM:** Diabetic nephropathy: A clinical and pathological study based on renal biopsies. Medicine (Baltimore) 38:321, 1959
150. **Olsen S:** Mesangial thickening and nodular glomerular sclerosis in diabetes mellitus and other diseases. Acta Pathol Microbiol Scand 233(suppl):203, 1972
151. **Schubert GE, Adam A:** Glomerular nodules and long-spacing collagen in kidneys of patients with multiple myeloma. J Clin Pathol 27:800, 1974
152. **Ramdall RE, Williamson WC, Mullinax F, et al.:** Manifestations of systemic light chain deposition. Am J Med 60:293, 1976
153. **Salinas-Madrigal L, Pirani CL, Pollak VE:** Glomerular and vascular "insudative" lesions of diabetic nephropathy: Electron microscopic observations. Am J Pathol 59:369, 1970
154. **Davies MJ, Woolf N, Carstairs KC:** Immunohistochemical studies in diabetic glomerulosclerosis. J Pathol Bacteriol 92:441, 1966

155. **Fahquahr A, MacDonald MK, Ireland JT:** The role of fibrin deposition in diabetic glomerulosclerosis: A light, electron, and immunofluorescence microscopic study. J Clin Pathol 25:657, 1972
156. **Cohen AS:** Amyloidosis. N Engl J Med 277:552, 574, 628, 1967
157. **Glener GG, Page DL:** Amyloid, amyloidosis and amyloidogenesis. Int Rev Exp Pathol 15:1, 1976
158. **Brownstein MH, Helwig EB:** Secondary systemic amyloidosis: Analysis of underlying disorders. South Med J 64:491, 1971
159. **Rukavina JG, Block WD, Jackson CE, et al.:** Primary systematic amyloidosis: A review of an experimental, genetic, and clinical study of 29 cases with particular emphasis on the familial form. Medicine (Baltimore) 35:239, 1956
160. **Briggs GW:** Amyloidosis. Ann Intern Med 55:943, 1961
161. **Hamburger J, Richet G, Cosnier J, et al.:** Nephrology. Philadelphia, W. B. Saunders, 1968, p 882
162. **Overbach O, Stemmerman MG:** Renal amyloidosis. Arch Intern Med 74:244, 1944
163. **Jao W, Pirani CL:** Renal amyloidosis: Electron microscopic observations. Acta Pathol Microbiol Scand [A] 233(suppl):217, 1972
164. **Ansell ID, Joekes AM:** Spicular arrangement of amyloid in renal biopsy. J Clin Pathol 25:1056, 1972
165. **Derosena R, Koss MN, Pirani CL:** Demonstration of amyloid fibrils in urinary sediment. N Engl J Med 293:1131, 1975
166. **Shirahama T, Skinner M, Cohen AS, Benson MD:** Uncertain value of urinary sediments in the diagnosis of amyloidosis. N Engl J Med 297:821, 1977
167. **DeChadarevian J-P, Kaplas BS:** The hemolytic uremic syndrome of childhood. Perspect Pediatr Pathol 4:465, 1978
168. **Gianantonio CA, Vitacco M, Mendiharzu F, et al.:** The hemolytic uremic syndrome. Nephron 11:174, 1973
169. **Clarkson AR, Lawrence JR, Meadows R, Seymour AE:** The haemolytic uraemic syndrome in adults. Q J Med 39:227, 1970
170. **Brown CB, Clarkson AR, Robson JS, et al.:** Haemolytic uraemic syndrome in women taking oral contraceptives. Lancet 1:1479, 1973
171. **Sun NCJ, Johnson WJ, Sung CTW, Woods JE:** Idiopathic renal failure: Review and case report of a successful renal transplantation. Mayo Clin Proc 50:395, 1975
172. **Shigematsu H, Dikman SH, Churg J, et al.:** Mesangial involvement of hemolytic uremic syndrome: A light and electron microscopic study. Am J Pathol 85:349, 1976
173. **Churg J:** Coagulation and the kidney. In Churg J, Spargo BH, Mostofi FK, Abell MR (eds): Kidney Disease: Present Status. Baltimore, Williams & Wilkins, 1979, p 140
174. **Bohle A, Helmchen U, Grund KE, et al.:** Malignant nephrosclerosis in patients with hemolytic uremic syndrome (primary malignant nephrosclerosis). Curr Top Pathol 65:81, 1977
175. **Clarkson AR, Meadows R, Lawrence JR:** Post-partum renal failure: The generalized Schwartzman reaction? Australas Ann Med 18:209, 1969
176. **Shehan HL, Lynch JB:** Pathology of Toxemia of Pregnancy. London, Churchill Livingstone, 1973, p 53
177. **Spargo BH, Lichtig C, Luger AM, et al.:** The renal lesion in pre-eclampsia. In Lindheimer MD, Katz AI, Zuspan FP (eds): Hypertension of Pregnancy. New York, John Wiley, 1976, p 129
178. **Brouet J-C, Clauvel J-P, Danon F, et al.:** Biological and clinical significance of cryoglobulins: A report of 86 cases. Am J Med 57:775, 1974
179. **Morel-Maroger L, Varroust P:** Glomerular lesions in dysproteinemias. Kidney Int 5:249, 1974

180. **Schubert GE, Veigel J, Lennert K:** Structure and function of the kidney in multiple myeloma. Virchows Arch [Pathol Anat] 355:135, 1972
181. **Sickel GW:** Crystalline glomerular deposits in multiple myeloma. Am J Med 27:354, 1959
182. **Callard P, Feldman G, Prandi D, et al.:** Immune complex glomerulonephritis in cirrhosis of the liver. Am J Pathol 80:329, 1975
183. **Berger RJ, Yaneva H, Nabarra B:** Glomerular changes in patients with cirrhosis of the liver. Adv Nephrol 7:3, 1978
184. **Bernstein J:** Hereditary renal disease. In Churg J, Spargo BH, Mostofi FK, Abell MR (eds): Kidney Disease: Present Status. Baltimore, Williams & Wilkins, 1979, p 295
185. **Kaufman DB, McIntosh RM, Smith FG, Vernier RL:** Diffuse familial nephropathy: A clinicopathological study. J Pediatr 77:37, 1970
186. **Gaboardi F, Edefonti A, Ibasciati E, et al.:** Alport's syndrome (progressive hereditary nephritis). Clin Nephrol 2:143, 1974
187. **Spear GS:** Alport's syndrome: A consideration of pathogenesis, editorial. Clin Nephrol 1:336, 1973
188. **Felts JH:** Hereditary nephritis with the nephrotic syndrome. Arch Intern Med 125:459, 1970
189. **Hinglais N, Grunfeld J-P, Bois E:** Characteristic ultrastructural lesion of the glomerular basement membrane in progressive hereditary nephritis (Alport's syndrome). Lab Invest 27:473, 1972
190. **Churg J, Sherman RL:** Pathologic characteristics of hereditary nephritis. Arch Pathol 95:374, 1973
191. **Hill GS, Jenis EH, Goodloe S Jr:** The nonspecificity of the ultrastructural alterations in hereditary nephritis with additional observations of benign familial hematuria. Lab Invest 31:516, 1974
192. **George CR, Hickman RO, Stricker GE:** Infantile nephrotic syndrome. Clin Nephrol 5:20, 1976
193. **Huttnen N-P, Savilahti E, Rapola J:** Selectivity of proteinuria in congenital nephrotic syndrome of Finnish type. Kidney Int 8:255, 1975
194. **Habib R, Bois E:** Heterogeneité des syndromes néphrotiques à debut precose du Hourrisson (syndrome nephrotique "infantile") étude anatomoclinique et gênétique de 37 observations. Helv Pediatr Acta 28:91, 1973
195. **Simila S, Vesa L, Wasz-Hockert O:** Hereditary onycho-osteodysplasia (the nail-patella syndrome) with nephrosis-like renal disease in a newborn boy. Pediatrics 46:61, 1970
196. **Bennett WM, Musgrave JE, Campbell RA, et al.:** The nephropathy of the nail-patella syndrome: Clinicopathologic analysis of 11 kindred. Am J Med 54:304, 1973
197. **Morita T, Laughlin LO, Kawano K, et al.:** Nail-patella syndrome. Light and electron microscopic studies in the kidney. Arch Intern Med 131:271, 1973
198. **Hoyer JR, Michael AF, Vernier RL:** Renal disease in nail-patella syndrome: Clinical and morphological studies. Kidney Int 2:231, 1972
199. **Savi M, Olivetti G, Neri T, Curtoni C:** Clinical, histopathological, and biochemical findings of Fabry's disease: A case report and family study. Arch Pathol Lab Med 101:536, 1977
200. **Gubler M-C, Lenoir G, Grunfeld JP, et al.:** Early changes in hemizygous and heterozygous patients with Fabry's disease. Kidney Int 13:223, 1978
201. **Scott CR, Lagunoff D, Pritzl P:** A mucopolysaccharide storage disease with involvement of the renal glomerular epithelium. Am J Med 54:549, 1973

6

ELECTRON MICROSCOPY OF SKELETAL MUSCLE AND PERIPHERAL NERVE BIOPSY SPECIMENS

SYDNEY S. SCHOCHET, JR.

Many of the ultrastructural alterations encountered in skeletal muscle and peripheral nerve biopsy specimens are, by themselves, relatively nonspecific. Therefore, the changes must be interpreted carefully in relation to the clinical data and the light microscopic findings. In this chapter, only selected ultrastructural abnormalities with distinctive morphologic features of diagnostic significance will be considered in detail.

Ideally, specimens of muscle for electron microscopy should be submitted in separate isometric clamps. However, most of our muscle biopsy specimens are obtained primarily for frozen sections and are submitted in a single clamp or without the use of an isometric clamp. Under these circumstances, we fasten a small fascicle of apparently untraumatized muscle to an applicator stick prior to fixation. Nerve biopsy specimens are also usually fixed while fastened to an applicator stick, although suspension with a small weight is a superior technique. Neither muscle nor nerve specimens should be cut into small cubes ("diced") until after partial aldehyde fixation. We routinely employ cacodylate-buffered glutaraldehyde followed by cacodylate-buffered osmium tetroxide for fixation. Our tissues are then block-stained with aqueous magnesium uranyl acetate prior to dehydration. Since this procedure tends to leech out glycogen, we occasionally process additional blocks omitting this step.

Thick sections are cut from all blocks of epoxy-embedded tissue and stained with alkaline toluidine blue for light microscopy. We do not perform electron microscopy until these and all other preparations have been examined. Sometimes useful information can be obtained by ultrastructural examination of nerve twigs included in muscle biopsy specimens. These nerves are often remarkably free of preparation artifacts and offer the investigator an opportunity to study motor nerves as well as the predominantly sensory nerves (e.g., sural) on which biopsies are routinely done for diagnostic purposes.

NORMAL SKELETAL MUSCLE

The periphery of the myofiber is surrounded by a relatively thick basement membrane. The surface of the muscle fiber itself, the sarcolemma or plasmalemma, contains multiple small invaginations, the caveolae. Myofiber nuclei are found normally in a subsarcolemmal position and usually appear irregularly infolded. At the poles of the nuclei are collections of glycogen granules, occasional dense bodies, mitochondria, and Golgi cisterns. Rough endoplasmic reticulum and free ribosomes are inconspicuous.

The highly ordered spatial orientation of the myofilaments imparts the characteristic banded appearance to the myofibrils. The intervals between successive Z-discs are designated as sarcomeres (Fig. 6-1). These vary in length according to the degree of contraction and are often quite short in routinely handled muscle biopsy specimens. The electron-dense Z-discs have an incompletely delineated fine structure and chemical composition. They display a quadratic configuration when viewed in cross-section. The thin filaments extend from the Z-discs through the I-bands and terminate by interdigitating among the thick filaments. The thin filaments measure approximately 1 μm in length and 6 nm in diameter. They contain actin, tropomyosin, and troponin. The A-bands contain thick myosin filaments and, toward either end, the interdigitating thin filaments. The myosin filaments measure approximately 1.5 μm in length and 16 nm in diameter. They display a regular hexagonal arrangement when viewed in cross-section. The central regions of the A-bands where there are no interdigitating thin filaments are designated as H-bands. In the middle of the thick filaments, the polarity of the myosin chains reverses, and the adjacent thick filaments are interconnected by cross-bridges. This region is relatively electron-dense and is designated as the M-band.

Myofibers contain two intersecting membranous systems. The transverse tubular system is continuous with the extracellular space and invaginates the myofiber more or less perpendicular to its long axis. The sarcoplasmic reticulum forms a fenestrated

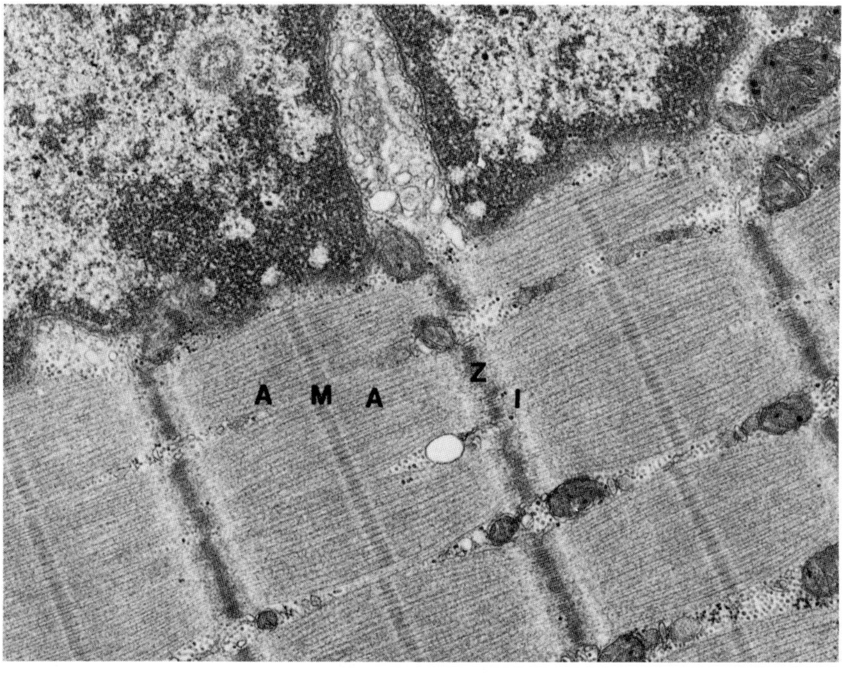

FIG. 6-1. Longitudinal section of normal skeletal muscle. The rigid spatial orientation of the myofilaments imparts the characteristic banded appearance to the myofibrils. ×26,250.

network about each of the myofibrils. Triads (Fig. 6-2) are specialized structures that arise where the two membranous systems intersect. Each triad consists of a central T-tubule of the transverse tubular system that is flanked by two terminal cisterns of the sarcoplasmic reticulum. The terminal cisterns contain an osmiophilic amorphous material and are more electron-dense than the T-tubules or the remainder of the sarcoplasmic reticulum. Small osmiophilic projections, the sarcoplasmic feet, impart a scalloped appearance to the junctions between the terminal cisterns and the T-tubules.

At the present time, there are no satisfactory ultrastructural criteria for distinguishing type 1 and type 2 myofibers in human muscle biopsy specimens. This has seriously impeded correlation of ultrastructural findings with histochemical data in the various disease states.

ABNORMALITIES OF MYOFIBER NUCLEI

A wide variety of abnormalities has been observed in myofiber nuclei.[1-5] Degenerating nuclei may contain vacuoles and deeply invaginated cytoplasm (nuclear "pseudoinclusions"). Intranuclear skeins of smooth filaments, measuring approximately 7 nm in diameter (Fig. 6-3), have been encountered in patients with various forms of myositis[6-9] and in some individuals who are carriers of Duchenne muscular dystrophy.[10] These thin, smooth filaments have been interpreted variously as deriva-

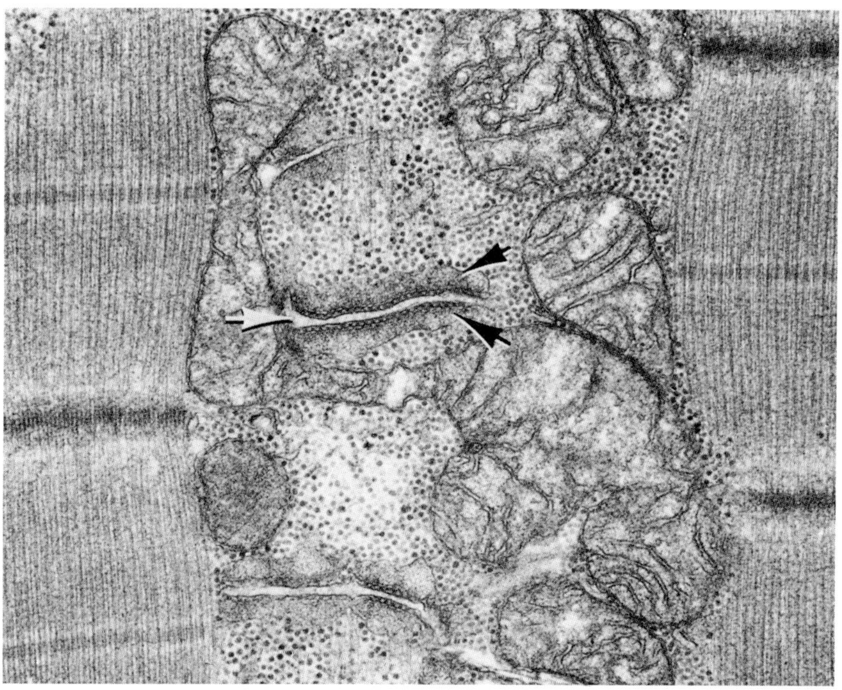

FIG. 6-2. Triads are composed of a T-tubule (white arrow) flanked by two terminal cisterns (black arrows) of the sarcoplasmic reticulum. ×38,438.

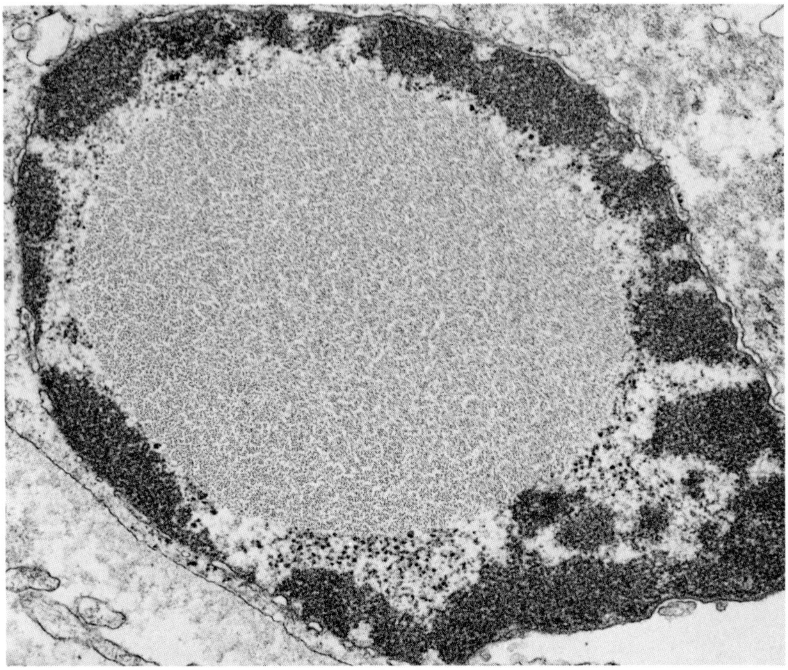

FIG. 6-3. Muscle from a patient with polymyositis. The nuclear inclusion is composed of fine filamentous material. ×30,000.

tives of myofilaments,[6] actin,[9] or nuclear rodlets.[8] Patients with so-called inclusion body myositis have masses of coarser fibrils measuring 15 to 18 nm in diameter in the cytoplasm and nuclei of affected myofibers.[6,9,11,12] Although the nature of the coarse fibrils is unknown, their presence is considered diagnostic of inclusion body myositis. Intranuclear nemaline rods have been observed in several cases of adult-onset[13] and in at least one case of infantile[14] nemaline myopathy. They also have been reported in a case of myositis,[15] but this may represent the simultaneous occurrence of two diseases.

ABNORMALITIES OF MYOFIBER MITOCHONDRIA

Diverse alterations in the distribution, number, size, and configuration of mitochrondria have been observed in diseased skeletal muscle.[1–5] Often the presence of myofibers containing large numbers of abnormal mitochondria can be suspected from the routine frozen sections. With the modified trichrome stain, the affected myofibers commonly display abnormal reddish granules and are termed *ragged red* fibers.[16] By far, the most spectacular of the mitochondrial abnormalities is the presence of rectangular paracrystalline inclusions (Fig. 6-4). These are found within the intracristal spaces and/or between the inner and outer mitochondrial membranes.[2,4,17,18] The inclusions are composed of sets of parallel osmiophilic lines with periodic punctate densities. In transverse section, each set appears to consist of

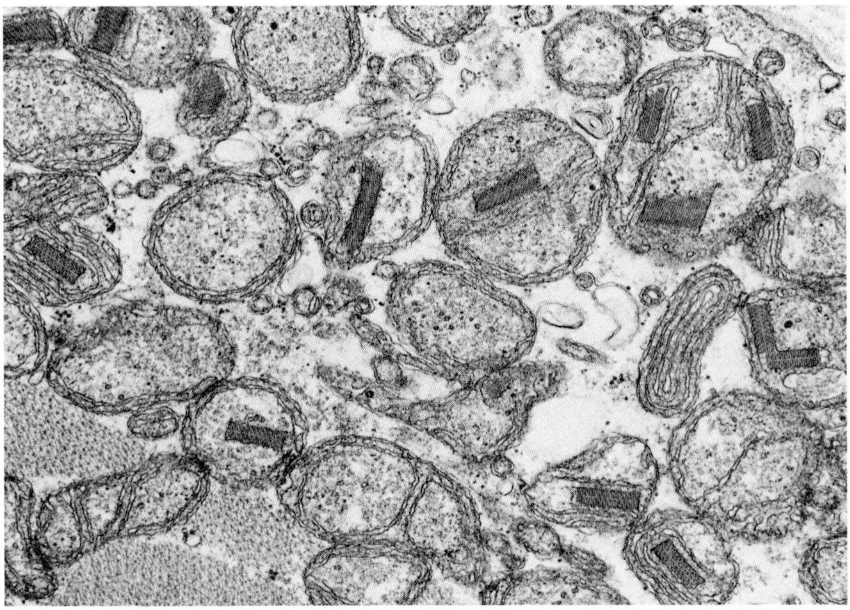

FIG. 6-4. Subsarcolemmal accumulation of abnormal mitochondria with paracrystalline inclusions; from a patient with an oculopharyngeal myopathy. ×30,000.

four parallel lines with the osmiophilic densities alternating between the first and second and between the third and fourth lines (Fig. 6-5). Relatively little is known about the chemical composition and pathogenesis of these paracrystalline inclusions. It has been suggested that they are biochemically inert, space-occupying structures.[18] Nevertheless, myofiber mitochondria isolated from certain patients in whom these inclusions were abundant have shown loosely coupled or uncoupled oxidative phosphorylation,[19,20] abnormal divalent cation transport,[21] or other abnormalities in respiratory control.[22,23] Comparable mitochondrial inclusions have been induced experimentally by the in vivo infusion of uncoupling agents or oleic acid.[24]

Although the presence of these paracrystalline mitochondrial inclusions does not delineate a specific mitrochondrial myopathy, they comprise the characteristic and predominant morphologic findings in Luft disease[19-21] and in the Kearns-Shy syndrome.[16,17,25] They are commonly present in association with lipid droplets in carnitine deficiency myopathy.[26-28] Lesser numbers have been encountered in a wide variety of disparate conditions, including polymyositis,[1,15] thyrotoxic periodic paralysis,[29] various glycogenoses,[30,31] and other incompletely defined myopathies.[2,22,32-35]

TUBULAR AGGREGATES IN MYOFIBERS

In certain pathologic conditions, type 2 myofibers contain randomly oriented fascicles of parallel tubules.[1,4,36] These tubular aggregates are most often just beneath the sarcolemma (Fig. 6-6) and about the poles of the myofiber nuclei. Glycogen granules and occasionally other organelles may be found in the interstices among the fascicles

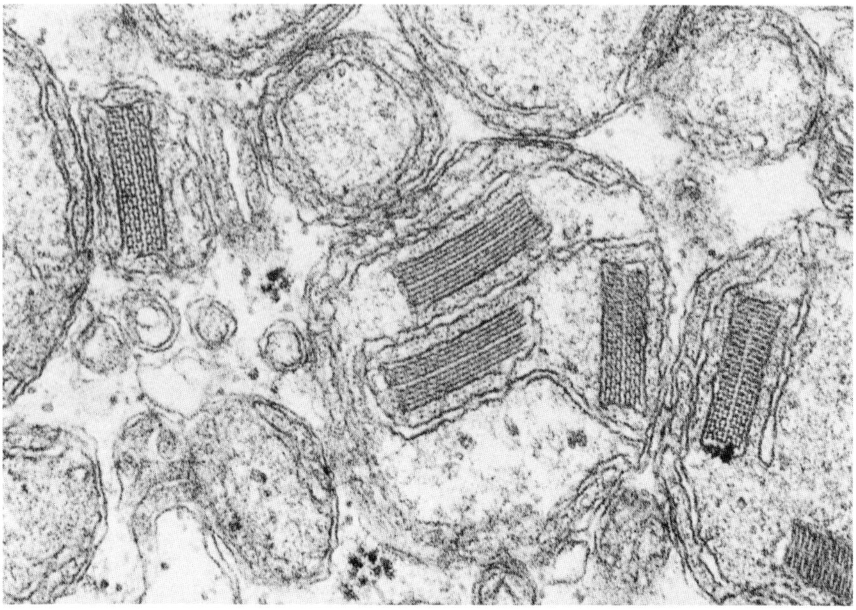

FIG. 6-5. The paracrystalline inclusions are composed of sets of four parallel lines with periodic densities. ×75,000.

of tubules. The individual tubules measure 60 to 90 nm in diameter and have smooth osmiophilic walls (Fig. 6-7). Occasionally tubules are fused, giving rise to dumbbell, trefoil, and even more complex cross-sectional configurations. At least some of the tubules are coextensive with and apparently derived from the dilated terminal cisterns of the sarcoplasmic reticulum. The lumina of the tubules often contain coaxial inner tubules, rod-shaped cores, or columns of spheroids. The lumenal contents and the contents of the terminal cisterns of the sarcoplasmic reticulum have a similar degree of granularity and osmiophilia.

Fascicles of another type of tubule are encountered less commonly. These tubules range from 50 to 80 nm in diameter and are characterized by the presence of regularly spaced, longitudinally oriented filaments applied to their exterior surfaces (Fig. 6-8). These striated tubules have either electron-lucent or granular lumenal contents.

Tubular aggregates have been described most often in patients with the periodic paralyses[36–41] but also have been encountered occasionally in other conditions.[1,4,36,42–44]

Z-DISC ALTERATIONS AND DERIVATIVES

Z-disc alterations are among the most common anomalies encountered in muscle biopsy specimens.[1–5] Malalignment or mild disarray of the Z-discs can be seen even in specimens from presumably normal individuals and cannot be regarded as evidence of neuromuscular disease.

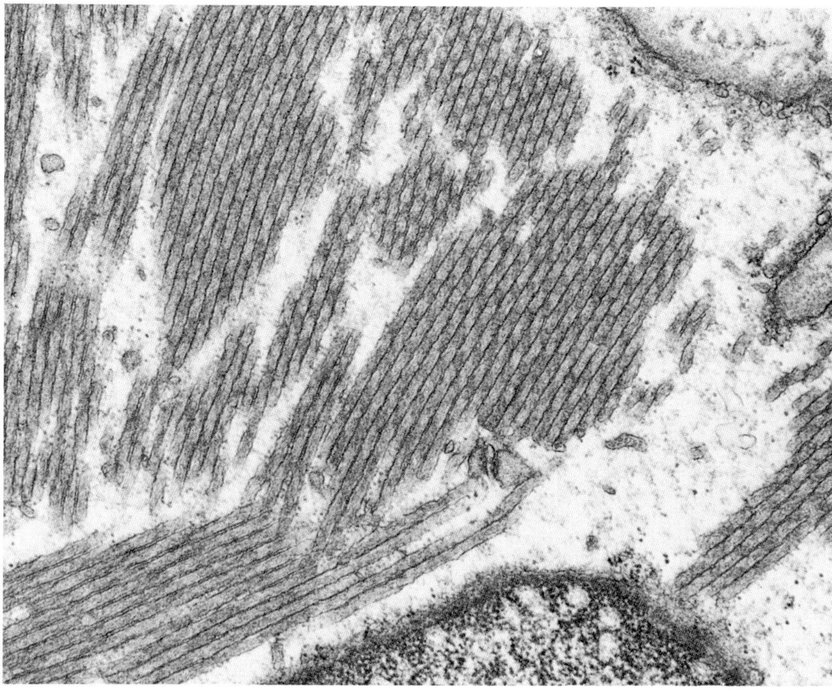

FIG. 6-6. Subsarcolemmal accumulation of smooth-walled tubules in a patient with periodic paralysis. ×28,125.

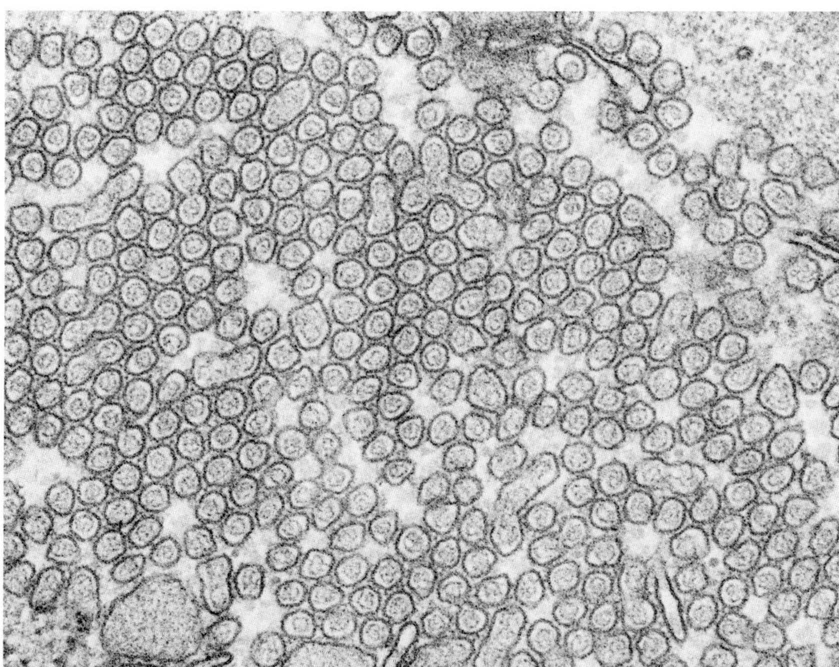

FIG. 6-7. Higher magnification showing the coaxial inner tubules. ×75,000.

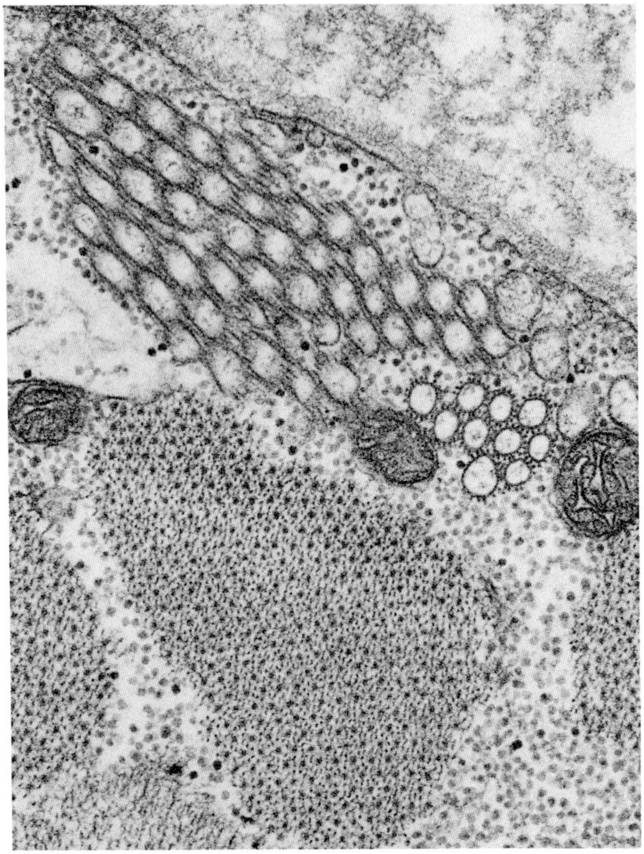

FIG. 6-8. Subsarcolemmal collection of striated tubules in a patient with neuronal ceroid-lipofuscinosis. Note the thin filaments applied to the outer walls of the tubules. ×60,000.

Another common but somewhat more significant alteration is Z-disc streaming. This abnormality occurs preferentially in type 1 myofibers and consists of longitudinal, band-like deposits of osmiophilic granular material that extend for a variable distance from the Z-disc. Details of the underlying myofilaments are obscured by the displaced Z-disc material. The more severe forms of Z-disc streaming are often accompanied by myofibrillar derangement and a reduction in the number of mitochondria (Fig. 6-9). These changes are most pronounced in myofibers undergoing target fiber formation,[4,45–47] in the unstructured cores of central core disease,[2,48] and in the small core-like structures of multicore disease.[49]

Cytoplasmic bodies are probably Z-disc derivatives and when small, tend to coincide with the location of Z-discs.[50] However, it also has been suggested that they may result from the contraction of degenerated myofilaments.[51] These two concepts are not necessarily mutually exclusive. The cytoplasmic bodies are two-zoned masses of variable size (Fig. 6-10). The electron-dense cores contain compact tangles of very fine filaments (less than 30 nm in diameter) and amorphous osmiophilic material. The less dense peripheral halos contain both the very fine filaments and radially oriented thick and thin myofilaments. They are found predominantly in type 2

FIG. 6-9. Muscle from a patient with neurogenic atrophy and target fiber formation. Note the reduction in mitochondria where the Z-disc streaming is most marked. ×6000.

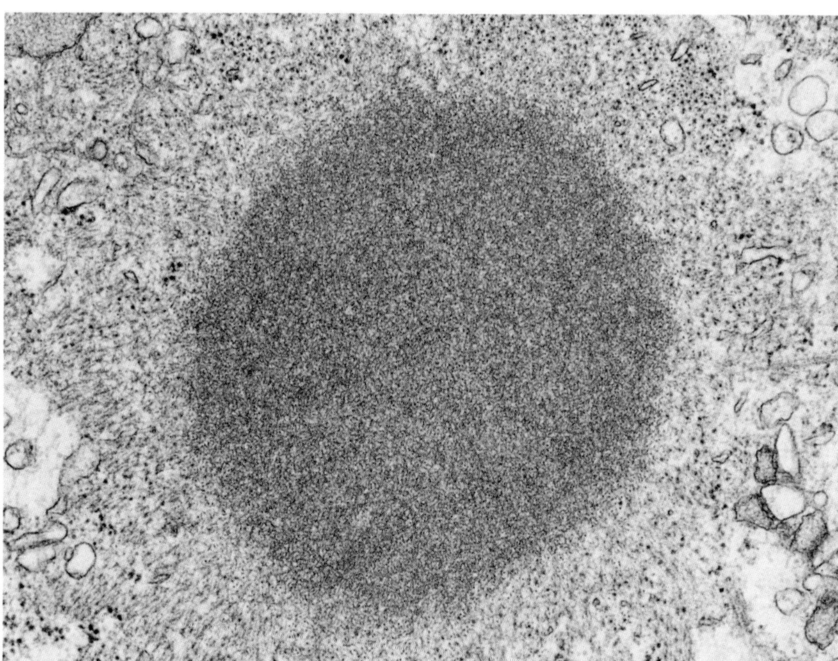

FIG. 6-10. Cytoplasmic body in muscle from a patient with polymyositis. ×40,000.

myofibers and when present, may be accompanied by a reduced number of mitochondria. Cytoplasmic bodies are encountered relatively infrequently but in a wide variety of disorders, including inflammatory myopathies,[6,50] muscular dystrophies,[1,50,51] periodic paralyses,[52,53] and certain other metabolic myopathies.[1,4,54,55]

Nemaline rods are highly ordered structures that are generally believed to be derived from components of the Z-discs.[56–58] In longitudinal sections, nemaline rods appear to consist of closely packed parallel filaments with periodic cross-striations at intervals of about 17 nm (Fig. 6-11). The ends of the rods appear to be in continuity with the thin myofilaments. In transverse sections, the nemaline rods display a quadratic latticework configuration that is very similar to comparably sectioned Z-discs (Fig. 6-12). The lattice squares measure approximately 10 nm on edge and appear to be formed by filaments that measure approximately 3 nm in diameter. Although nemaline rods have been the subject of intense investigation, their precise fine structure and chemical composition remain unsettled.

Nemaline rods are not entirely disease specific. They were first described in, and regarded characteristic of, a unique nonprogressive congenital myopathy.[59] However, subsequent cases have shown varying patterns of inheritance and a spectrum of clinical manifestations.[58,60,61] In some cases the rods selectively involve type 1 myofibers, while in others both fiber types are involved. In many cases, there is concomitant type 1 myofiber predominance and/or atrophy. Some authors regard these alterations of greater significance than the presence of the nemaline rods.[60,62] The extent of rod formation in an individual muscle biopsy specimen does not reflect the

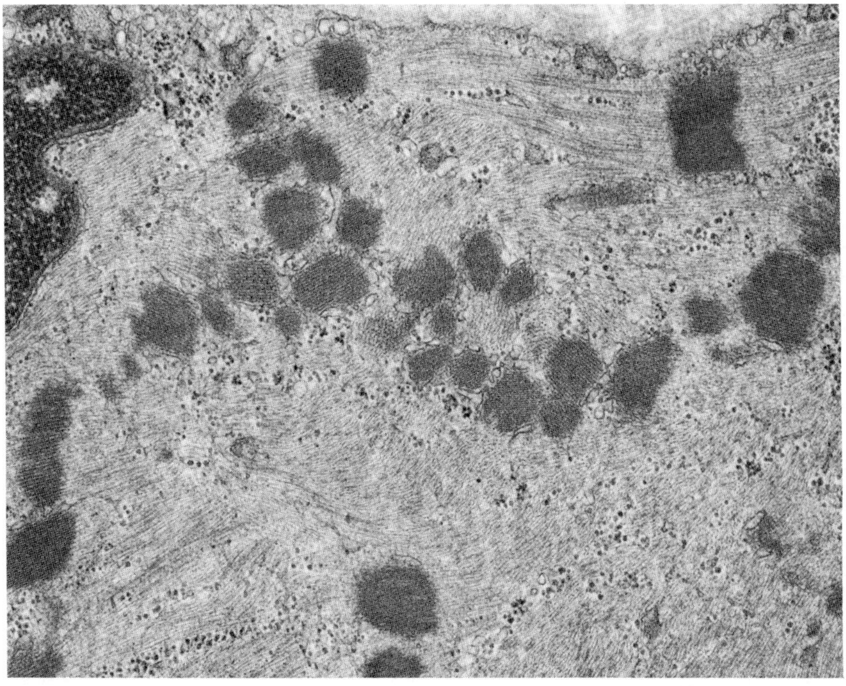

FIG. 6-11. Nemaline rods in muscle from a patient with centronuclear myopathy. ×30,000.

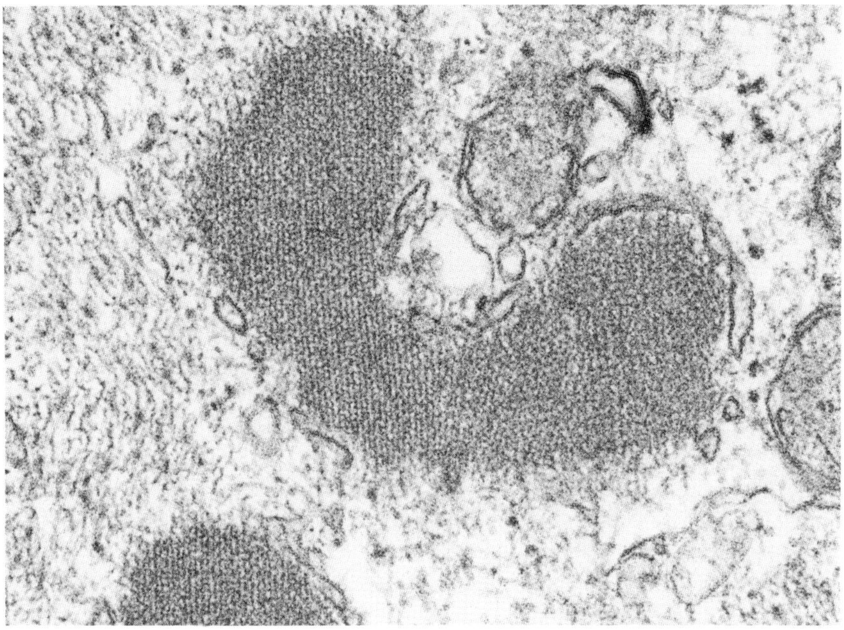

FIG. 6-12. Higher magnification of nemaline rods; muscle from a patient with oculopharyngeal myopathy. ×90,000.

course of the clinical disease, and there is some evidence that the number of rods increases with the passage of time. Sreter et al. demonstrated specific deficiencies among the light chains of fast myosin extracted from muscle of patients with nemaline myopathy.[63] These alterations are not explained by the disproportion of fiber types and may reflect the fundamental biochemical abnormality.

Cases of so-called adult-onset rod disease have been described.[13,58,64] Some of these cases are further characterized by the presence of intranuclear nemaline rods that have virtually the same ultrastructural features as their accompanying intracytoplasmic counterparts.[13] At least one case of congenital nemaline myopathy with both intracytoplasmic and intranuclear nemaline rods has been reported.[14]

Occasional nemaline rods, usually in association with other myofiber abnormalities, are observed in a wide variety of disparate conditions. These include muscular dystrophies,[1] inflammatory myopathies,[15,58] and other congenital myopathies.[2,4,60] They even may be seen in extraocular muscles[65] and in the vicinity of myotendinous insertions from apparently normal individuals.[58]

GLYCOGEN AND CARBOHYDRATE DEPOSITS IN MYOFIBERS

The glycogen content of skeletal muscle varies somewhat from individual to individual and is normally greater in type 2 myofibers than in type 1 myofibers. A moderate increase in the number of glycogen granules is seen commonly as a nonspecific reaction in a wide variety of muscle diseases.[1,2,4] Several of the glycogenoses prominently involve skeletal muscle; however, caution must be exercised in suggesting these diseases on the basis of ultrastructural findings. Definitive diagnosis can be made only by appropriate biochemical studies.

Skeletal muscle biopsy specimens from patients with the infantile form of type 2 glycogenosis (Pompe's disease) contain very large quantities of glycogen that are both membrane-bound and free in sarcoplasm.[66–69] By contrast, specimens from individuals with the adult form of type 2 glycogenosis show far less glycogen accumulation. There are focal areas of myofiber degeneration and autophagic vacuoles with remarkably heterogeneous contents (Fig. 6-13). These include glycogen granules, membranous whorls, and intensely osmiophilic granular material.[30,67,70–72] Occasionally this diagnosis is suspected initially from the ultrastructural findings, since the clinical manifestations may mimic adult-onset dystrophy or inflammatory myopathy, and the light microscopic alterations may be minimal. It must be emphasized, however, that occasional membrane-bound (presumably intralysosomal) collections of glycogen granules can be found in diverse conditions when glycogen is especially abundant. The precise biochemical basis for type 2 glycogenosis and its varying patterns of expression remains controversial.[72]

Some patients with type 3 glycogenosis have prominent neuromuscular involvement. Muscle biopsy specimens from these individuals have shown large quantities of morphologically normal glycogen, much of which is subsarcolemmal in location

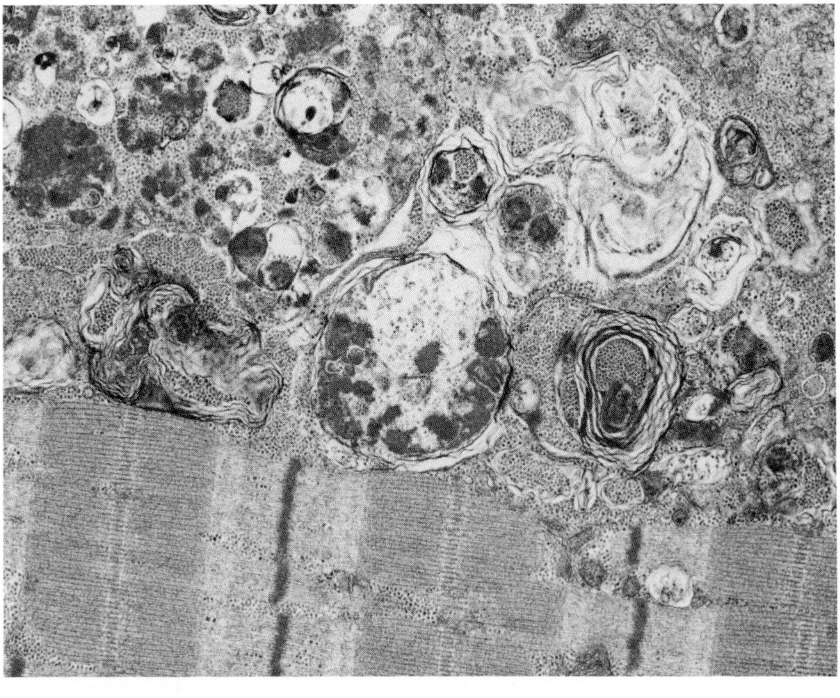

FIG. 6-13. Muscle biopsy specimen from a patient with the adult form of type 2 glycogenosis. Note the glycogen granules and autophagic vacuoles with heterogeneous contents. ×18,750.

(Fig. 6-14).[73,74] Similar ultrastructural findings have been reported in types 5, 7, and 10 glycogenoses.[68,69,75,76]

Type 4 glycogenosis is characterized by an accumulation of an amylopectin-like polyglucosan in various tissues of the body. Conspicuous, morphologically distinctive deposits with a granulofibrillar ultrastructure can be found in skeletal muscle.[77,78] Deposits with a similar ultrastructure but different light microscopic appearance have been reported in muscle biopsy specimens from patients with Lafora disease.[79–81]

LIPID AND LIPOPIGMENT DEPOSITS IN MYOFIBERS

Droplets of neutral lipid are normally more abundant in type 1 myofibers than in type 2 myofibers. Ultrastructurally, the lipid droplets appear as faintly osmiophilic or partially empty, unbounded vacuoles that are usually adjacent to mitochondria. A mild increase in lipid content is seen in a variety of muscle disorders. However, lipid droplets are especially numerous, particularly in type 1 myofibers, in patients with the myopathic and systemic forms of carnitine deficiency.[26–28,82–85] In some of these cases, mitochondria are abnormal and may contain intracristal paracrystalline inclusions.[26–28] In addition, lipid droplets have been observed in Schwann cells, pericytes, endothelial cells, and leukocytes from some of these patients.[28,82,84]

FIG. 6-14. Subsarcolemmal and intermyofibrillary accumulation of glycogen in a patient with type 3 glycogenosis. ×14,000.

Electron microscopy of muscle biopsy specimens from patients with Fabry disease has disclosed occasional lamellar bodies within the myofibers as well as the expected pleomorphic lipid inclusions in capillary endothelial cells.[86,87] Several reports have documented the ultrastructural findings in skeletal muscle from patients with the various forms of neuronal ceroid-lipofuscinosis (Batten-Vogt syndrome). Skeletal muscle has been reported to harbor deposits of osmiophilic granular material in the infantile (Haltia-Santavuori) form of this syndrome.[88] In the late infantile (Jansky-Bielschowsky) form, myofibers contain large numbers of membrane-bound deposits of curvilinear profiles beneath the sarcolemma (Fig. 6-15) and between the myofibrils.[4,89,90] Similar deposits may be found in the endothelial cells of the intramuscular capillaries. In the juvenile (Spielmeyer-Sjögren) form, muscle fibers may contain deposits of identical curvilinear profiles[90] or cytosomes with so-called rectilinear profiles.[89] Recently, cytosomes containing curvilinear profiles have been observed in muscle from a patient with the adult (Kufs) form of this disease.[91] While muscle biopsy specimens are suitable for the diagnosis of this syndrome, they may not be satisfactory for distinguishing the various forms. (See further discussion in section on peripheral nerve.)

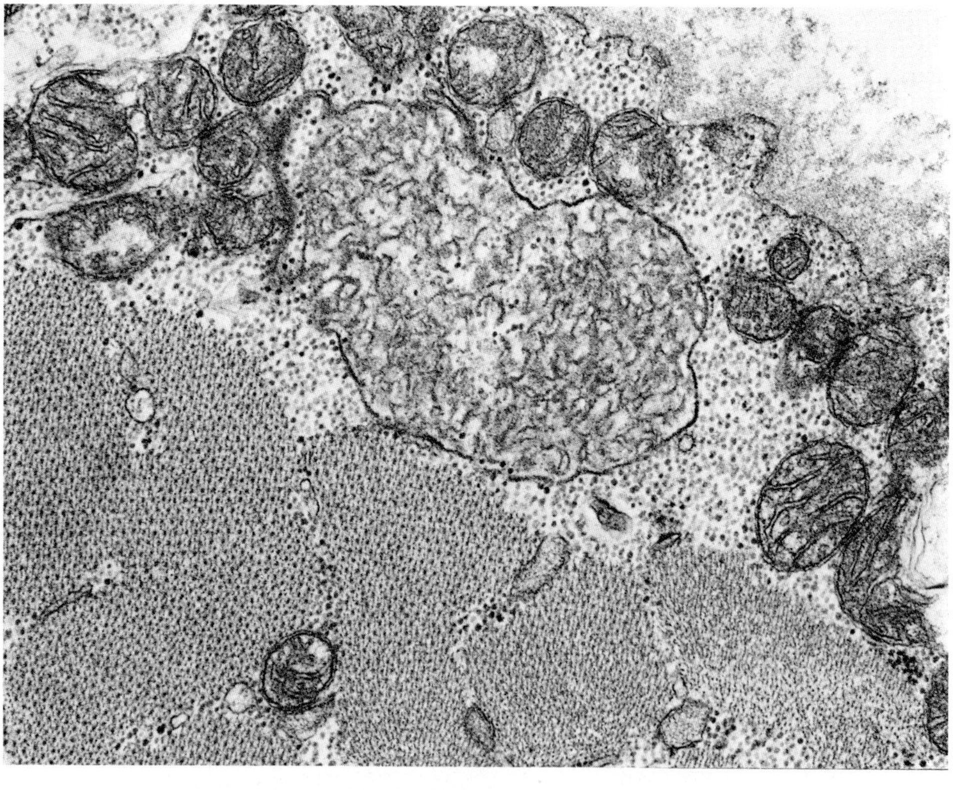

FIG. 6-15. Muscle biopsy specimen from a child with the Jansky-Bielschowsky form of neuronal ceroid-lipofuscinosis. Note the membrane-bounded deposit of curvilinear profiles. ×40,000.

MYOFIBER STRUCTURES OF UNKNOWN DERIVATION AND SIGNIFICANCE

In the course of ultrastructural examination of muscle biopsy specimens, one occasionally encounters morphologically striking structures of unknown derivation and significance. When first described, they often have been thought to be characteristic of a particular disease process. Later, when seen in diverse conditions, they are recognized as nonspecific but may remain of unknown derivation. Some examples are discussed below.

Concentric laminated bodies are hollow cylinders with walls of fibrillary laminae (Fig. 6-16). Each lamina appears to be formed from a series of paired parallel circumferential fibrils and intersecting longitudinal fibrils. The precise structural organization of the mural fibrils has been interpreted somewhat differently by various observers. The cores of these cylinders may contain glycogen granules, amorphous granular material, or small vesicles, or may appear to be empty. Concentric laminated bodies usually occur in small groups in a subsarcolemmal location. They have been described in a wide variety of unrelated conditions.[4,76,92]

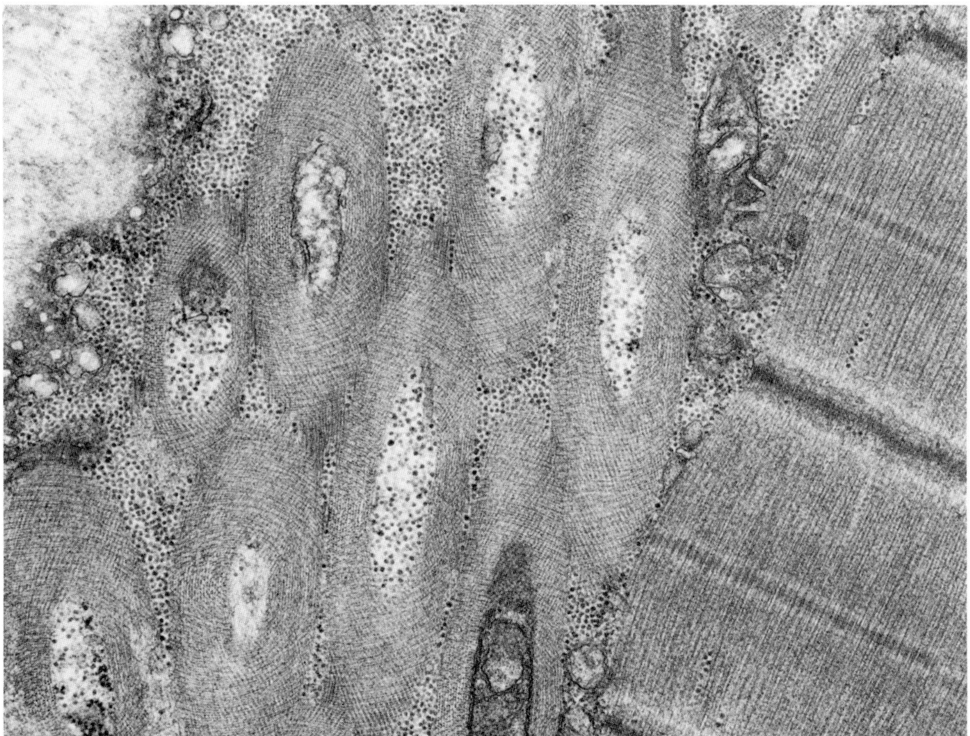

FIG. 6-16. Subsarcolemmal accumulation of concentric laminated bodies encountered in a muscle biopsy specimen from a child with the Willi-Prader syndrome. ×30,000.

Filamentous bodies are relatively large, compact skeins of tightly packed filaments (Fig. 6-17) that measure 6 to 7 nm in diameter. They are usually found just beneath the sarcolemma and are often juxtanuclear. It is not clear whether the filamentous bodies are derived from the thin myofilaments or some other fibrous protein. The filaments are thicker than the central components of cytoplasmic bodies. Filamentous bodies have been described in diverse conditions,[1,4] but we have found them most often in various forms of periodic paralysis.

Zebra-striped bodies or leptomeres are composed of roughly parallel fine filaments that are periodically traversed by narrow bands of osmiophilic granular material. They have been observed in various myopathies.[4,72] Very similar but somewhat more regular structures have been seen in normal and strabismic extraocular muscles (Fig. 6-18), in intrafusal fibers of muscle spindles, and near myotendinous insertions.[1,65]

Fingerprint inclusion bodies are composed of complex arrays of parallel straight and curved osmiophilic lamellae. The individual lamellae are serrated and spaced at intervals of 30 to 36 nm. The inclusion bodies are usually subsarcolemmal in location and may merge with masses of filaments that resemble filamentous bodies. These structures have been described in nonprogressive childhood myopathies,[93,94] myotonic dystrophy,[95] and dermatomyositis.[95] They are thought to affect type 1 myofibers selectively.

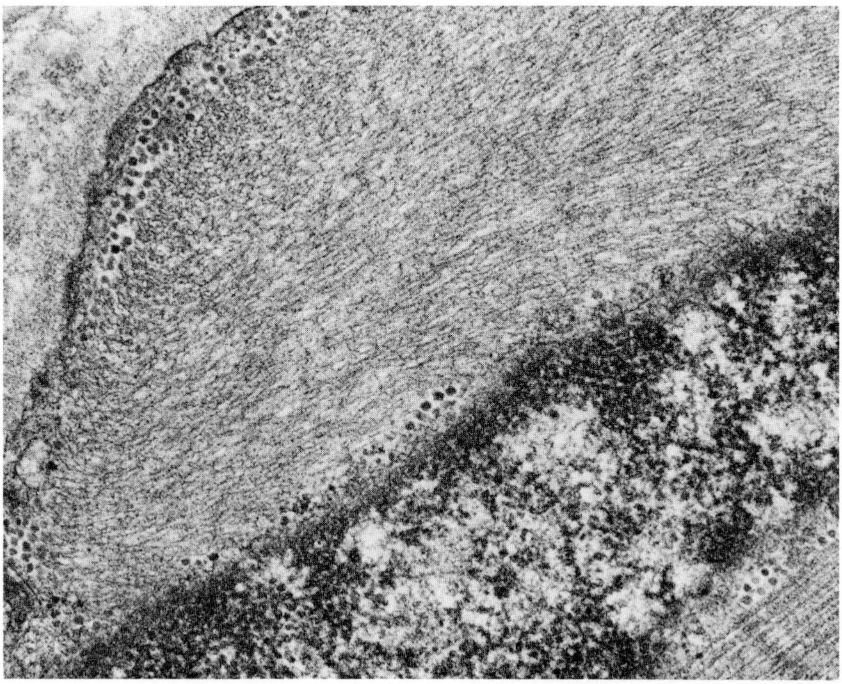

FIG. 6-17. Subsarcolemmal filamentous body in muscle from a patient with the adult form of type 2 glycogenosis. ×52,500.

FIG. 6-18. Zebra-striped body in the inferior oblique extraocular muscle. ×30,000.

Reducing bodies are roughly ovoid masses of coarse osmiophilic granules. The inclusions contain numerous cavities in which glycogen granules and cytoplasmic organelles are sequestered. Reducing bodies are thought to be composed of ribosomal or viral RNA. They were described initially in two children with a severe, progessive neuromuscular disease.[2,55] Subsequently, similar if not identical structures have been reported in another child with a milder myopathy.[96]

NORMAL PERIPHERAL NERVE

The nerve fascicles in a multifascicular nerve are bound together by the epineurium. This is composed of connective tissue and contains longitudinally oriented arteries, veins, and lymphatics. Each nerve fascicle is ensheathed by the perineurium (Fig. 6-19). This is a highly specialized structure composed of one or more layers of modified Schwann cells. These cells contain numerous pinocytotic vesicles and are bound to one another by tight junctions. Each layer is invested by relatively prominent basal laminae. The perineurium is functionally important as one component of the blood-nerve barrier.[97] The endoneurium consists of the interstitial connective tissues within the nerve fascicle. The endothelial cells of the endoneurial capillaries are joined to one another by tight junctions and function as the other component of the blood-nerve barrier.[98]

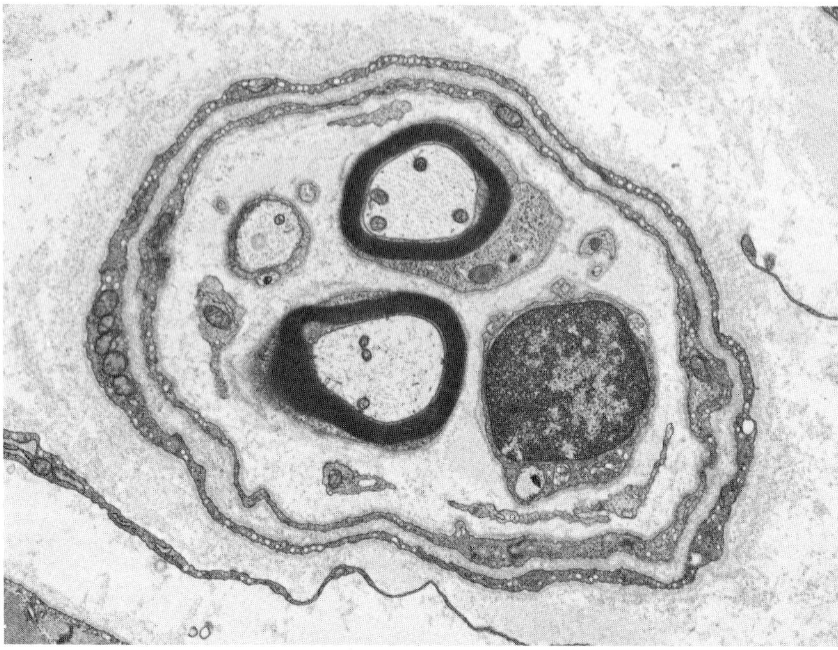

FIG. 6-19. Small unifascicular intramuscular nerve invested by a perineurium composed of two laminae of modified Schwann cells. ×9000.

In the adult, myelinated nerve fibers normally range from about 2 to 18 μm in diameter, with bimodal peaks at 4 and 11 μm. Each myelin internode is formed and maintained by a single Schwann cell. The nucleus is located toward the middle of the internode, where the cytoplasm is most abundant. Elsewhere, at least a thin layer of Schwann cell cytoplasm covers the myelin and separates it from the surrounding basal lamina (Fig. 6-20). The thickness of the myelin sheath is roughly proportional to the diameter of the ensheathed axon.

Peripheral nerve myelin displays a laminar periodicity of about 15 nm. The major dense lines arise from fusion of the specialized inner surfaces of Schwann cell membranes, while the minor dense lines arise from apposition of the outer surfaces of these membranes. Schmidt-Lantermann incisures are wedge-shaped discontinuities within the myelin internodes, where the major dense lines split and enclose Schwann cell cytoplasm. Similar splitting occurs adjacent to the nodes of Ranvier, where the internodal myelin sheath terminates in overlapping tongues of Schwann cell cytoplasm. The cytoplasmic tongues associated with each more peripheral dense line extend further toward the center of the node (Fig. 6-21). The peripheral layer of Schwann cell cytoplasm forms a collar that interdigitates with a similar collar formed by the next successive Schwann cell. Thus, even the node of Ranvier is covered by Schwann cell cytoplasm and a basal lamina.

Normally, most unmyelinated nerve fibers range from 1 to 2 μm in diameter and have a unimodal size distribution. They outnumber the myelinated fibers by a ratio

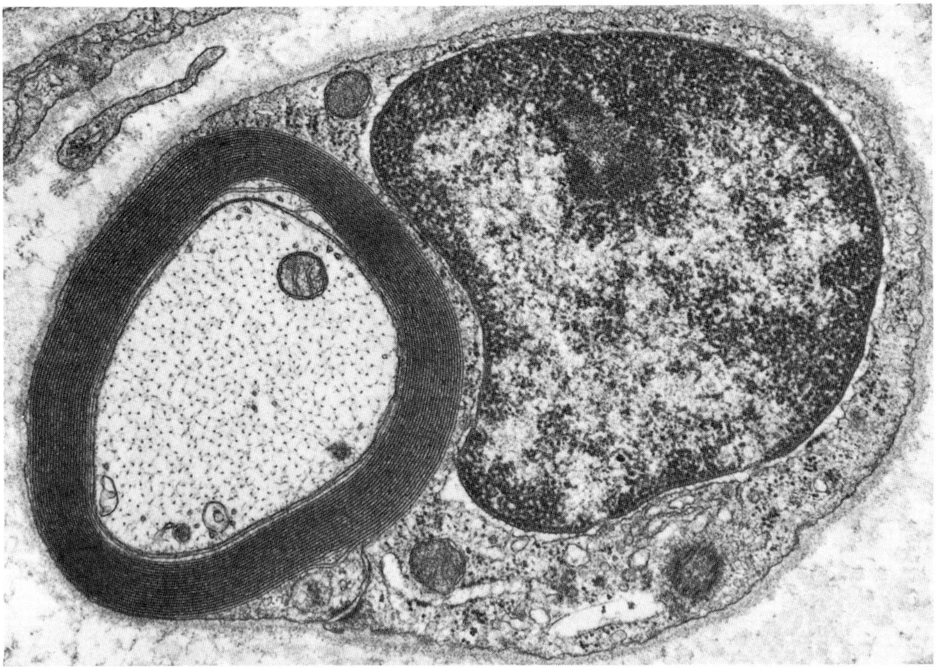

FIG. 6-20. Myelinated nerve fiber. The Schwann cell invests a single axon. ×25,500.

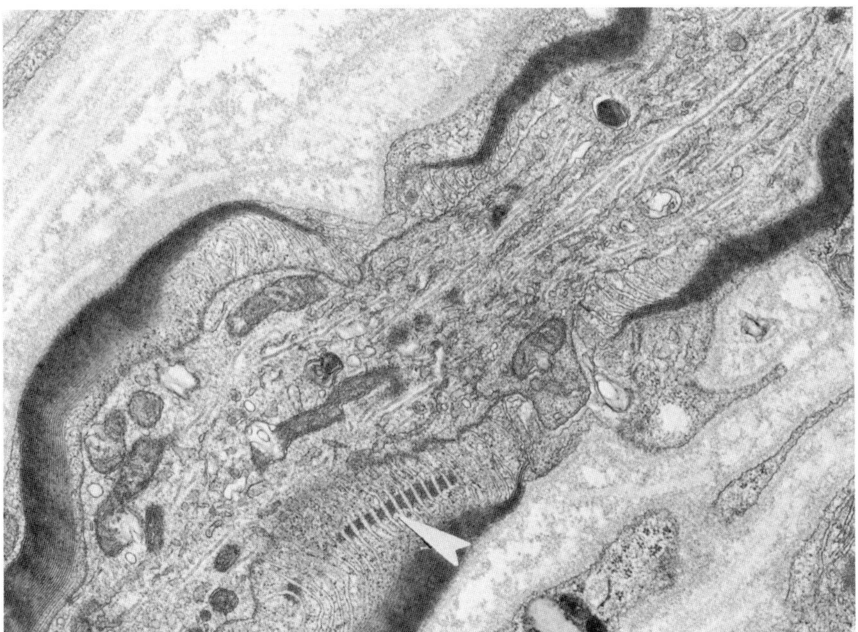

FIG. 6-21. Node of Ranvier. Note the termination of the major dense lines in overlapping tongues of cytoplasm. These may contain a series of regularly spaced osmiophilic deposits that is designated a vinculum (arrow). ×20,000.

of approximately 4:1. The associated Schwann cells form overlapping and interconnecting rows with several unmyelinated axons invested by the perikaryon and processes of each Schwann cell (Fig. 6-22). The unmyelinated somatic and accompanying unmyelinated autonomic nerve fibers cannot be distinguished morphologically in peripheral nerve biopsy specimens.

AXONAL ALTERATIONS

Reactive axonal enlargements ("reactive axons") are bulbous swellings that develop rapidly at or near the site of focal, severe nerve injury. The evolution of these structures has been studied extensively in experimental animal models following nerve transection, crush, and ligation.[99,100] Although acute post-traumatic lesions are rarely seen in human diagnostic material, similar axonal changes may be seen as the result of occlusive or inflammatory vasculitides and in certain severe inflammatory neuropathies. The axonal enlargements are filled with myriads of intact and degenerating axoplasmic organelles. Less pronounced accumulations of intact and degenerating organelles extend for a variable distance both proximally and distally from the site of severe injury. Thus, similar axoplasmic changes may be seen even where the axons are still invested with apparently normal myelin (Fig. 6-23). The distal segment of a transected nerve undergoes wallerian degeneration. This process con-

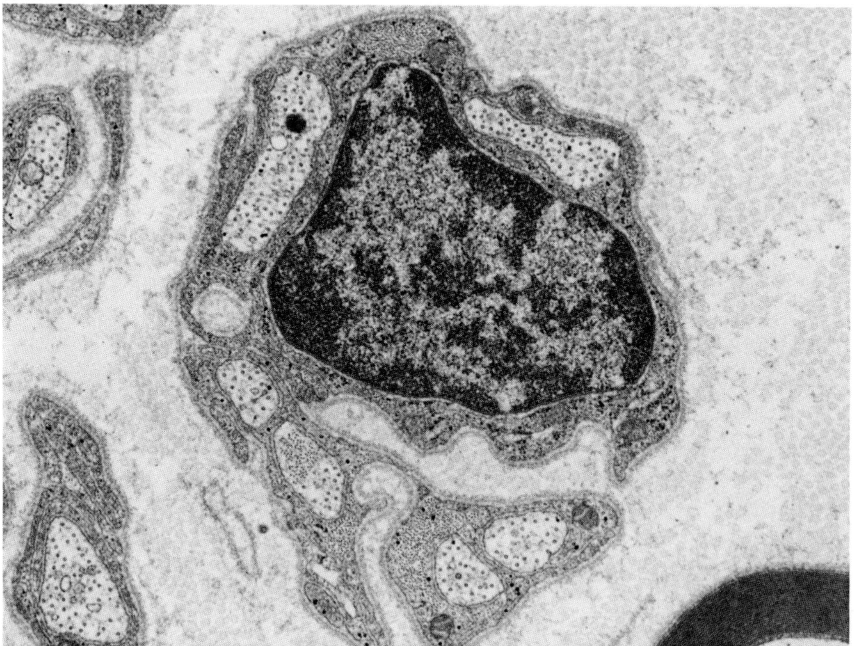

FIG. 6-22. Unmyelinated nerve fibers. A Schwann cell and its processes invest multiple axons. ×20,000.

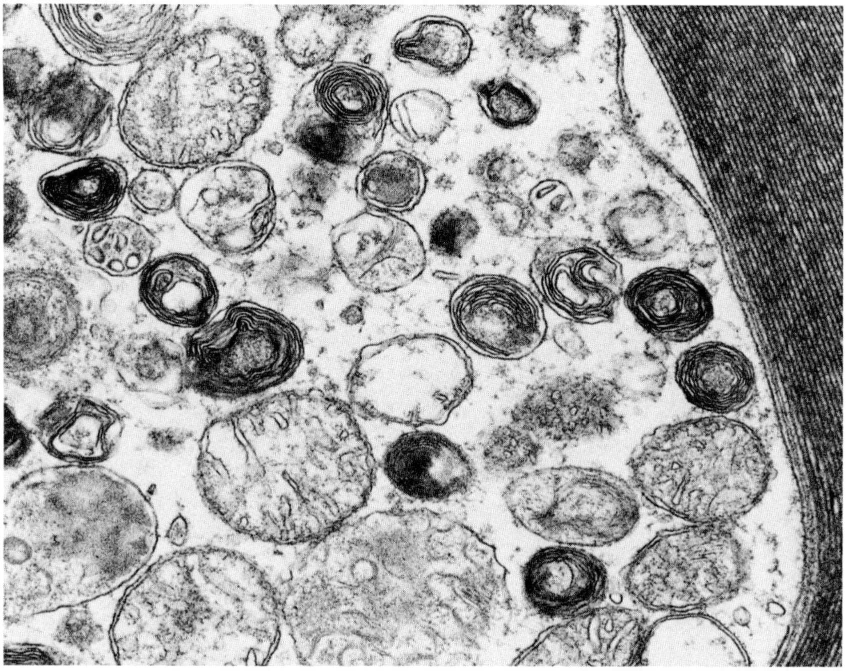

FIG. 6-23. Degenerating axoplasmic organelles including lamellar dense bodies from a patient with porphyric neuropathy. ×45,000.

sists of a relatively stereotyped sequence of concurrent degradative changes involving the axoplasm, myelin sheaths, and Schwann cells.[101, 102] Hematogenous macrophages are thought to participate in the process.[103]

Neuropathies due to a wide variety of toxic agents, metabolic derangements, and nutritional deficiencies have been designated variously as dying-back neuropathies, axonal degenerations or axonal dystrophies.[104–107] These are distinguished from wallerian degeneration by their slower course, tendency to affect specific populations of nerve fibers, tendency to be more severe distally, and, most significantly, by the absence of any focal, severe nerve injury or transection. The morphologic changes in the affected nerve fibers may resemble those of early wallerian degeneration; however, only scattered individual fibers are involved. Furthermore, depending upon the cause, certain axoplasmic organelles may be affected preferentially.

Among the most striking examples of selective organelle involvement encountered in human diagnostic material are the hereditary[108–112] and acquired [113–117] forms of giant axonal neuropathy. In these conditions, there are focal, often paranodal, axonal enlargements containing massive accumulations of neurofilaments (Fig. 6-24). Other organelles, such as microtubules, mitochondria, and vesicles of endoplasmic reticulum, are confined largely to small channels traversing the accumulations of neurofilaments. Especially in the hereditary forms, there may also be deposits of osmiophilic granular material (Fig. 6-25) and sheaves of coarse osmiophilic fibrils scattered among the neurofilaments. Very similar intra-axonal sheaves of osmiophi-

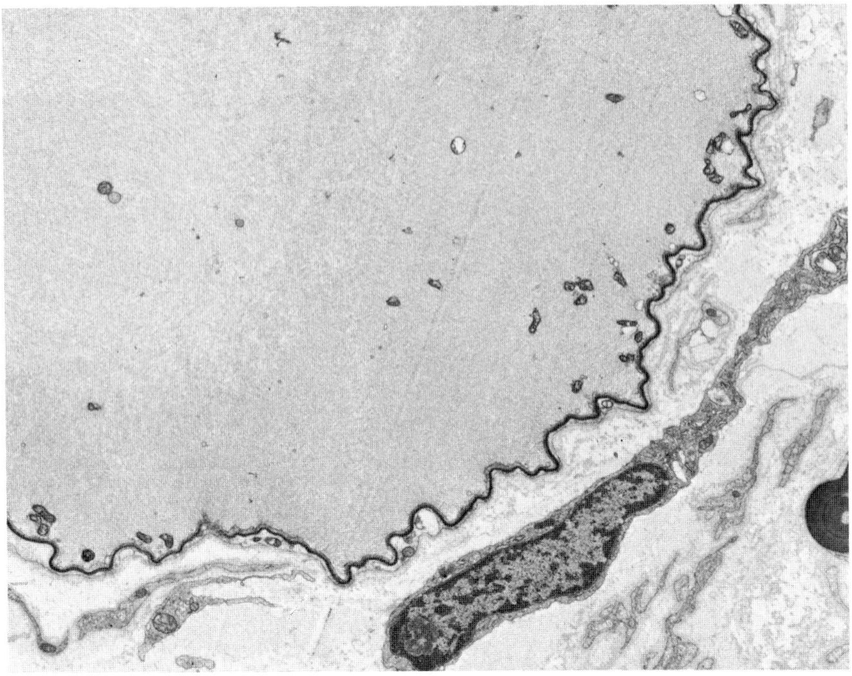

FIG. 6-24. Acquired giant axonal neuropathy. Note the enormous axonal enlargement with neurofilaments. ×6000.

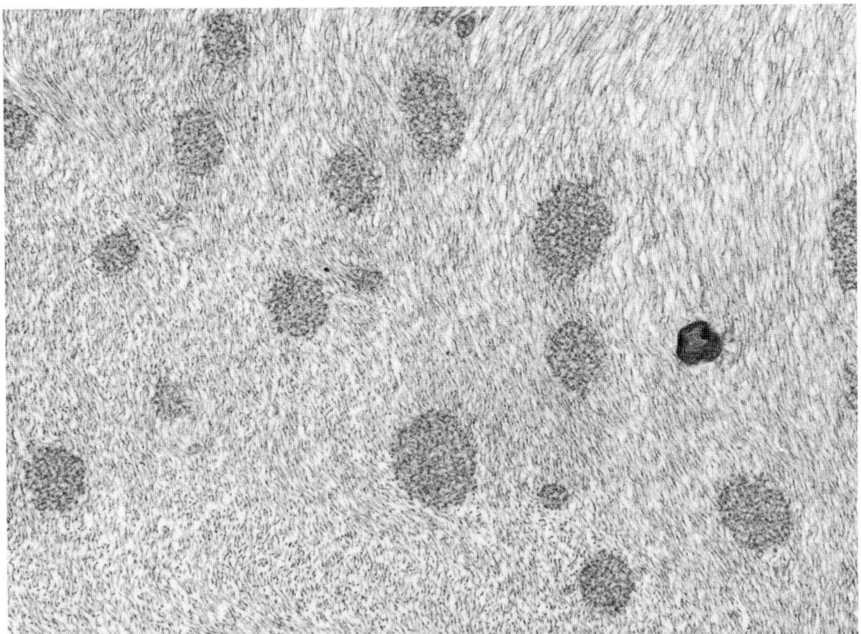

FIG. 6-25. Hereditary giant axonal neuropathy. Note the deposits of osmiophilic granular material scattered among the neurofilaments. ×30,000.

lic fibrils have been observed in sural nerve biopsy specimens from a patient with Fabry disease[118] and from a presumably normal, older man.[119] Intra-axonal deposits of glycogen—free, membrane-bound, or within mitochondria—have been described in uremic neuropathy,[120] hypothyroid neuropathy,[121] Fabry disease,[118] and infantile neuroaxonal dystrophy.[122] In the last condition, the axons also contain elaborate arrays of smooth membranes, tubules, and vesicles. During the course of axonal degeneration, portions of the abnormal axoplasm may be sequestered and degraded by invaginations of the adaxonal Schwann cell cytoplasm. Although this phagocytic process has been studied primarily in experimental models,[123] it has been observed in human material (Fig. 6-26).

Axonal atrophy may be a distinct pathologic process or a very indolent form of axonal degeneration. Axonal atrophy occurs predominantly in the very chronic hereditary neuropathies.[104] The light microscopic findings are dominated by a reduction in the number of myelinated nerve fibers with active axonal degeneration rarely demonstrated. Ultrastructurally, the process is characterized by the presence of small, i.e., atrophic, axons surrounded by disproportionately thick myelin sheaths (Fig. 6-27).

Regardless of the antecedent process, wallerian degeneration, axonal degeneration, or axonal atrophy, the degenerated myelinated nerve fibers are reduced eventually to bands of Büngner. These are columns of Schwann cells and Schwann cell processes enclosed by basal lamina (Fig. 6-28). Ultimately, these degenerate too un-

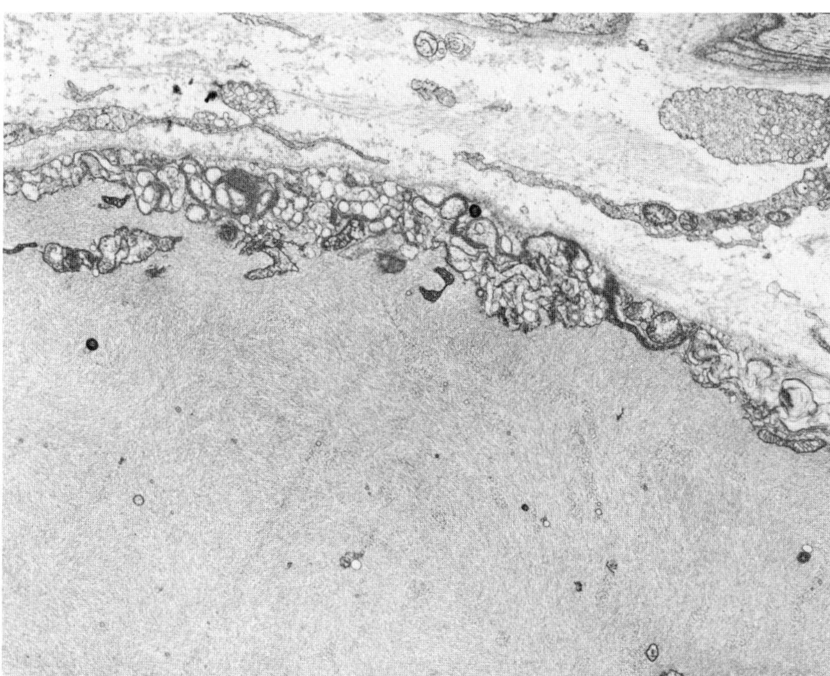

FIG. 6-26. Acquired giant axonal neuropathy. Abnormal axoplasm is being sequestered and degraded by invaginations of the adaxonal Schwann cell cytoplasm. ×6000.

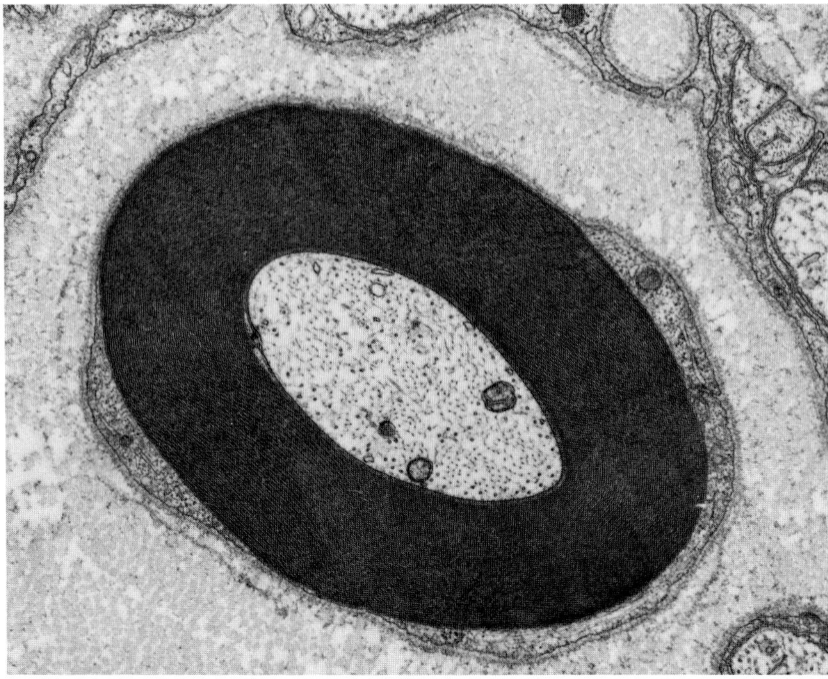

FIG. 6-27. Axonal atrophy in a patient with a variant of type 3 mucolipidosis. Note the small atrophic axon surrounded by a disproportionately thick myelin sheath. ×18,750.

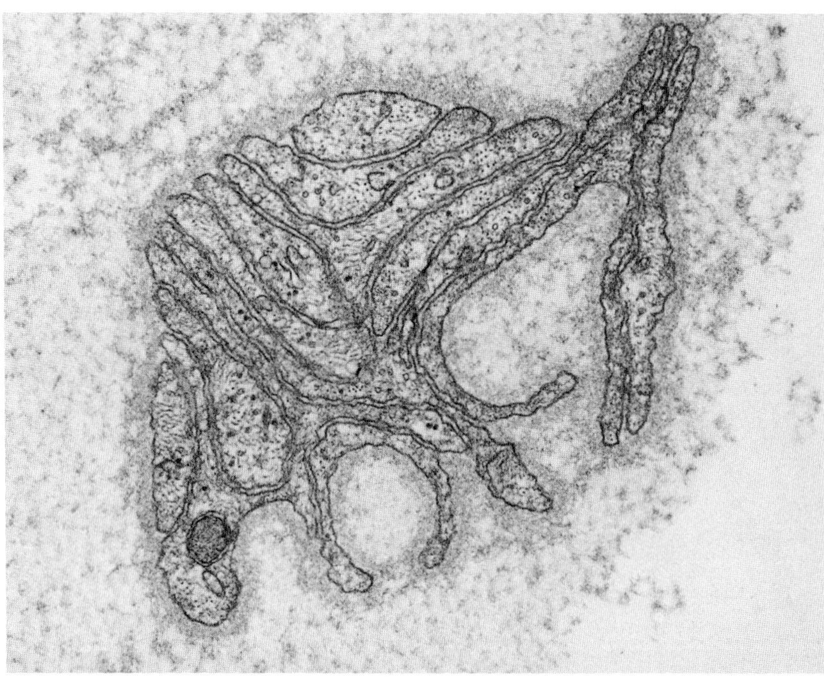

FIG. 6-28. Band of Büngner in a patient with the neuronal form of Charcot-Marie-Tooth disease (HMSN—II). ×40,000.

less invaded by regenerating axonal sprouts. This occurs more readily following nondisruptive processes than nerve transection. Axonal regeneration may even occur concurrently with axonal degeneration. The axonal sprouts are difficult to identify with certainty. Distally, they may contain abundant filaments, mitochondria, dense bodies, and complex arrays derived from the endoplasmic reticulum (Fig. 6-29). Eventually, the regenerating axons enlarge and may become remyelinated. However, the remyelinated internodes are shorter and their myelin sheaths are thinner than the internodes of the original intact nerve fiber. In transverse section, the regenerating axons may appear as clusters of small unmyelinated or thinly myelinated axons bounded peripherally by persistent basement membrane (Fig. 6-30).[119]

In many neuropathies, unmyelinated nerve fibers also are affected. The pathologic alterations involving these fibers are more difficult to assess. Recognized changes include hydropic degeneration, distended growth cones, filamentous accumulations (in giant axonal neuropathy), and atrophy.[124] Small stacks of flattened Schwann cell processes, analogous to bands of Büngner, persist following the degeneration of unmyelinated axons. Collagen pockets, which are small bundles of collagen fibers encircled by Schwann cell cytoplasm, also have been interpreted by some authors as sequelae of degeneration of unmyelinated axons. The extent to which unmyelinated fibers participate in "onion bulb" formations remains controversial.

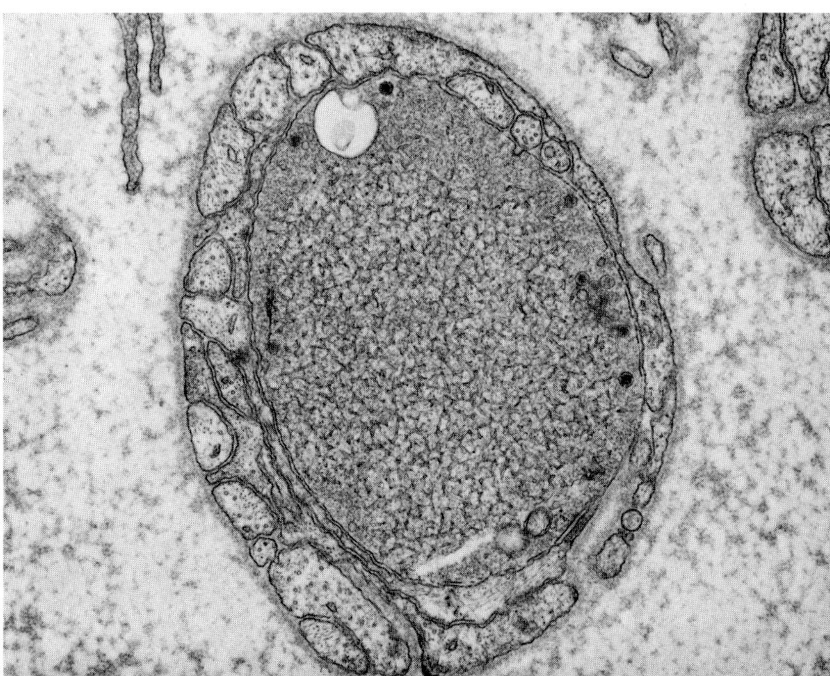

FIG. 6-29. Regenerating axonal sprout containing an elaborate tubular array derived from the endoplasmic reticulum. ×22,500.

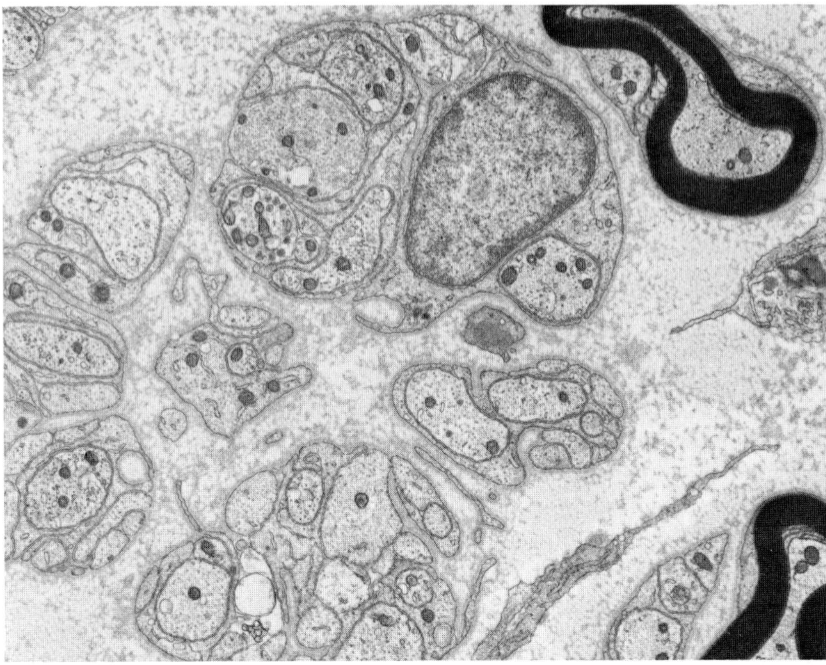

FIG. 6-30. Clusters of regenerating axonal sprouts invested by persistent basement membrane. ×9000.

SEGMENTAL DEMYELINATION

Segmental demyelination is the selective demyelination of certain internodes. Traditionally, this process has been attributed to disorders that directly and selectively damage Schwann cells and/or their associated myelin sheaths. More recent studies have shown the occurrence of segmental demyelination in association with axonal degeneration and axonal atrophy. In fact, Dyck has suggested that most examples of segmental demyelination may be accompanied by some degree of axonal dysfunction.[104] The presence of segmental demyelination in biopsy material is best appreciated by examination of teased nerve fiber preparations.

"Onion bulb" formations are important morphologic features of many nerve biopsy specimens.[125] They may be sufficiently prominent to be seen readily by light microscopy of paraffin-embedded sections, or they may be seen clearly only in epoxy-embedded thick sections. Occasionally they can be demonstrated and distinguished from clusters of regenerating axons only by electron microscopy. Onion bulbs consist of a demyelinated or remyelinated axon surrounded by concentric layers of Schwann cells, basal laminae, and longitudinally oriented bundles of collagen fibrils (Fig. 6-31). Occasional unmyelinated axons may be incorporated in the Schwann cell lamellae of the onion bulb formations. Ochoa has suggested that paucity of unmyelinated fibers between onion bulbs in some hypertrophic neuropathies reflects the par-

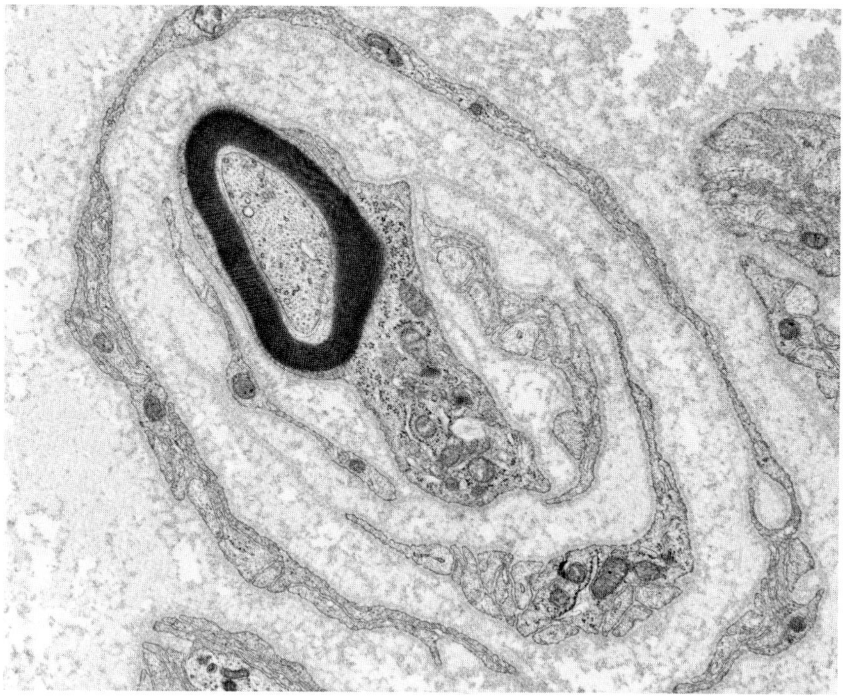

FIG. 6-31. Onion bulb formation in a patient with an acquired giant axonal neuropathy. ×14,000.

ticipation of unmyelinated fibers in onion bulb formation.[124] Nevertheless, these structures are generally accepted as morphologic evidence of repeated waves of demyelination and remyelination. Although they are not disease specific, conditions such as Dejerine-Sottas disease, hypertrophic Charcot-Marie-Tooth disease (HMSN-I), Refsum's disease, and chronic relapsing polyneuropathy should be considered when they are especially prominent. Smaller, less conspicuous onion bulb formations have been described in a wide variety of other conditions with a demyelinative component.

OTHER SCHWANN CELL ALTERATIONS

Altered Schwann cell organelles and/or intracytoplasmic deposits of adventitious material have been observed in a variety of conditions. Some of these alterations and deposits are sufficiently distinctive to suggest specific diagnoses. In at least some cases of hereditary giant axonal neuropathy, the Schwann cells along with endoneurial endothelial cells may contain prominent accumulations of microfilaments.[110–112] This feature may be useful for distinguishing acquired giant axonal neuropathy from hereditary giant axonal neuropathy or may even delineate a distinct form with generalized microfilament overproduction. Membrane-bound paracrystalline arrays thought to be derived from mitochondria have been observed in the Schwann cells of patients with Refsum's disease.[126] When seen, clinical data and other morphologic features must be evaluated carefully, since similar, if not identical, inclusions have been observed in other neuropathies (Fig. 6-32), including diabetes and metachromatic leukodystrophy.[126]

Peripheral nerve specimens from patients with metachromatic leukodystrophy contain a spectrum of pleomorphic membrane-bound deposits in both the Schwann cells and the endoneurial macrophages. Morphologically, the most distinctive deposits consist of stacks of osmiophilic and electron-lucent discs (Fig. 6-33). These have been designated as tuff-stone deposits.[127]

Nerve biopsy specimens from patients with Krabbe's disease have pleomorphic electron-lucent lipid deposits in Schwann cells and endoneurial macrophages.[128] However, the multiangular and twisted tubular configurations that are characteristic of this condition are found predominantly in the endoneurial macrophages (Figs. 6-34 and 6-35). Extrapolating from studies on Gaucher's disease,[129] it would seem likely that these "hollow" multiangular and tubular profiles are actually stacks of flat cerebroside crystals that were largely leeched out during tissue processing. Somewhat similar acicular lipid lamellae have been observed in Schwann cell cytoplasm in a few cases of adrenoleukodystrophy.[130]

Patients with Fabry's disease have pleomorphic lipid deposits that are most conspicuous in endothelial cells and in the modified Schwann cells of the perineurium.[86, 87, 118] The dual localization of the lipid deposits is highly characteristic,

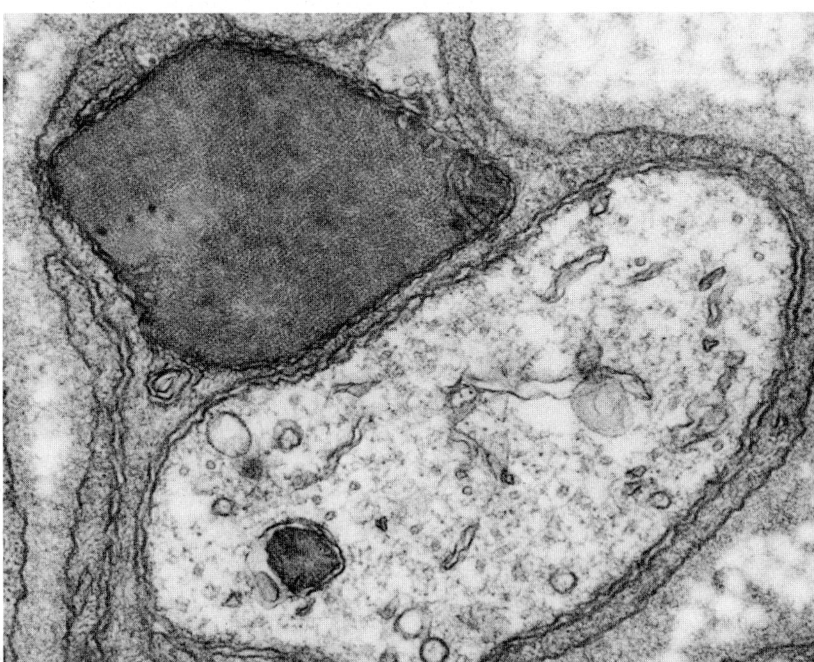

FIG. 6-32. Paracrystalline structure thought to be derived from Schwann cell mitochondrion. ×60,000.

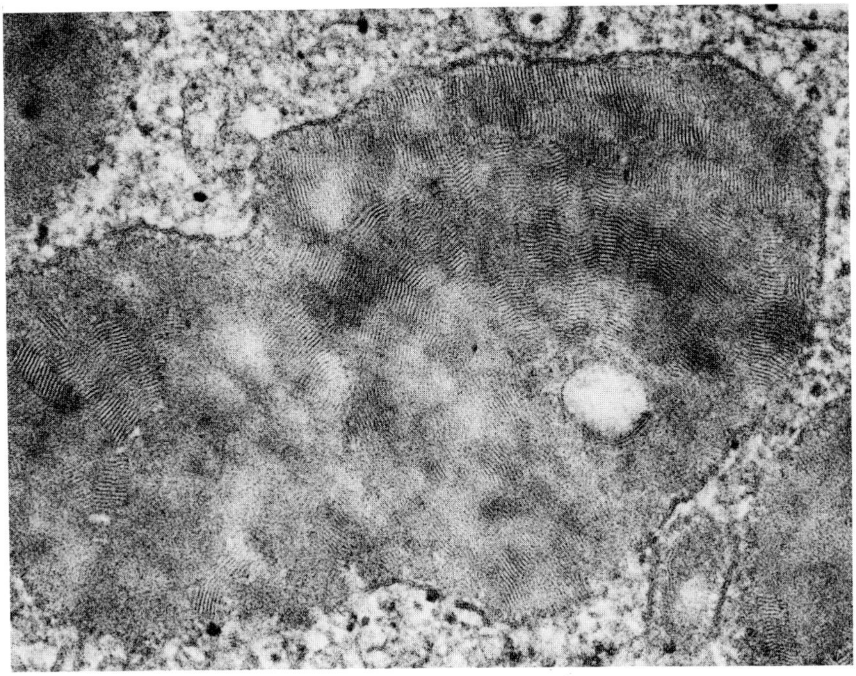

FIG. 6-33. So-called tuff-stone deposit in a Schwann cell from a patient with late infantile metachromatic leukodystrophy. ×100,625.

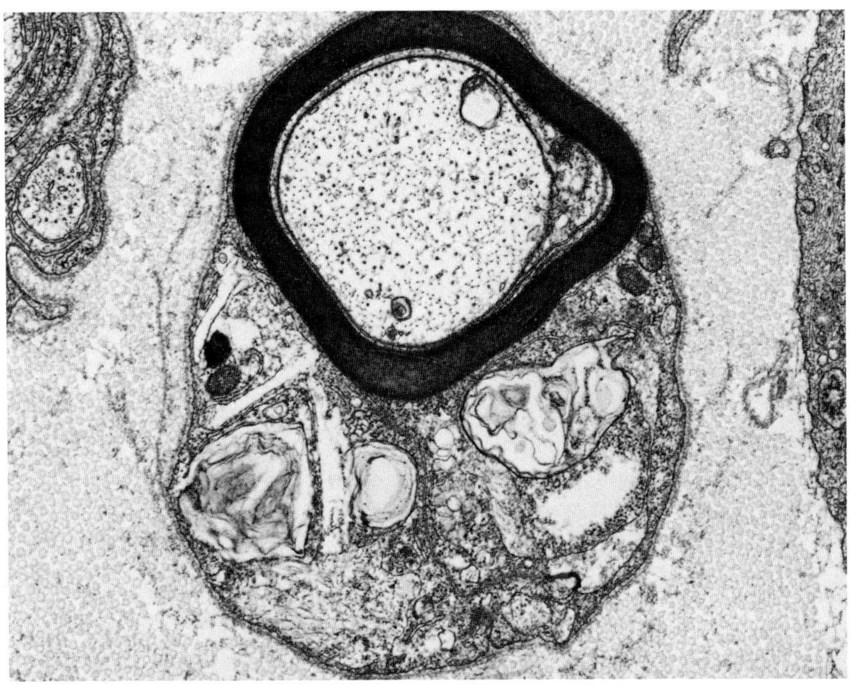

FIG. 6-34. Sural nerve biopsy specimen from a patient with Krabbe's disease. Note the pleomorphic lipid deposits in the Schwann cell cytoplasm. ×18,750.

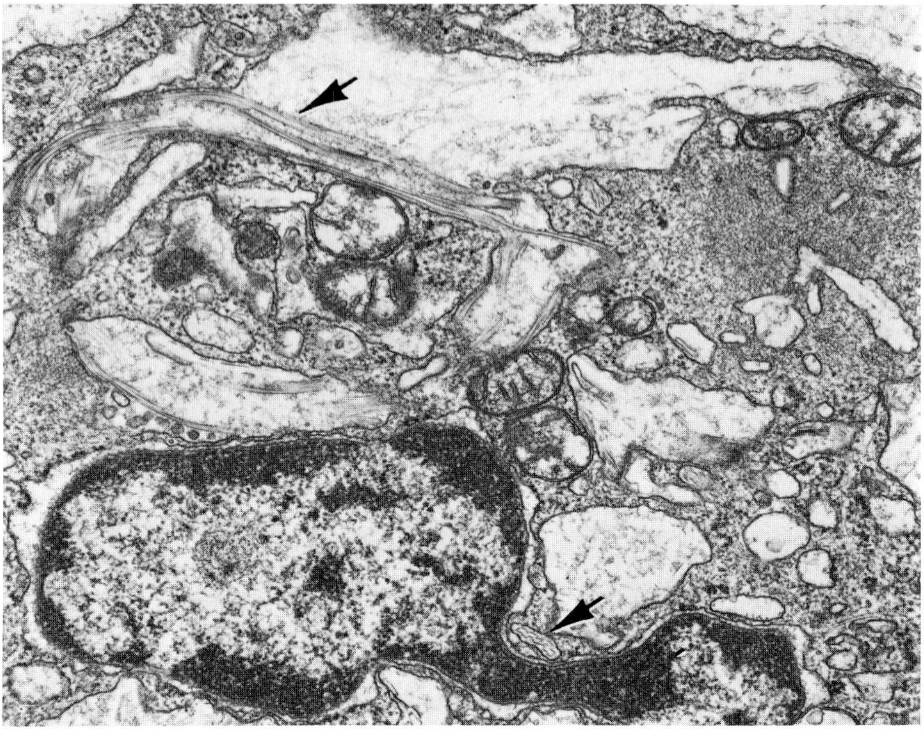

FIG. 6-35. Sural nerve biopsy specimen from a patient with Krabbe's disease. Note the characteristic cerebroside deposits (arrows) in the endoneurial macrophage. ×28,125.

even though the configuration of the individual deposits (alternating osmiophilic and electron-lucent lamellae) is not unique. The peripheral nerves also show preferential loss of small myelinated and unmyelinated fibers.[118]

Peripheral nerves are among the tissues most suitable for the diagnosis of neuronal ceroid-lipofuscinosis.[107,131] The distinctive membrane-bound deposits of curvilinear profiles (Fig. 6-36) can be recognized readily and with certainty in Schwann cells associated with both myelinated and unmyelinated nerve fibers. Similar deposits also may be found in the endothelial cells of the endoneurial capillaries.

We have observed prominent deposits of an unidentified lipid and possible mucopolysaccharide in Schwann cells and endoneurial macrophages in a sural nerve biopsy specimen from a child with a variant of type 3 mucolipidosis (Fig. 6-37). Morphologically similar deposits also have been described, especially in endoneurial macrophages, in other forms of mucolipidoses and mucopolysaccharidoses.[132–134]

Membrane-bound deposits of glycogen have been described in the Schwann cells of children with the infantile form of type 2 glycogenosis (Pompe's disease).[135] However, these deposits are more extensive in skeletal muscle. Recently, McMaster et al. have reported deposits of a branched fibrillar and granular material in peripheral nerve from a patient with type 4 glycogenosis.[78]

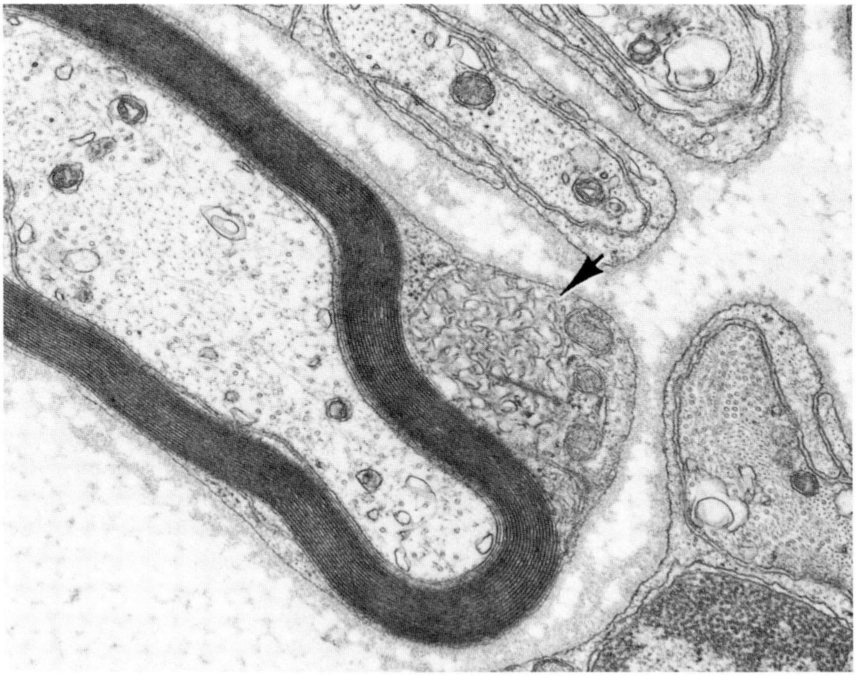

FIG. 6-36. Sural nerve biopsy specimen from a child with the Jansky-Bielschowsky form of neuronal ceroid-lipofuscinosis. Note the membrane-bound deposit of curvilinear profiles (arrow). ×30,000.

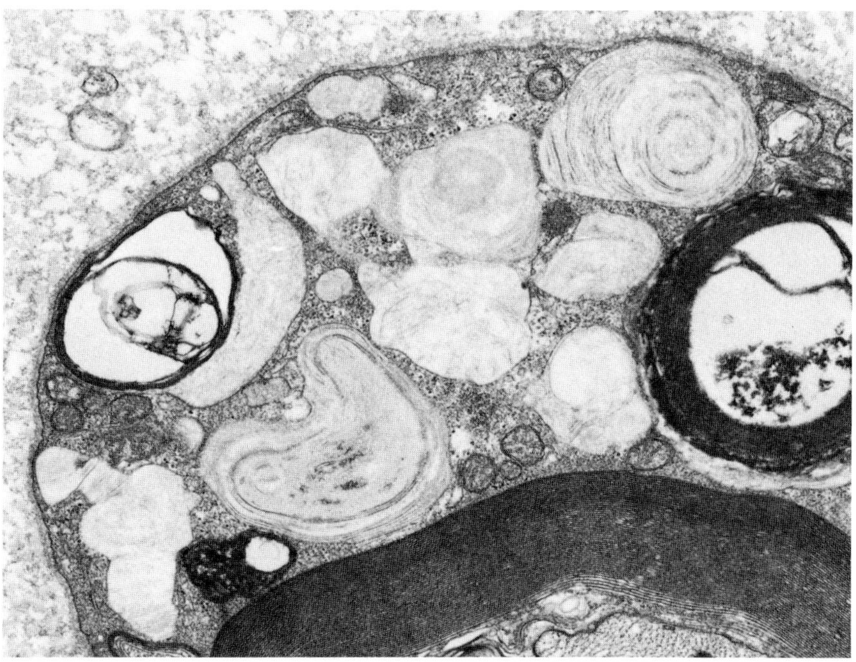

FIG. 6-37. Sural nerve biopsy from a child with a variant of type 3 mucolipidosis. Note the pleomorphic Schwann cell deposits. ×18,750.

ENDONEURIAL, PERINEURIAL, AND EPINEURIAL ALTERATIONS

Among these alterations, deposition of amyloid is of the greatest diagnostic significance.[136] Amyloid is initially sought by light microscopy with polarized light after Congo red staining. Unfortunately, the deposits may be small, and the characteristic apple-green, birefringent color may not be as obvious as desired. Therefore, electron microscopy is an extremely useful means for confirming the presence of this adventitious material. The amyloid appears as stiff fibrils about 7 nm in diameter. The deposits may be encountered in the epineurium, beneath the perineurium, as masses within the endoneurium, or about endoneurial blood vessels (Fig. 6-38). It is especially important to distinguish the subperineurial deposits from Renaut corpuscles, which have a somewhat similar appearance especially in epoxy sections.[137]

Prominent vacuolated endoneurial fibroblasts were described initially in hereditary sensory neuropathy.[138] Subsequently, the vacuolated fibroblasts have been encountered in other chronic neuropathies, especially hypertrophic neuropathies with abundant interstitial myxoid material.[139]

Basal laminae of endoneurial blood vessels thicken with increasing age and as a nonspecific response to various chronic neuropathies. A more prominent thickening and duplication of the vascular basal laminae have been observed in patients with diabetes.[140] Recently, Johnson et al. reported quantitatively significant thickening of the perineural basal laminae in this disease.[141] Although the thickening of the base-

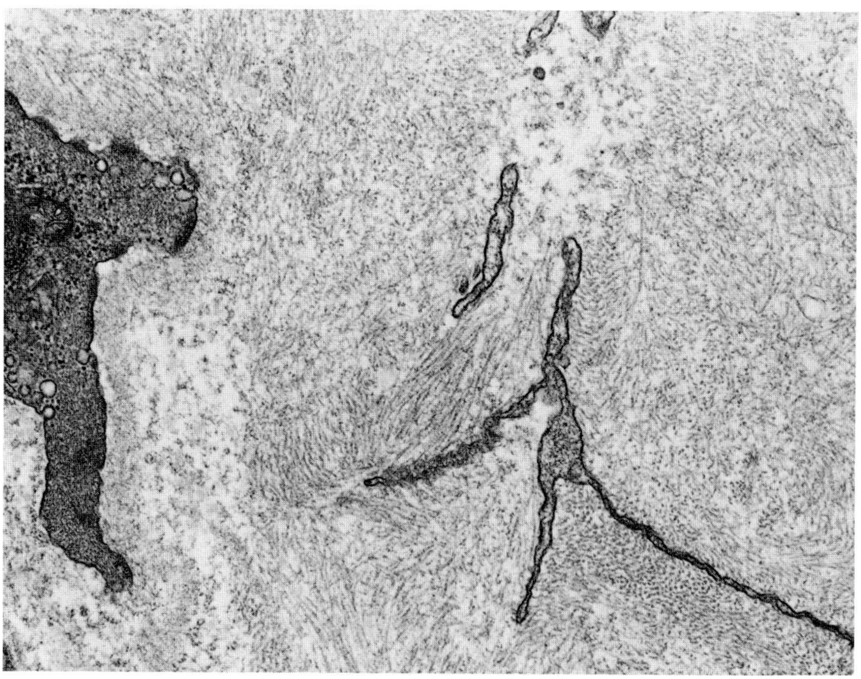

FIG. 6-38. Perivascular amyloid deposit. ×30,000.

ment membranes may reflect important aspects of the pathogenesis of diabetic neuropathy, the marked variation in magnitude precludes its use for diagnosis in individual cases.

ACKNOWLEDGMENTS

Special thanks to Ms. Gail Stewart for typing and retyping the manuscript and to Ms. Sally Anderson for preparing the illustrations.

REFERENCES

1. **Mair WGP, Tomé FMS:** Atlas of the Ultrastructure of Diseased Human Muscle. London, Churchill Livingstone, 1972
2. **Neville HE:** Ultrastructural changes in muscle disease. In Dubowitz V, Brooke MH: Muscle Biopsy: A Modern Approach. London, Saunders, 1973, pp 383–444
3. **Price HM:** Ultrastructural pathologic characteristics of the skeletal muscle fiber: An introductory survey. In Pearson CM, Mostofi FK (eds): The Striated Muscle. Baltimore, Williams & Wilkins, 1973, pp 144–184
4. **Schochet SS Jr, Lampert PW:** Diagnostic electron microscopy of skeletal muscle. In Trump BF, Jones RT (eds): Diagnostic Electron Microscopy. New York, Wiley, 1978, vol 1, pp 209–251
5. **Engel AG, Macdonald RD:** Ultrastructural reactions in muscle disease and their light microscopic correlates. In Walton JN, Canal N, Scarlato G (eds): Muscle Diseases. Amsterdam, Excerpta Medica, 1970, pp 71–89
6. **Yunis EJ, Samaha FJ:** Inclusion body myositis. Lab Invest 25:240–248, 1971
7. **Schochet SS Jr, McCormick WF:** Polymyositis with intranuclear inclusions. Arch Neurol 28:280–283, 1973
8. **Chou SM:** Prospects of viral etiology in polymyositis. In Kakulas BA (ed): Clinical Studies in Myology. Amsterdam, Excerpta Medica, 1973, pt 2, pp 17–28
9. **Carpenter S, Karpati G, Heller I, et al.:** Inclusion body myositis: A distinct variety of idiopathic inflammatory myopathy. Neurology (Minneap) 28:8–17, 1978
10. **Ionescu V, Radu H, Nicolescu P:** Identification of Duchenne muscular dystrophy carriers. Arch Pathol 99:436–441, 1975
11. **Hughes JT, Esiri MM:** Ultrastructural studies in human polymyositis. J Neurol Sci 25:347–360, 1975
12. **Oteruelo FT:** Intranuclear inclusions in a myopathy of late onset. Virchows Arch [Cell Pathol] 20:319–324, 1976
13. **Engel WK, Oberc MA:** Abundant nuclear rods in adult-onset rod disease. J Neuropathol Exp Neurol 34:119–132, 1975
14. **Jenis EH, Lindquist RR, Lister RC:** New congenital myopathy with crystalline intranuclear inclusions. Arch Neurol 20:281–287, 1969
15. **Sato T, Walker DL, Peters HA, et al.:** Chronic polymyositis and myxovirus-like inclusions. Arch Neurol 24:409–418, 1971
16. **Olson W, Engel WK, Walsh GO, et al.:** Oculocraniosomatic neuromuscular disease with "ragged-red" fibers. Arch Neurol 26:193–211, 1972
17. **Karpati G, Carpenter S, Larbrisseau A, et al.:** The Kearns-Shy syndrome. J Neurol Sci 19:133–151, 1973
18. **Bonilla E, Schotland DL, DiMauro S, et al.:** Electron cytochemistry of crystalline inclusions in human skeletal muscle mitochondria. J Ultrastruct Res 51:404–408, 1975

19. **Luft R, Ikkos D, Palmieri G, et al.:** A case of severe hypermetabolism of nonthyroid origin with a defect in the maintenance of mitochondrial respiratory control: A correlated clinical, biochemical and morphological study. J Clin Invest 41:1776–1804, 1962
20. **Afifi AK, Ibrahim ZB, Bergman RA, et al.:** Morphologic features of hypermetabolic mitochondrial disease: A light microscopic, histochemical and electron microscopic study. J Neurol Sci 15:271–290, 1972
21. **DiMauro S, Bonilla E, Lee CP, et al.:** Luft's disease: Further biochemical and ultrastructural studies of skeletal muscle in the second case. J Neurol Sci 27:217–232, 1976
22. **DiMauro S, Schotland DL, Bonilla E, et al.:** Progressive ophthalmoplegia, glycogen storage and abnormal mitochondria. Arch Neurol 29:170–179, 1973
23. **Schotland DL, DiMauro S, Bonilla E, et al.:** Neuromuscular disorder associated with a defect in mitochondrial energy supply. Arch Neurol 33:475–479, 1976
24. **Melmed C, Karpati G, Carpenter S:** Experimental mitochondrial myopathy produced by *in vivo* uncoupling of oxidative phosphorylation. J Neurol Sci 26:305–318, 1975
25. **Berenberg RA, Pellock JM, DiMauro S, et al.:** Lumping or splitting? "Ophthalmoplegia-plus" or Kearns-Sayre syndrome? Ann Neurol 1:37–54, 1977
26. **Karpati G, Carpenter S, Engel AG, et al.:** The syndrome of systemic carnitine deficiency. Neurology (Minneap) 25:16–24, 1975
27. **Isaacs H, Heffron JJA, Badenhorst M, et al.:** Weakness associated with the pathological presence of lipid in skeletal muscle: A detailed study of a patient with carnitine deficiency. J Neurol Neurosurg Psychiatry 39:1114–1123, 1976
28. **Engel AG, Banker BQ, Eiben RM:** Carnitine deficiency: Clinical, morphological, and biochemical observations in a fatal case. J Neurol Neurosurg Psychiatry 40:313–322, 1977
29. **Takagi A, Schotland DL, DiMauro S, et al.:** Thyrotoxic periodic paralysis. Neurology (Minneap) 23:1008–1016, 1973
30. **Engel AG, Dale AJD:** Autophagic glycogenosis of late onset with mitochondrial abnormalities: Light and electron microscopic observations. Mayo Clin Proc 43:233–279, 1968
31. **Torvik A, Dietrichson P, Svaar H, et al.:** Myopathy with tremor and dementia: A metabolic disorder? J Neurol Sci 21:181–190, 1974
32. **Hudgson P, Bradley WG, Jenkison M:** Familial "mitochondrial" myopathy: A myopathy associated with disordered oxidative metabolism in muscle fibres. J Neurol Sci 16:343–370, 1972
33. **Iannaccone ST, Griggs RC, Markesbery WR, et al.:** Familial progressive external ophthalmoplegia and ragged-red fibers. Neurology (Minneap) 24:1033–1038, 1972
34. **Kamineniecka Z:** Myopathies with abnormal mitochondria. Acta Neurol Scand 55:57–75, 1976
35. **Kamieniecka Z, Schmalbruch H:** Myopathies with abnormal mitochondria: A clinicopathologic classification. Muscle Nerve 1:413–415, 1978
36. **Engel WK, Bishop DW, Cunningham GG:** Tubular aggregates in type II muscle fibers: Ultrastructural and histochemical correlation. J Ultrastruct Res 31:507–525, 1970
37. **Bradley WG:** Ultrastructural changes in adynamia episodica hereditaria and normokalemic familial periodic paralysis. Brain 92:379–390, 1969
38. **Ionasescu V, Zellweger H, Schochet SS Jr, et al.:** Biochemical abnormalities of muscle ribosomes during attacks of hyperkalemic periodic paralysis. J Neurol Sci 19:389–398, 1973
39. **Meyers KR, Gilden DH, Rinaldi CF, et al.:** Periodic muscle weakness, normokalemia, and tubular aggregates. Neurology (Minneap) 22:269–279, 1972
40. **Schiffer D, Giordana MT, Monga G, et al.:** Histochemistry and electron microscopy of muscle fibres in a case of congenital paramyotonia. J Neurol 211:125–133, 1976

41. **Chesson AL, Schochet SS Jr, Peters BH:** Biphasic periodic paralysis. Arch Neurol 36:700–704, 1979
42. **Morgan-Hughes JA, Mair WGP, Lascelles PT:** A disorder of skeletal muscle associated with tubular aggregates. Brain 93:873–880, 1970
43. **Chui LA, Neustein H, Munsat TL:** Tubular aggregates in subclinical alcoholic myopathy. Neurology (Minneap) 25:405–412, 1975
44. **Fardeau M, Tomé FMS, Simon P:** Muscle and nerve changes induced by perhexiline maleate in man and mice. Muscle Nerve 2:24–39, 1979
45. **Schotland DL:** An electron microscopic study of target fibers, target-like fibers and related abnormalities in human muscle. J Neuropathol Exp Neurol 28:214–228, 1969
46. **Kovarsky J, Schochet SS Jr, McCormick WF:** The significance of target fibers. A clinicopathologic review of 100 patients with neurogenic atrophy. Am J Clin Pathol 59:790–797, 1973
47. **DeCoster W, DeReuck J, Vander Eecken H:** The target phenomenon in human muscle. Acta Neuropathol (Berl) 34:329–338, 1976
48. **Neville HE, Brooke MH:** Central core fibers: Structured and unstructured. In Kakulas BA (ed): Basic Research in Myology. Amsterdam, Excerpta Medica, 1973, pt 1, pp 497–511
49. **Engel AG, Gomez MR, Groover RV:** Multicore disease. Mayo Clin Proc 46:666–681, 1971
50. **Macdonald RD, Engel AG:** The cytoplasmic body: Another structural anomaly of the Z disk. Acta Neuropathol (Berl) 14:99–107, 1969
51. **Cullen MJ, Fulthorpe JJ:** Stages in fibre breakdown in Duchenne muscular dystrophy. J Neurol Sci 24:179–200, 1975
52. **Macdonald RD, Rewcastle NB, Humphrey JG:** The myopathy of hyperkalemic periodic paralysis. Arch Neurol 19:274–283, 1968
53. **Howes EL Jr, Price HM, Pearson CM, et al.:** Hypokalemic periodic paralysis: Electromicroscopic changes in the sarcoplasm. Neurology (Minneap) 16:242–256, 1966
54. **D'Agostino AN, Ziter FA, Rallison ML, et al.:** Familial myopathy with abnormal muscle mitochondria. Arch Neurol 18:388–401, 1968
55. **Brooke MH, Neville HE:** Reducing body myopathy. Neurology (Minneap) 22:829–840, 1972
56. **Engel AG, Gomez MR:** Nemaline (Z disk) myopathy: Observations on the origin, structure and solubility properties of the nemaline structures. J Neuropathol Exp Neurol 26:601–619, 1967
57. **Stromer MH, Tabatabai LB, Robson RM, et al.:** Nemaline myopathy, an integrated study: Selective extraction. Exp Neurol 50:402–521, 1976
58. **Engel WK:** Rod (nemaline) disease. In Goldensohn ES, Appel SH (eds): Scientific Approaches to Clinical Neurology. Philadelphia, Lea & Febiger, 1977, pp 1667–1691
59. **Shy GM, Engel WK, Sommers JE, et al.:** Nemaline myopathy. A new congenital myopathy. Brain 86:793–810, 1963
60. **Brooke MH:** A Clinician's View of Neuromuscular Diseases. Baltimore, Williams & Wilkins, 1977, pp 209–211
61. **Bender AN, Willner JP:** Nemaline (rod) myopathy: The need for histochemical evaluation of affected families. Ann Neurol 4:37–42, 1978
62. **Kinoshita M, Satoyoshi E:** Type I fiber atrophy and nemaline bodies. Arch Neurol 31:423–425, 1974
63. **Sreter FA, Aström K-E, Romanul FCA, et al.:** Characteristics of myosin in nemaline myopathy. J Neurol Sci 27:99–116, 1976
64. **Brownell AKW, Gilbert JJ, Shaw DT, et al.:** Adult onset nemaline myopathy. Neurology (Minneap) 28:1306–1309, 1978

65. **Martinez AJ, Hay S, McNeer KW:** Extraocular muscles: Light microscopy and ultrastructural features. Acta Neuropathol (Berl) 34:237–253, 1976
66. **Garancis JC:** Type II glycogenosis: Biochemical and electron microscopic study. Am J Med 44:289–300, 1968
67. **Engel AG:** Vacuolar myopathies: Multiple etiologies and sequential structural studies. In Pearson CM, Mostofi FK (eds): The Striated Muscle. Baltimore, Williams & Wilkins, 1973, pp 301–341
68. **Schotland DL:** Ultrastructure of muscle in glycogen storage diseases. In Pearson CM, Mostofi FK (eds): The Striated Muscle. Baltimore, Williams & Wilkins, 1973, pp 410–426
69. **McAdams AJ, Hug G, Bove KE:** Glycogen storage disease, types I to X. Hum Pathol 5:463–487, 1974
70. **Schlenska GK, Heene R, Spalke G, et al.:** The symptomatology, morphology and biochemistry of glycogenosis type II (Pompe) in the adult. J Neurol 212:237–252, 1976
71. **Martin JJ, deBarsy Th, den Tandt WR:** Acid maltase deficiency in non-identical adult twins. J Neurol 213:105–118, 1976
72. **DiMauro S, Stern LZ, Mehler M, et al.:** Adult-onset acid maltase deficiency: A postmortem study. Muscle Nerve 1:27–36, 1978
73. **Brunberg JA, McCormick WF, Schochet SS Jr:** Type III glycogenosis: An adult with diffuse weakness and muscle wasting. Arch Neurol 25:171–178, 1971
74. **Murase T, Ikeda H, Muro T, et al.:** Myopathy associated with type III glycogenosis. J Neurol Sci 20:287–295, 1973
75. **Tobin WE, Huijing F, Porro BS, et al.:** Muscle phosphofructokinase deficiency. Arch Neurol 28:128–130, 1973
76. **Korényi-Both A, Smith BH, Baruah JK:** McArdle's syndrome. Fine structural changes in muscle. Acta Neuropathol (Berl) 40:11–19, 1977
77. **Schochet SS Jr, McCormick WF, Kovarsky JL:** Light and electron microscopy of skeletal muscle in type IV glycogenosis. Acta Neuropathol (Berl) 19:137–144, 1971
78. **McMaster KR, Powers JM, Hennigar GR Jr, et al.:** Nervous system involvement in type IV glycogenosis. Arch Pathol Lab Med 103:105–111, 1979
79. **Carpenter S, Karpati G, Andermann F, et al.:** Lafora's disease: Peroxisomal storage in skeletal muscle. Neurology (Minneap) 24:531–536, 1974
80. **Neville HE, Brooke MH, Austin JH:** Studies in myoclonus epilepsy (Lafora body form), pt 4. Arch Neurol 30:466–474, 1974
81. **Coleman DL, Gambetti P, DiMauro S, et al.:** Muscle in Lafora disease. Arch Neurol 31:396–406, 1974
82. **Markesbery WR, Quillen MP, Procopis PQ, et al.:** Muscle carnitine deficiency. Arch Neurol 31:320–324, 1974
83. **VanDyke DH, Griggs RC, Markesbery WR, et al.:** Hereditary carnitine deficiency of muscle. Neurology (Minneap) 25:154–159, 1975
84. **Angelini C, Lücke S, Cantarutti F:** Carnitine deficiency of skeletal muscle: Report of a treated case. Neurology (Minneap) 28:633–637, 1976
85. **Hart ZH, Chang C-H, DiMauro S, et al.:** Muscle carnitine deficiency and fatal cardiomyopathy. Neurology (Minneap) 28:147–151, 1978
86. **Tomé FMS, Fardeau M, Lenoir G:** Ultrastructure of muscle and sensory nerve in Fabry's disease. Acta Neuropathol (Berl) 38:187–194, 1977
87. **Sima AAF, Robertson DM:** Involvement of peripheral nerve and muscle in Fabry's disease. Arch Neurol 35:291–301, 1978
88. **Ceuterick CH, Martin JJ, Casaer P, et al.:** The diagnosis of infantile generalized ceroid lipofuscinosis (type Hagberg-Santavuori) using skin biopsy. Neuropaediatrie 7:250–260, 1976

89. **Carpenter S, Karpati G, Andermann F:** Specific involvement of muscle, nerve, and skin in late infantile and juvenile amaurotic idiocy. Neurology (Minneap) 22:170–186, 1972
90. **Goebel HH, Zeman W, Pilz H:** Significance of muscle biopsies in neuronal ceroid-lipofuscinosis. J Neurol Neurosurg Psychiatry 38:985–993, 1975
91. **Dom R, Brucher JM, Ceuterick C, et al.:** Adult ceroid-lipofuscinosis (Kufs' disease) in two brothers. Acta Neuropathol (Berl) 45:67–72, 1979
92. **Payne CM, Curless RG:** Concentric laminated bodies. J Neurol Sci 29:311–322, 1976
93. **Engel AG, Angelini C, Gomez MR:** Fingerprint body myopathy. Mayo Clin Proc 47:377–388, 1972
94. **Payne CM, Curless RG:** Fingerprint inclusions. J Neurol Sci 31:379–386, 1977
95. **Tomé FMS, Fardeau M:** "Fingerprint inclusions" in muscle fibres in dystrophia myotonica. Acta Neuropathol (Berl) 24:62–67, 1973
96. **Tomé FMS, Fardeau M:** Congenital myopathy with "reducing bodies" in muscle fibres. Acta Neuropathol (Berl) 31:207–217, 1975
97. **Thomas PK, Olsson Y:** Microscopic anatomy and function of the connective tissue components of peripheral nerve. In Dyck PJ, Thomas PK, Lambert EH (eds): Peripheral Neuropathy. Philadelphia, Saunders, 1975, vol 1, pp 168–189
98. **Olsson Y:** Vascular permeability in the peripheral nervous system. In Dyck PJ, Thomas PK, Lambert EH (eds): Peripheral Neuropathy. Philadelphia, Saunders, 1975, vol 1, pp 190–200
99. **Blümcke S, Niedorf HR, Rode J:** Axoplasmic alterations in the proximal and distal stumps of transected nerves. Acta Neuropathol (Berl) 7:44–61, 1966
100. **Zelená J, Lubinska L, Gutmann E:** Accumulation of organelles at the ends of interrupted axons. Z Zellforsch 91:200–219, 1968
101. **Singer M, Steinberg MC:** Wallerian degeneration: A reevaluation based on transected and colchicine-poisoned nerves in the amphibian, triturus. Am J Anat 133:51–84, 1972
102. **Donat JR, Wisniewski HM:** The spatio-temporal pattern of wallerian degeneration in mammalian peripheral nerves. Brain Res 53:41–53, 1973
103. **Asbury AK:** Biology of Schwann cells. In Dyck PJ, Thomas PK, Lambert EH (eds): Peripheral Neuropathy. Philadelphia, Saunders, 1975, vol 1, pp 201–212
104. **Dyck PJ:** Alterations of the peripheral nervous system of man. In Dyck PJ, et al. (eds): Peripheral Neuropathy. Philadelphia, Saunders, 1975, vol 1, pp 296–336
105. **Weller RO, Cervós-Navarro J:** Pathology of Peripheral Nerves. London, Butterworths, 1977, pp 61–89
106. **Asbury AK, Johnson PC:** Pathology of Peripheral Nerve. Philadelphia, Saunders, 1978, pp 50–95
107. **Lampert PW, Schochet SS Jr:** Ultrastructural changes of peripheral nerve. In Trump BF, Jones RT (eds): Diagnostic Electron Microscopy. New York, John Wiley, 1979, vol 2, pp 307–350
108. **Asbury AK, Gale MK, Cox SC, et al.:** Giant axonal neuropathy—A unique case with segmental neurofilamentous masses. Acta Neuropathol (Berl) 20:237–247, 1972
109. **Carpenter S, Karpati G, Andermann F, et al.:** Giant axonal neuropathy: A clinically and morphologically distinct neurological disease. Arch Neurol 31:312–316, 1974
110. **Prineas JW, Ouvrier RA, Wright RG, et al.:** Giant axonal neuropathy: A generalized disorder of cytoplasmic microfilament formation. J Neuropathol Exp Neurol 35:458–470, 1976
111. **Koch T, Schultz P, Williams R:** Giant axonal neuropathy: A childhood disorder of microfilaments. Ann Neurol 1:438–451, 1977
112. **Gambarelli D, Hassoun J, Pellisaier JF, et al.:** Giant axonal neuropathy: Involvement

of peripheral nerve, myenteric plexus and extra-neuronal area. Acta Neuropathol (Berl) 39:261–269, 1977
113. **Korobkin R, Asbury AK, Summer AF, et al.:** Glue-sniffing neuropathy. Arch Neurol 32:158–162, 1975
114. **Towfighi J, Gonatas NK, Pleasure D, et al.:** Glue sniffer's neuropathy. Neurology (Minneap) 26:238–243, 1976
115. **Davenport JG, Farrell DF, Sumi SM:** Giant axonal neuropathy caused by industrial chemicals. Neurology (Minneap) 26: 919–923, 1976
116. **Schochet SS Jr, Chesson AL Jr:** Giant axonal neuropathy: Possibly secondary to vitamin B_{12} malabsorption. Acta Neuropathol (Berl) 40:79–83, 1977
117. **Rizzuto N, Terzian H, Galiazzo-Rizzuto S:** Toxic polyneuropathies in Italy due to leather cement poisoning in shoe industries. J Neurol Sci 31:343–354, 1977
118. **Ohnishi A, Dyck PJ:** Loss of small peripheral sensory neurons in Fabry's disease. Arch Neurol 31:120–127, 1974
119. **Ochoa J, Mair WGP:** The normal sural nerve in man: II. Changes in the axons and Schwann cells due to aging. Acta Neuropathol (Berl) 13:217–239, 1969
120. **Dyck PJ, Johnson WJ, Lambert EH, et al.:** Segmental demyelination secondary to axonal degeneration in uremic neuropathy. Mayo Clin Proc 46:400–431, 1971
121. **Dyck PJ, Lambert EH:** Polyneuropathy associated with hypothyroidism. J Neuropathol Exp Neurol 29:631–658, 1970
122. **Berard-Badier M, Gambarelli D, Pinsard N, et al.:** Infantile neuroaxonal dystrophy or Seitelberger's disease: II. Peripheral nerve involvement: Electron microscopic study in one case. Acta Neuropathol [Suppl] (Berl) V: 30–39, 1971
123. **Spencer PS, Thomas PK:** Ultrastructural studies of the dying-back process: II. The sequestration and removal by Schwann cells and oligodendrocytes of organelles from normal and diseased axons. J Neurocytol 3:763–783, 1974
124. **Ochoa J:** Recognition of unmyelinated fiber disease: Morphologic criteria. Muscle Nerve 1:375–386, 1978
125. **Asbury AK, Johnson PC:** Pathology of Peripheral Nerve. Philadelphia. Saunders, 1978, pp 136–147
126. **Fardeau M:** Pathology of Refsum's disease. In Dyck PJ, Thomas PK, Lambert EH (eds): Peripheral Neuropathy. Philadelphia, Saunders, 1975, vol 2, pp 881–890
127. **Thomas PK, King RHM, Kocen RS, et al.:** Comparative ultrastructural observations on peripheral nerve abnormalities in the late infantile, juvenile and late onset forms of metachromatic leukodystrophy. Acta Neuropathol (Berl) 39:237–245, 1977
128. **Schochet SS Jr, McCormick WF, Powell GF:** Krabbe's disease: A light and electron microscopic study. Acta Neuropathol (Berl) 36:153–160, 1976
129. **Lee RE, Peters SP, Glew RH:** Gaucher's disease: Clinical, morphologic, and pathogenetic considerations. In Sommers SC, Rosen PP (eds): Pathology Annual. New York, Appleton, 1977, pt 2, pp 309–339
130. **Powers JM, Schaumburg HH:** Adreno-leukodystrophy (sex-linked Schilder's disease). Am J Pathol 76:481–500, 1974
131. **Goebel HH, Zeman W, Pilz H:** Ultrastructural investigations of peripheral nerves in neuronal ceroid-lipofuscinosis (NCL). J Neurol 213:295–303, 1976
132. **Martin JJ, Leron JG, Farriaux JP, et al.:** I-cell disease (mucolipidosis II). Acta Neuropathol (Berl) 33:285–305, 1975
133. **Kornfeld M, Snyder RD, Wenger DA:** Fucosidosis with angiokeratoma. Arch Pathol Lab Med 101:478–485, 1977
134. **Swift TR, McDonald TF:** Peripheral nerve involvement in Hunter syndrome (mucopolysaccharidosis II). Arch Neurol 33:845–846, 1976
135. **Goebel HH, Lenard HG, Kohlschütter A, et al.:** The ultrastructure of the sural nerve in Pompe's disease. Ann Neurol 2:111–115, 1977

136. **Asbury AK, Johnson PC:** Pathology of Peripheral Nerve. Philadelphia, Saunders, 1978, pp 148–155
137. **Asbury AK:** Renaut bodies: A forgotten endoneurial structure. J Neuropathol Exp Neurol 32:334–343, 1973
138. **Schoene WC, Asbury AK, Aström KE, et al.:** Hereditary sensory neuropathy. J Neurol Sci 11:463–487, 1970
139. **Asbury AK, Cox SC, Baringer JR:** The significance of giant vacuolation of endoneurial fibroblasts. Acta Neuropathol (Berl) 18:123–131, 1971
140. **Thomas PK, Eliasson SG:** Diabetic neuropathy. In Dyck PJ, Thomas PK, Lambert EH (eds): Peripheral Neuropathy. Philadelphia, Saunders, 1975, vol 2, pp 956–981
141. **Johnson PC, Yoshimura MP, Gaines JA:** Thickening of non-vascular basement membranes in diabetic peripheral nerves. J Neuropathol Exp Neurol 37:637, 1978

7

HEMATOLOGIC DISORDERS

BARBARA M. OSBORNE

The contribution of electron microscopy to the diagnostic evaluation of lymphoid and hematopoietic proliferations and neoplasms varies. Ultrastructural studies may be diagnostic by differentiating certain lymphomas or leukemias from histologically similar tumors that may mimic them. Electron microscopy can also be useful in conjunction with light microscopy in establishing a precise subclassification of these neoplasms. Because of the small size of samples for electron microscopy, the ultrastructural studies must always be correlated with the light microscopy and clinical information.

LEUKEMIAS

Electron microscopy is useful for the accurate identification and classification of leukemias when there is a problem in the differential diagnosis, as may occur when they present in extramedullary sites. Familiarity with the ultrastructural characteristics of the normal bone marrow cells and their development is helpful, although the malignant counterparts often deviate from normal. In the normal bone marrow, there are undifferentiated cells that are agranular and have many free ribosomes and few cytoplasmic elements. These are thought to be stem cells from which cells of at least the granulocytic, monocytic, and erythroid series arise. The first recognizable granulocytic precursor is the myeloblast, which ultrastructurally possesses characteristics of a stem cell but shows early primary granule formation with a positive peroxidase reaction that will also stain the perinuclear cisternae and Golgi bodies (where the granules originate).[4] In the granulocytic series, the mature neutrophil is distinguished from eosinophils and basophils by its specific cytoplasmic granules. During maturation of the granulocytic series, one of two types of granules predominates, depending on the stage of development. Promyelocytes contain mostly primary lysosomal granules, which are azurophilic. With the myelocyte stage, a second population of granules appears. These specific granules are larger and less dense than the primary granules, and ultimately comprise 80% of the mature granulocyte granule population. They differ from primary granules in their enzymatic composition (containing alkaline phosphatase), and they have antibacterial functions. In addition, as maturation proceeds, the nucleus loses its nucleoli and becomes indented and then lobulated with increasing chromatin condensation. The cytoplasm loses rough endoplasmic reticulum and mitochondria. The size of the Golgi complex decreases, because at the metamyelocyte or band stage there is no further granule production. Maturation of basophils and eosinophils is similar, except that the specific granule morphology and function are different.

By contrast, maturation of monocytes is not as clearly defined. There is fibrillar formation in the perinuclear cytoplasm, which is said to be less marked than in leukemic monocytes.[5] In addition to the small primary lysosomal granules, features of monocytes include a U-shaped nucleus, vesicles near the Golgi region and cell surface, small mitochondria, and short profiles of rough endoplasmic reticulum. Further development of monocytes occurs after migration to tissues.

Lymphocytes are characterized by their agranularity, free ribosomes, and a few long single strands of rough endoplasmic reticulum.[29] Occasional lysosomal granules may be present. In their resting (unstimulated) stage, it is not possible to distinguish T from B lymphocytes ultrastructurally.

It is felt that at least 90% of blastic leukemias may be subclassified on the basis of ultrastructural studies, in combination with smears and cytochemistry (which may be performed at the ultrastructural level).[4] Another frequently encountered problem is the examination of bone marrows, particularly in the pediatric age group, to differentiate between metastatic tumors and acute leukemia. Electron microscopy is particularly helpful in identifying metastatic small round cell tumors.

The cells of chronic granulocytic leukemia (Fig. 7-1) show a variation in number of granules from cell to cell. Usually the number of granules in the cells is decreased

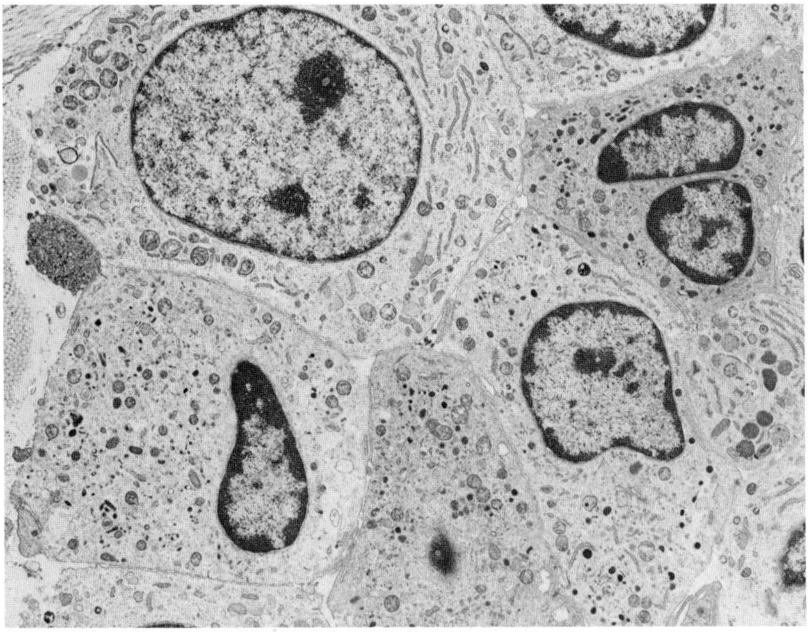

FIG. 7-1. Granulocytic leukemia. This 45-year-old male presented with multiple subcutaneous nodules. Most of the cells demonstrate a profusion of cytoplasmic granules. The nuclei show varying stages of maturation, indicating that this is subacute or chronic granulocytic leukemia. X4400.

compared with normal cells. The granules themselves are often larger or smaller than normal and lack a crystalline substructure.

The changes described above are more pronounced in acute granulocytic leukemia (Fig. 7-2). Approximately 5% of cases have a small percentage of cells containing Auer rods. These bodies are formed by fusion of granules, resulting in a parallel substructure with variable periodicity.[4]

In promyelocytic leukemia, the cells contain numerous atypical rod and splinter forms of granules, some of which are distinctive Auer rods showing a hexagonal pattern with 260 A periodicity. It is felt by some that acute promyelocytic leukemia should be separated from the other cases of acute granulocytic leukemia, as clinically this entity may be associated with a hemorrhagic diathesis. Most cases of acute granulocytic leukemia will show a number of blast cells without distinguishing characteristics. In about 10% of cases, it is not possible to classify the blastic leukemia, as most of the cells lack distinctive cytoplasmic features.[4]

In acute monomyelocytic leukemia, most studies show varying proportions of two cell lines—one granulocytic and one monocytic. It is not entirely certain whether both are neoplastic, and one theory is that the monocytic line is leukemic, while the granulocytic line is hyperplastic. Pure monocytic leukemia is rare, and, as in the illustrated case (Figs. 7-3, 7-4, 7-5), frequently presents with extramedullary involvement. Ultrastructurally, the nuclei are reniform, the small (primary lysosomal) granules are of uniform size, and some small vesicles are present.[14] A feature in this case

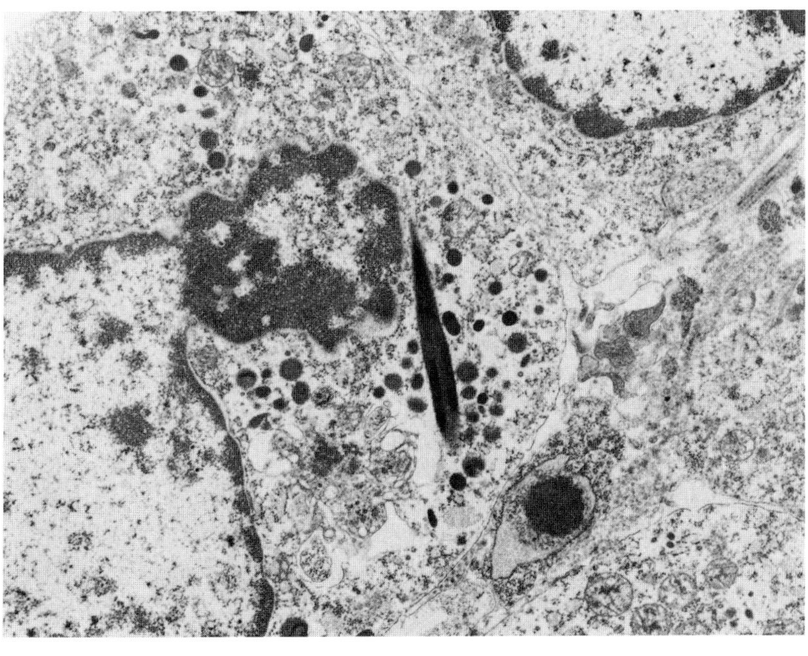

FIG. 7-2. Granulocytic leukemia. This 25-year-old male had a soft tissue mass involving the anterior chest wall, with a normal bone marrow. Light microscopy was suggestive of alveolar rhabdomyosarcoma. Although formalin-fixed, the cytoplasm of the neoplastic cells has many granules, and one Auer rod (linear structure), which at high power demonstrated periodicity. X9300.

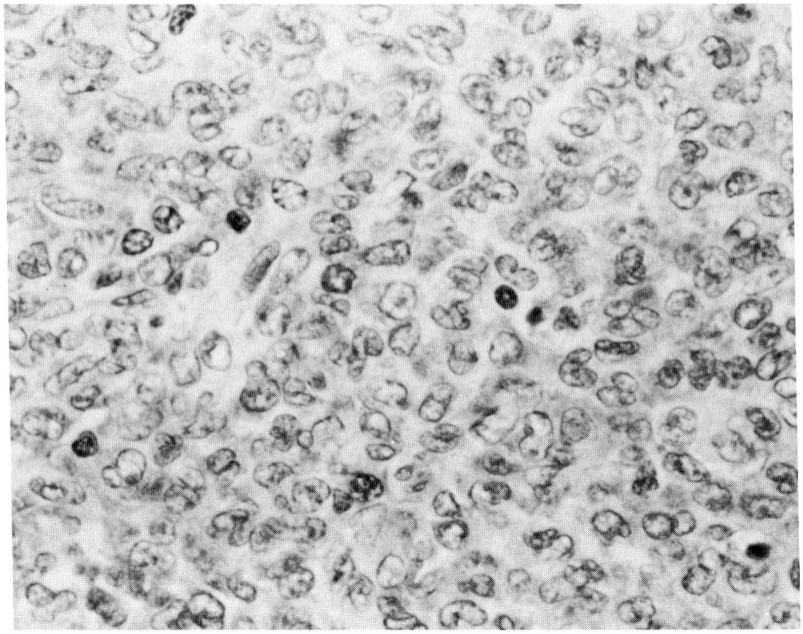

FIG. 7-3. Monocytic leukemia-lymphoma. This 55-year-old female presented with multiple subcutaneous nodules and enlarged lymph nodes that were diagnosed as large cell (histiocytic) lymphoma. Light microscopy of the lymph node shows a monomorphic infiltration of cytoplasmic cells with oval or reniform nuclei. The naphthol-ASD-chloroacetate esterase stain was negative, while nonspecific esterase on imprints was positive. X600.

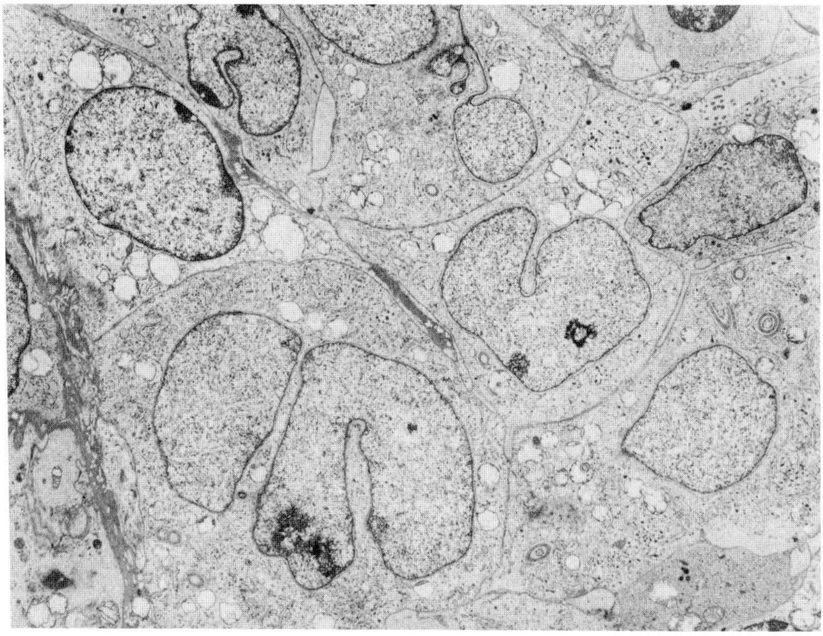

FIG. 7-4. Monocytic leukemia-lymphoma. The nuclei are uniformly reniform, while the cytoplasm shows small (primary lysosomal) granules of uniform size with some small vesicles. X3800.

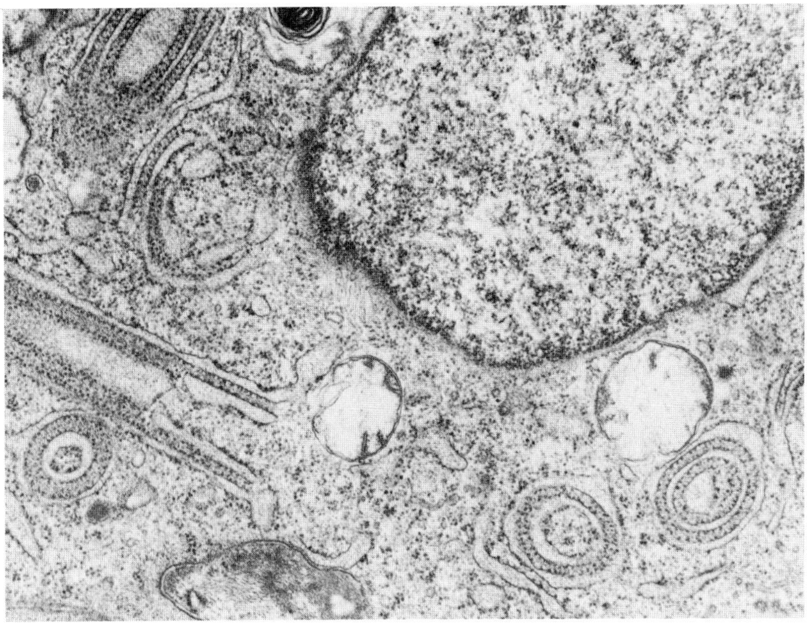

FIG. 7-5. Monocytic leukemia-lymphoma. Most of the cells show one or more cytoplasmic inclusions formed from rough endoplasmic reticulum. It is unclear whether the inclusions are specific, or whether they represent an unique finding in this case. Their formation may be related to effects of therapy. X20,000.

is the presence in most of the cells of one or more previously undescribed cylindrical conformations of rough endoplasmic reticulum that differ from ribosome-lamellar complexes. It is uncertain whether this is a common finding in all cases of monocytic (Schilling type) leukemia. In this case, the NASD stain was negative, while the nonspecific esterase stain on imprints was positive.

Acute lymphocytic leukemia is composed of cells with irregular nuclear contours only superficially resembling lymphocytes.[5] Most cells are 6 to 8 μm in diameter (Fig. 7-6), have marginated chromatin, and are agranular with the exception of occasional primary lysosomal granules. Some of the cells are 12 μm or greater and appear more active, or "transformed" (Fig. 7-7). It is possible that a large percentage of these cells in a leukemia may indicate a poor prognosis. In addition, there may be accumulation of glycogen within the leukemic lymphocytes.

Hairy cell leukemia (leukemic reticuloendotheliosis) is a disease predominantly involving males. It is clinically characterized by a chronic course with marked splenomegaly as well as pancytopenia due to infiltration by bland-appearing mononuclear cells within the bone marrow and spleen. It is not uncommon for a bone marrow tap to be "dry." The peripheral blood will show "hairy cells." Electron microscopic examination[6,9] shows distinctive cells, often with extensive cytoplasmic projections (Fig. 7-8), or if in tissue, complex, serpentine extensive cytoplasmic interdigitations due to molding of the projections as cells are compressed against each

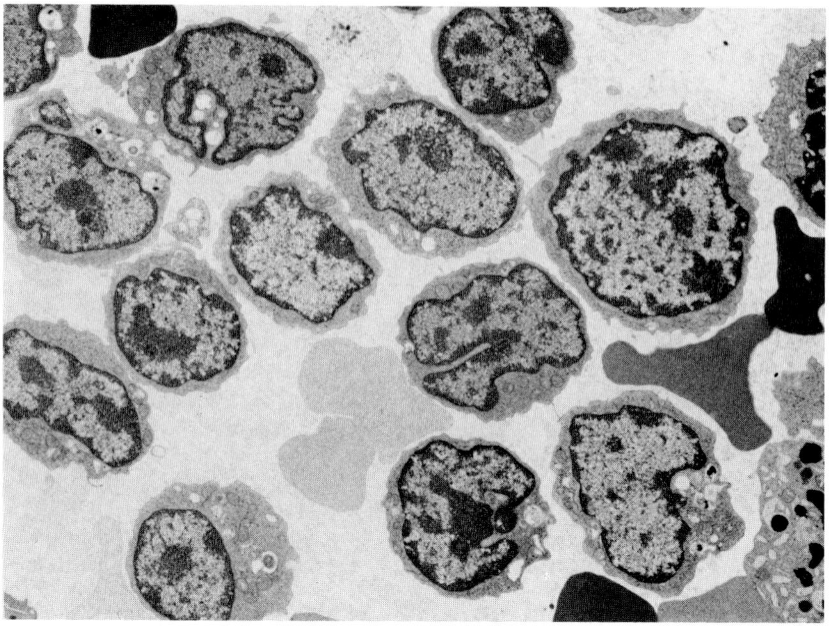

FIG. 7-6. Acute lymphocytic leukemia in peripheral blood. The leukemic cells have a high nuclear-cytoplasmic ratio, with nuclei varying from round to extremely irregular. The chromatin is finely clumped in the center of the nucleus, although at times, the nuclei may show considerable marginated dense chromatin. X3400.

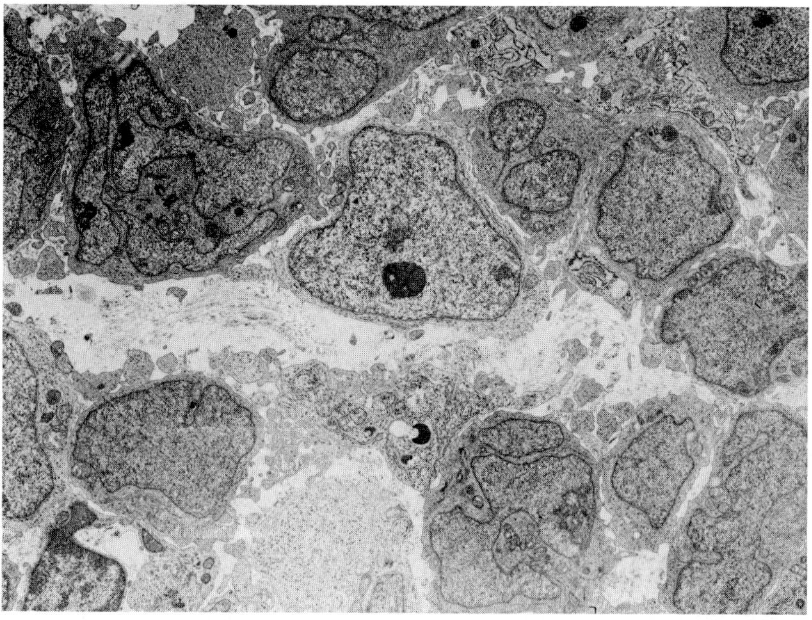

FIG. 7-7. Acute lymphocytic leukemia. This 7-year-old female with known acute lymphocytic leukemia in remission developed bilaterally enlarged kidneys. The renal biopsy specimen shows primitive blast cells within the renal interstitium. Most of these cells are larger and appear more "transformed" than is usual in acute lymphocytic leukemia. X6200.

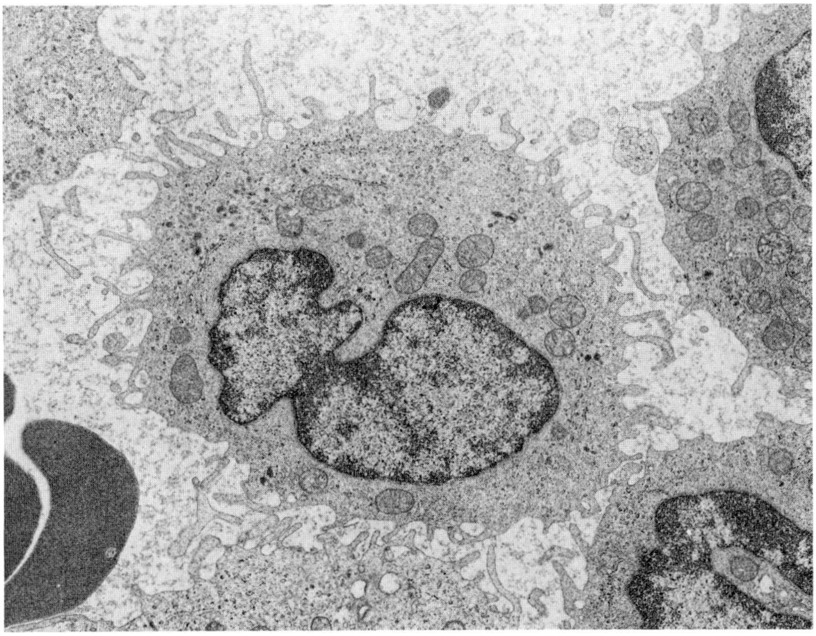

FIG. 7-8. Hairy cell leukemia in peripheral blood. The cytoplasm characteristically shows many long microvilli. X7000.

other (Fig. 7-9). The cytoplasm may also contain ribosome-lamellar complexes (Fig. 7-10), which are structures not entirely unique to hairy cell leukemia, as they may also be seen in other conditions and were first described in chronic lymphocytic leukemia.[48] The cell of origin is as yet uncertain. Ultrastructurally, the cytoplasm is not typical of either lymphocytes or monocytes, but shows features intermediate between the two. Electron microscopy may be helpful in making the diagnosis of hairy cell leukemia particularly when the bone marrow biopsy specimen is not classic, as may be the case with concurrent atypical mycobacterial disease, or in patients who have received previous therapy for other presumed diagnoses. The number of cytoplasmic projections on the leukemic cells varies from case to case. Leukemic cells from bone marrow aspirates that have been centrifuged before fixation may show fewer hairy projections.

LYMPHOMAS

The individual normal cells of lymph nodes from which lymphomas may arise include lymphocytes at various developmental stages in either a paracortical location (T-derived lymphocytes), or, in the case of B lymphocytes, in follicular centers where they undergo transformation. Immunoblastic sarcoma presumably arises from the large plasmacytoid transformed lymphocytes that have exited from the follicle (and thus, by definition, are always diffuse), and are arrested further along in their devel-

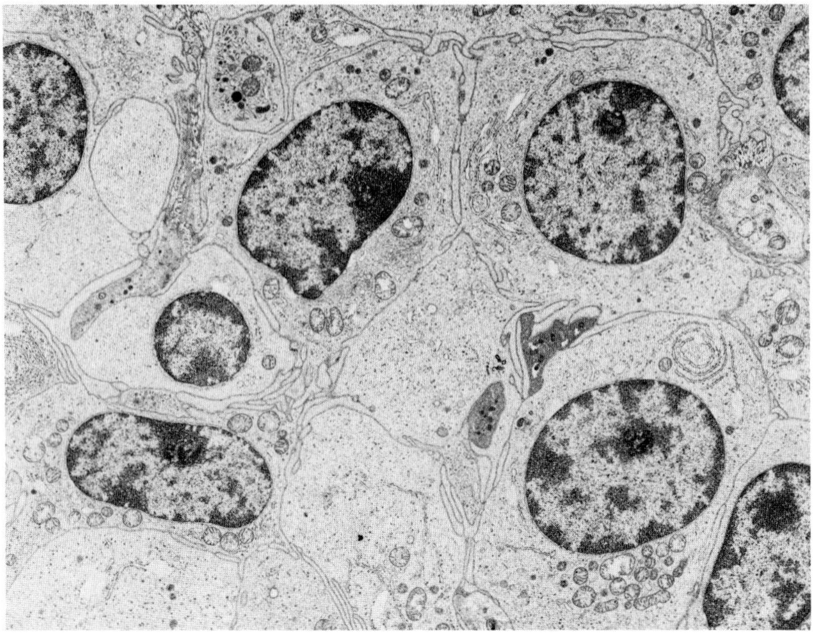

FIG. 7-9. Hairy cell leukemia in spleen. The cytoplasmic villi appear complexly interdigitated when the leukemic cells are examined in tissue. X3800.

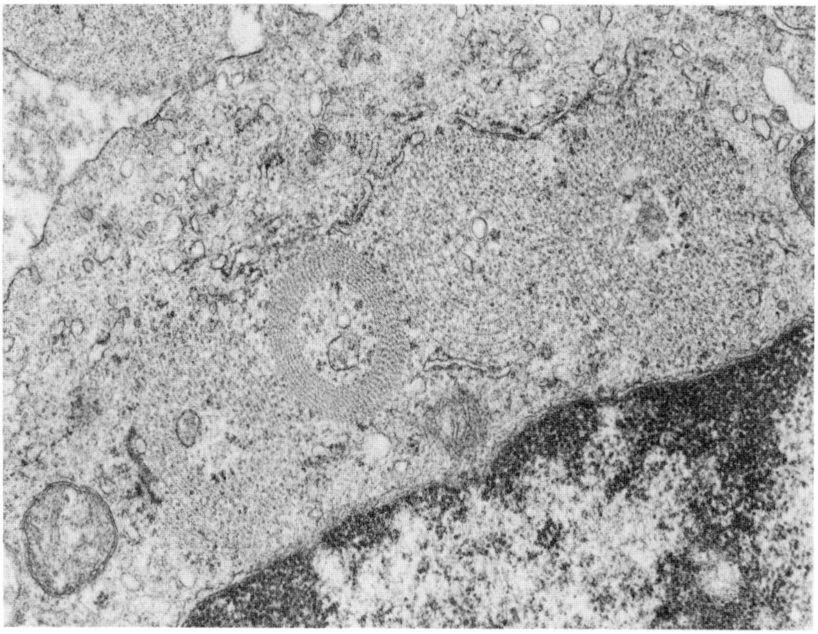

FIG. 7-10. Hairy cell leukemia. Approximately one half of cases of hairy cell leukemia will contain cytoplasmic ribosome-lamellar complexes. Although they are apparently most prevalent in this disease, similar structures have been described in chronic lymphocytic leukemia and in adenoma cells of the adrenal cortex. X23,000.

opment toward mature immunoglobulin-producing plasma cells. On rare occasions, true histiocytes may give rise to malignant histiocytosis. It is unclear whether the occasional diffuse, true histiocytic malignancy that presents in the absence of the usual clinical features of malignant histiocytosis is a separate entity (i.e., true diffuse histiocytic lymphoma or just a forme fruste of malignant histiocytosis). A monoclonal proliferation of plasma cells results in a plasmacytoma that more often occurs in extranodal sites, and most often in bone marrow.

Lymphocytes may show varying degrees of transformation depending normally on stimulation by antigen exposure to elicit either a humoral B-cell response with antibody production or a cell-mediated response reflected by T-lymphocyte transformation responsible for graft versus host reactions and cytotoxic reactions. The stimulated or transformed lymphocyte is termed an immunoblast. It is not as yet possible to distinguish accurately T and B lymphocytes or immunoblasts on an ultrastructural level, although some[8] have claimed that the nuclear chromatin pattern of the T immunoblast characteristically shows finely clustered aggregates of chromatin dispersed throughout the nucleus.[27] A small untransformed lymphocyte is approximately 7 μm in diameter, with peripherally clumped dense chromatin. By contrast, B lymphocytes that are undergoing transformation following antigen stimulation show increasing nuclear size with dispersion of chromatin, and increasingly prominent nucleoli. The larger amount of cytoplasm is characterized by many free polyribosomes, but other organelles are confined to some mitochondria and occasional long strands of rough endoplasmic reticulum.

Plasma cells are circular or oval with an eccentrically placed nucleus showing peripherally condensed aggregates of chromatin. The cytoplasm contains a few free ribosomes and, most characteristic, a large amount of concentrically arranged rough endoplasmic reticulum.

Macrophages or true lymph node histiocytes are larger cells that are distinguished ultrastructurally by the presence in their cytoplasm of secondary lysosomes, which are large dense bodies containing phagocytosed materials such as remnants of red cells, lymphocytes, or plasma cells. Cytoplasmic processes may be prominent. Dendritic macrophages concerned with antigen uptake and presentation are found in the lymph node follicle or germinal center, where their long dendritic processes interdigitate to form a labyrinth. They are supported on a framework of reticulum cells, which are modified fibroblasts connected by cell junctions.

It must be emphasized that the diagnosis of lymphoma versus reactive lymph node hyperplasia cannot be made on the basis of ultrastructural studies alone. It cannot be overstated that the best approach in the accurate diagnosis of lymph node problems is careful handling of tissues, good fixation in a medium with which the pathologist is familiar, and thin light microscopic sections that have been carefully processed and stained. The 1-μm sections prepared from the EM blocks are particularly useful in the study of cellular morphology and nuclear configuration.

At The M. D. Anderson Hospital, the Rappaport classification for non-Hodgkin lymphomas has been employed, although, the corresponding designation according to the Lukes-Collins nomenclature is indicated[26] because it more accurately reflects the cell type and development stage from which the particular lymphoma arose.

In well-differentiated lymphocytic lymphoma, which may be a tissue manifestation of chronic lymphocytic leukemia, there is a proliferation of small, round lymphocytes with round or oval nuclei, containing dense peripheralized chromatin (Fig. 7-11). In every case, a certain number of lymphocytes show some degree of transformation, and when these larger cells form aggregates, they may produce a pseudonodular pattern that can be mistaken for nodular L&H (lymphocyte predominance) Hodgkin's disease or, more commonly, for nodular, poorly differentiated lymphocytic lymphoma. An article by Evans et al.[10] has shown that this pattern is a variant of well-differentiated lymphocytic lymphoma, except when the mitotic rate exceeds 30 mitoses per 20 high power field, in which case the clinical course is more aggressive. Ultrastructural studies show what would be expected from the light microscopic examination; namely, in some cells in which the nuclei have become prominent, the chromatin pattern has become more dispersed, and a greater amount of cytoplasm is present. The size of these larger cells and their degree of transformation are generally less than that of the cells of histiocytic lymphoma, although if the grid were composed entirely of these larger cells, a mistaken impression of histiocytic lymphoma would be possible.

Poorly differentiated lymphocytic lymphoma (Fig. 7-12) generally presents with a nodular pattern that may ultimately become diffuse and/or show progression in histology towards large noncleaved cells. Morphologically, these poorly differentiated lymphocytes (small cleaved cells) show pronounced nuclear infoldings (cleavage)

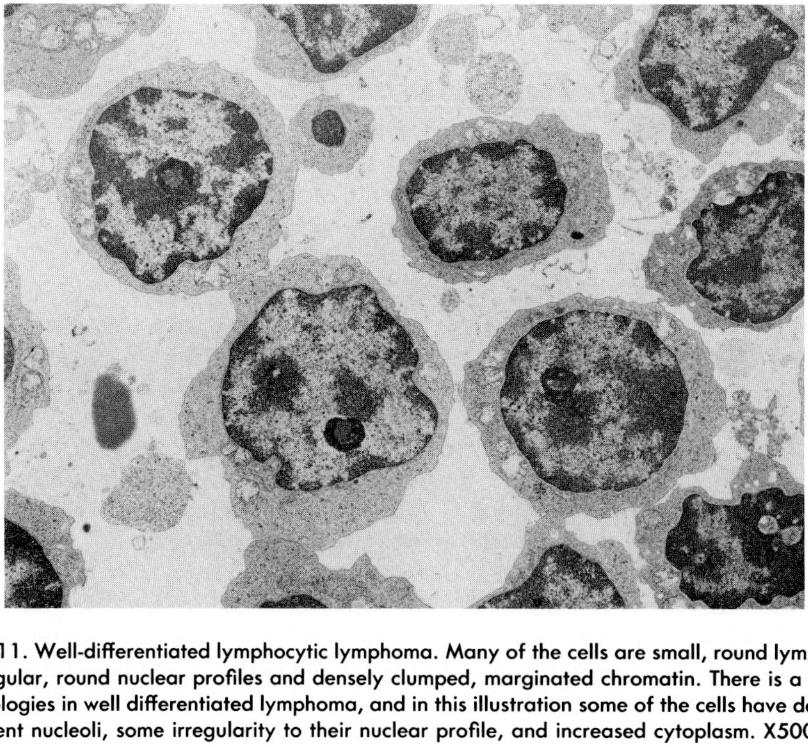

FIG. 7-11. Well-differentiated lymphocytic lymphoma. Many of the cells are small, round lymphocytes with regular, round nuclear profiles and densely clumped, marginated chromatin. There is a range of morphologies in well differentiated lymphoma, and in this illustration some of the cells have developed prominent nucleoli, some irregularity to their nuclear profile, and increased cytoplasm. X5000.

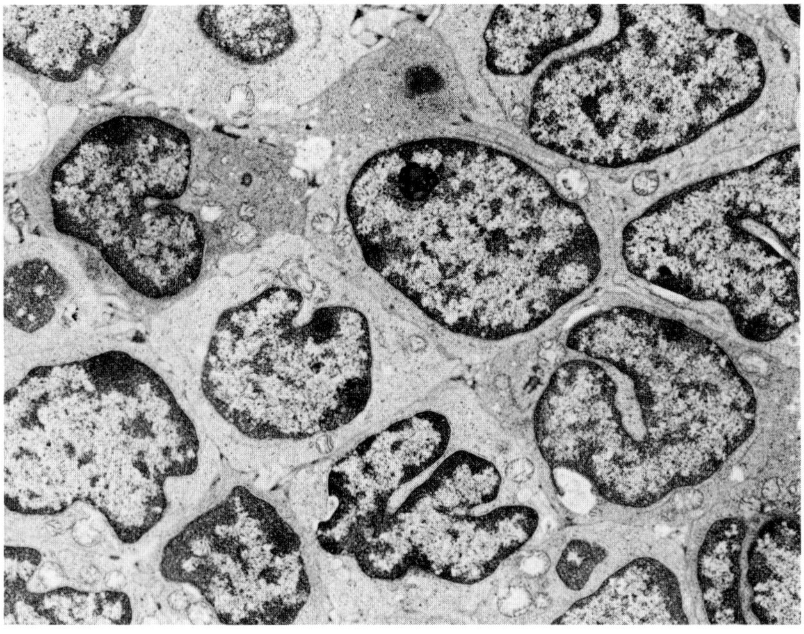

FIG. 7-12. Poorly differentiated lymphocytic lymphoma. The cells demonstrate a high nuclear cytoplasmic ratio, with irregular, deeply indented nuclei whose chromatin is prominently marginated. Nucleoli, when present, are small. X3600.

with peripherally coarsely clumped chromatin, and fairly scant cytoplasm. Nucleoli are improminent. To be emphasized in all lymphomas is the lack of attachments (desmosomes) between neoplastic cells.

An unusual morphologic variant of nodular lymphoma has been described,[24,43] which has been called signet ring lymphoma. Five out of seven of the initial cases were nodular, poorly differentiated lymphocytic lymphoma. Ultrastructural studies confirm that this is indeed lymphoma with a nucleus that has the nuclear chromatin pattern of small cleaved cell lymphoma but that is indented by a membrane-bound, large vacuole that may contain a variable number of microvesicles (Figs. 7-13 and 7-14). These vacuoles stain for monoclonal immunoglobulins. In the illustrated case, this was shown to be IgG/kappa.

Lymphoblastic lymphoma is a lymphoma which generally marks as T cell,[1,31] although morphologically it is indistinguishable from null cell acute lymphocytic leukemia appearing in an extramedullary location. It predominates in adolescent male children who frequently present with a mediastinal mass. This lymphoma has a high incidence of leukemic transformation. Conversely, cases of acute lymphocytic leukemia that mark as T cell rather than null cell may have an increased incidence of extramedullary involvement. By electron microscopy (Fig. 7-15), the cells may have extremely irregular, convoluted nuclei, although there are some cases in which the more classic convoluted cells are in the minority. The nuclear-cytoplasmic ratio is high. The nucleus frequently has a reduplicated nuclear membrane, which is a non-

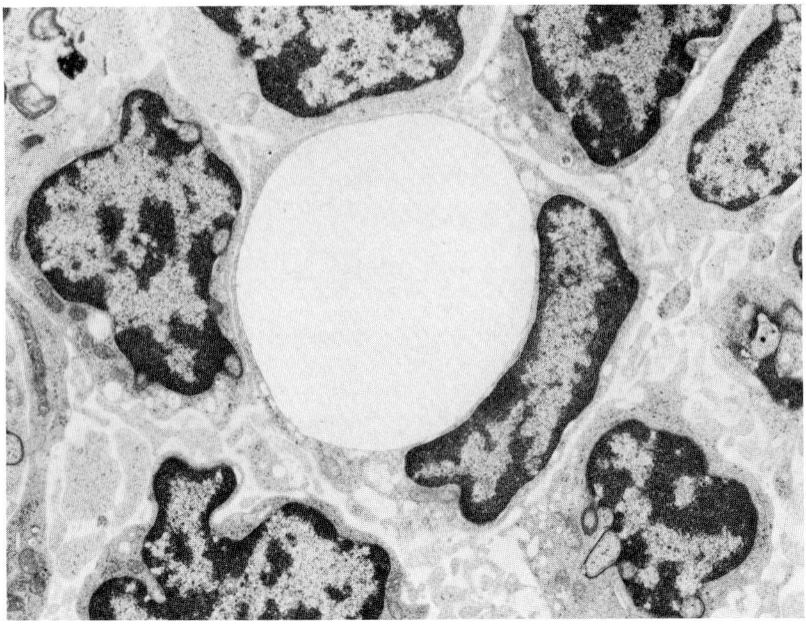

FIG. 7-13. Poorly differentiated lymphocytic lymphoma, signet ring cell type. This interesting variant of follicular center cell lymphoma demonstrates large vacuoles that compress the nucleus, imparting a signet ring appearance. X6700.

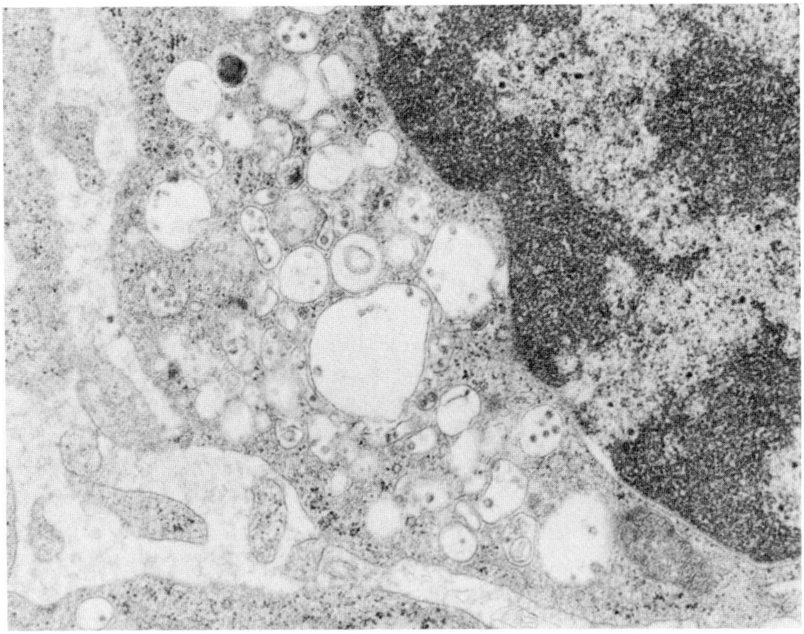

FIG. 7-14. Poorly differentiated lymphocytic lymphoma, signet ring cell variant. This high-power view demonstrates multiple small membrane-bound sacs containing small vesicles. These sacs presumably fuse to form the characteristic large membrane-bound vacuoles, which ultimately compress the nucleus and which in some cases have been shown to contain immunoglobulin. X21,200.

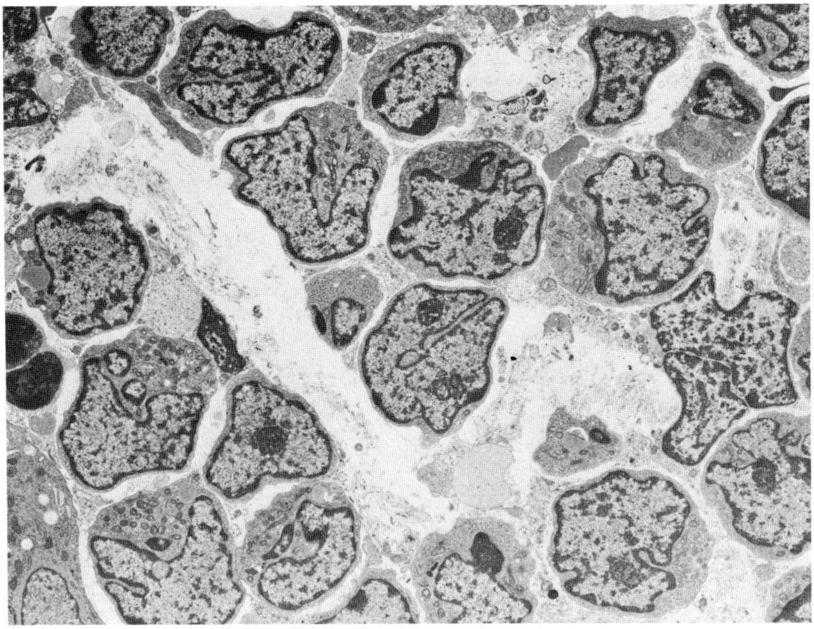

FIG. 7-15. Lymphoblastic lymphoma. Note that many of the cells are similar to the cells demonstrated in Fig. 7-4, commensurate with the fact that lymphoblastic lymphoma is morphologically indistinguishable from an extramedullary manifestation of acute lymphocytic leukemia. The cells have a high nuclear-cytoplasmic ratio, with a variable number of highly irregular, convoluted nuclei whose chromatin is uniformly dispersed in fine clumps with a generally thin peripheral margin of dense chromatin. Small nucleoli often adjacent to the nuclear membrane are uniformly present, although improminent. X3200.

specific finding. The chromatin pattern reflects the blastic homogeneous appearance by light microscopy, although it is greatly magnified and shows fine clumps of dense chromatin evenly distributed throughout the nucleus. An improminent nucleolus is present and is often adherent to the nuclear membrane. In less optimally fixed tissue, the chromatin may be more coarsely clumped, becoming indistinguishable from that of B cell, poorly differentiated lymphocytic lymphoma. In fact, the convoluted cell of lymphoblastic lymphoma cannot always be distinguished from a small cleaved cell of a (follicular center cell) poorly differentiated lymphocytic lymphoma using ultrastructural study alone.

Mixed cell lymphoma is composed of nodules of small cleaved cells (poorly differentiated lymphocytes) and large noncleaved follicular center cells, with the minority group comprising at least 25% of the total. Because of the mixture of cells, electron microscopy is not helpful in making the diagnosis, although ultrastructural studies do demonstrate the two types of cells. One population of cells has an angular nucleus and condensed chromatin (small cleaved). The features of the other cells are those of a more transformed lymphocytic cell: a large noncleaved nucleus with dispersed chromatin and a fairly prominent nucleus, abundant cytoplasmic free polyribosomes, and occasional long strands of rough endoplasmic reticulum.

Undifferentiated lymphomas (small noncleaved) include classic Burkitt's lymphomas and undifferentiated non-Burkitt lymphomas. The latter are characterized by an absent or less obvious starry-sky appearance, somewhat more pleomorphic nuclei along with larger nucleoli, less than the expected cytoplasmic lipid, and more than the usual small plasmacytoid rim of cytoplasm. At present, the strict morphologic criteria for the diagnosis of Burkitt's lymphoma by light microscopy[2] include a monotonous proliferation of round, regular cells with a narrow rim of cytoplasm containing lipid vacuoles. The nucleus contains finely clumped chromatin with one or more small nucleoli. The size of the neoplastic lymphoid cells is approximately the same as the nucleus of the benign phagocytosing histiocytes that are responsible for the starry-sky appearance.[2] Ultrastructural studies (Fig. 7-16) show predominantly round cells approximately 12 μm in diameter with rounded nuclei containing one or more small, but prominent, nucleoli and a moderately coarsely clumped chromatin pattern. The cytoplasm is clearly outlined and often electron-dense. It contains the characteristic lipid vacuoles that are more numerous in the classic Burkitt's undifferentiated lymphomas. The cytoplasm is packed with polyribosomes, and a few mitochondria may be grouped. The benign histiocytes are approximately 20 μm in diameter and show many digestive vacuoles filled with cellular debris. There is a wide range of nuclear shape and lipid content in these small noncleaved cell lymphomas. The so-called undifferentiated non-Burkitt's lymphoma (Fig. 7-17) basically has similar features, although with greater irregularity in nuclear profiles and less cytoplasmic lipid. It is unclear whether there are indeed two clinically justifiable separate entities.

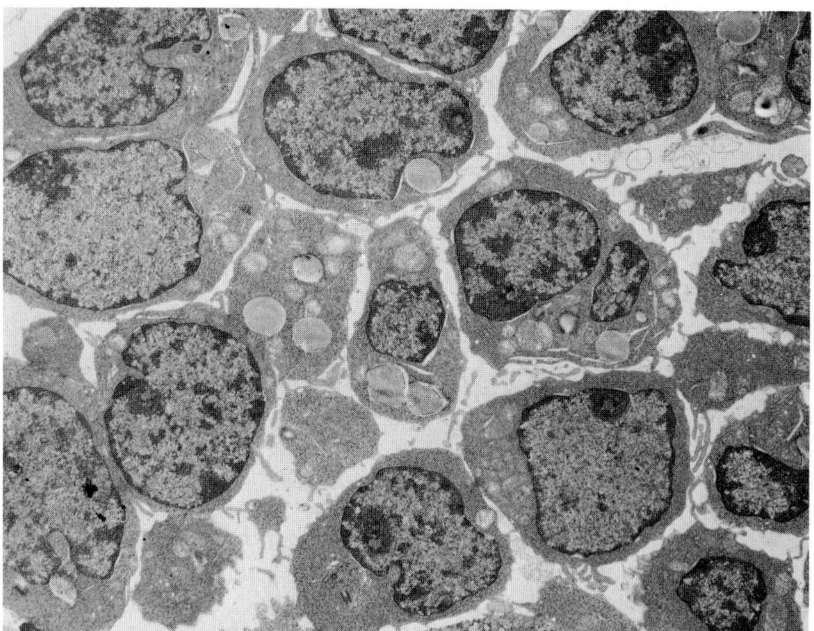

FIG. 7-16. Undifferentiated lymphoma, Burkitt type. The rounded nuclei are fairly uniform, with one to four prominent nucleoli. The clearly outlined cytoplasm is often electron-dense and characteristically contains a variable number of lipid vacuoles. X2000.

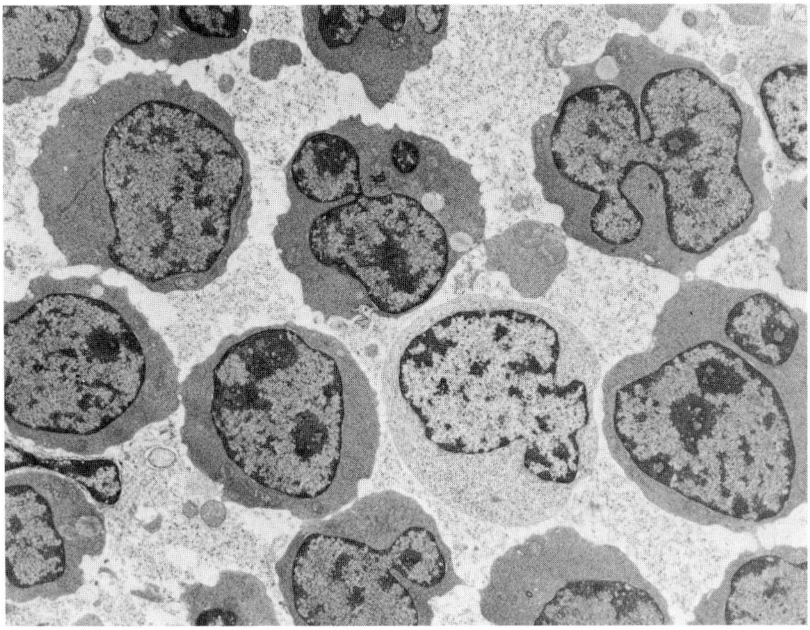

FIG. 7-17. Undifferentiated lymphoma, non-Burkitt's type. By comparison with acceptable cases of Burkitt's lymphoma, the nuclear profile is more irregular. The cytoplasm in most cells is still electron-dense, may be more plentiful, and shows generally fewer lipid vacuoles. Some cells appear to be in transition toward transformed lymphocytes. X3300.

The contribution of ultrastructural studies to the diagnosis of large cell (histiocytic) lymphoma is one of the most important in the clinical application of electron microscopy to reticuloendothelial problems. Cases are not infrequently diagnosed as poorly differentiated carcinoma, metastatic malanoma, malignant histiocytosis, liposarcoma, or malignant fibrohistiocytoma until the diagnosis of large cell lymphoma ultimately is confirmed by ultrastructural studies. It is now known that most histiocytic lymphomas derive from transformed lymphocytes at some stage of development either within the follicle or following their exit as immunoblasts from the follicular center. The morphologic appearance of large cell lymphomas covers a wide range that has only recently been recognized.[12,15] The diffuse large cell lymphomas of follicular center cell origin may be large cleaved, mixed follicular center cell, large noncleaved, or immunoblastic sarcoma (sometimes termed *pleomorphic pyroninophilic*).[40] Strauchen et al. also included the term *blastic lymphoma* in their study.[40] However, the description of blastic lymphoma is identical to that of undifferentiated non-Burkitt lymphoma. The large cleaved cell (Fig. 7-18) shows a high nuclear cytoplasmic ratio with a large, irregularly outlined nucleus containing frequent nuclear blebs and a dispersed chromatin pattern with improminent nuclei often abutting on the nuclear membrane. The cytoplasm is characterized by a moderate number of polyribosomes and few other organelles except for some mitochondria. The cytoplasmic periphery may show numerous undulations that should not be mistaken for

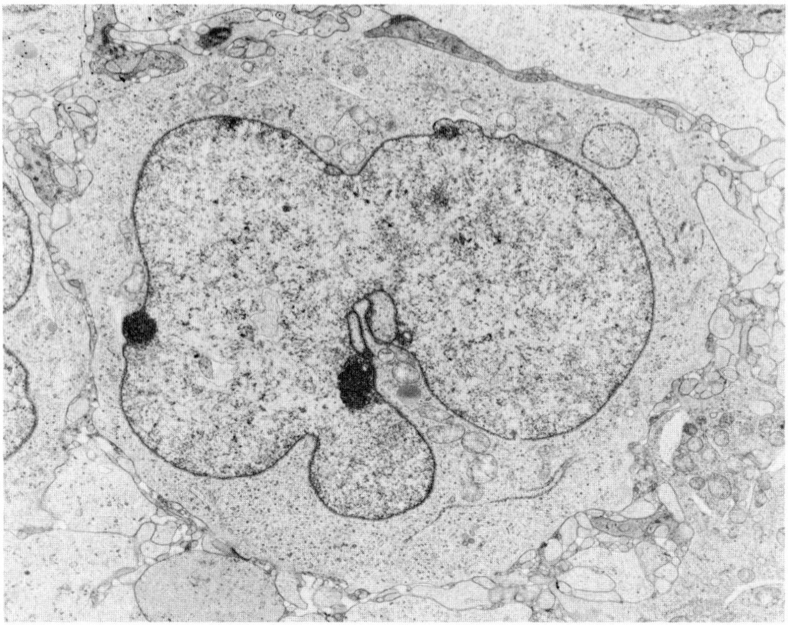

FIG. 7-18. Large cell (histiocytic) lymphoma. This is a large cleaved, transformed lymphocyte of histiocytic lymphoma. The large nucleus is irregular, with a totally dispersed chromatin pattern. One or more nucleoli may be present. The cytoplasm shows some polyribosomes, occasional small strands of rough endoplasmic reticulum, and a characteristically undulating profile of the cytoplasmic borders. X2900.

microvilli. To be reiterated is the fact that the neoplastic cells do not show any cell junctions between their borders. The large noncleaved cell (Fig. 7-19) appears to be, morphologically, somewhere between the large cleaved cell and the immunoblast (Fig. 7-20). It has a progressively rounder nuclear profile with a fairly prominent, more centrally placed nucleus. Again, the cytoplasmic borders, which show no attachments, may be irregular in outline. The cells of immunoblastic sarcoma will usually show an even larger, centrally placed nucleolus. It is not clear whether in all cases of immunoblastic sarcoma the degree of plasmacytoid differentiation appreciated by light microscopy can be correlated ultrastructurally, allowing for an unequivocal diagnosis of the immunoblastic sarcoma subclassification on the basis of electron microscopy alone.

Not infrequently, large cell lymphomas may be confused with other malignancies. A group of patients was studied in whom initial lymph node specimens demonstrated a predominantly sinusoidal pattern of large cell lymphoma (Fig. 7-21), which can be easily misinterpreted as malignant histiocytosis, metastatic melanoma, or metastatic poorly differentiated carcinoma.[32] In this group of patients, when electron microscopy was performed, the malignant sinusoidal proliferation was confirmed to be immunoblastic (Figs. 7-22 and 7-23). In two cases autopsy also confirmed that only large cell lymphoma (immunoblastic sarcoma) was present.

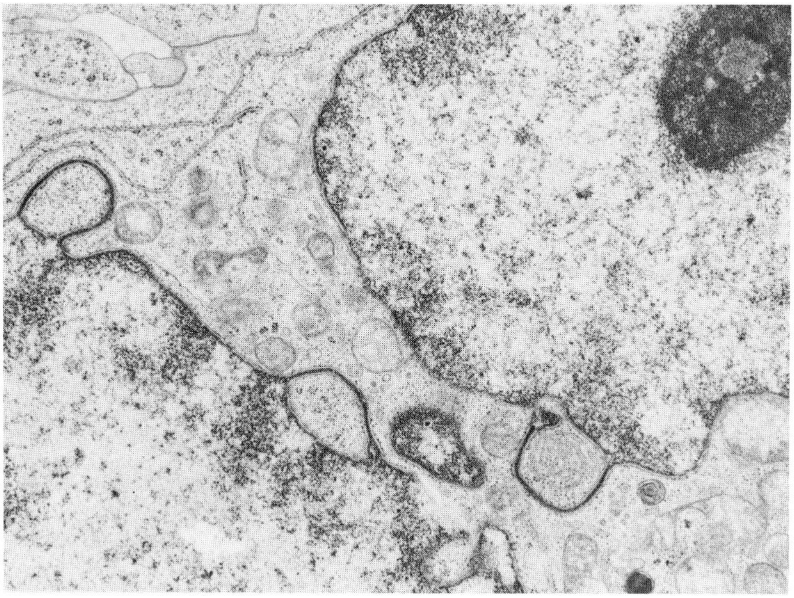

FIG. 7-19. Large cell (histiocytic) lymphoma. This is a high-power view demonstrating numerous nuclear blebs, which although more frequent in reticuloendothelial neoplasms, are not diagnostic, as they may be found in other tumors. X6200.

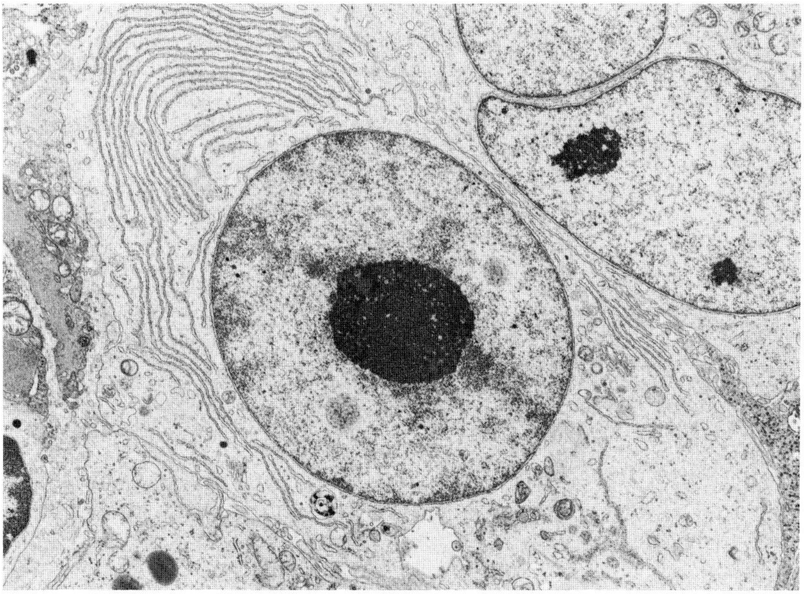

FIG. 7-20. Large cell (histiocytic) lymphoma, immunoblastic sarcoma subclassification. In this subgroup of large cell lymphoma, the plasmacytoid features appreciated by light microscopy are reflected in the parallel arrays of rough endoplasmic reticulum seen in the cytoplasm on ultrastructural examination. The large nucleus, with its dispersed chromatin pattern, shows a very prominent, often centrally placed, nucleolus. X4700.

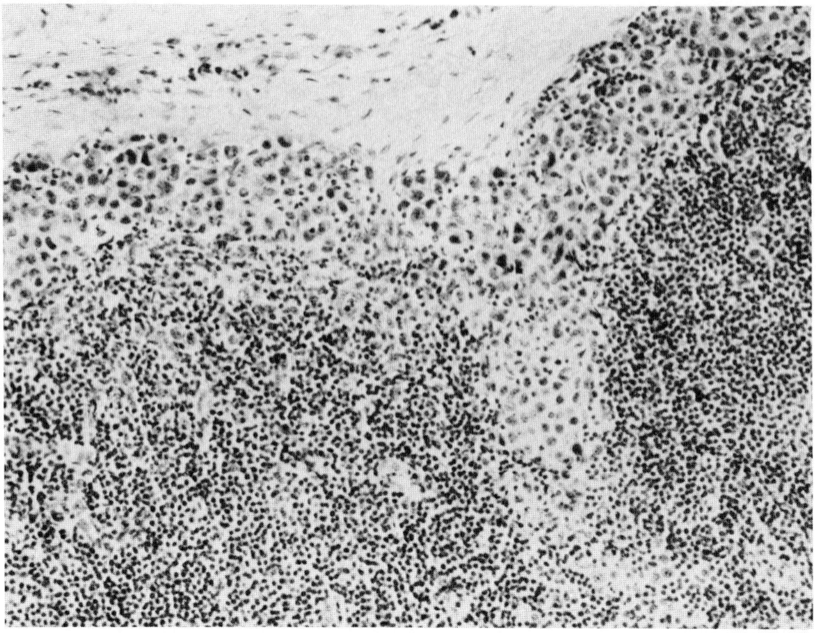

FIG. 7-21. Large cell (histiocytic) lymphoma, sinusoidal pattern. The light micrograph demonstrates a malignant, monomorphic proliferation of large cells, often with plasmacytoid differentiation characteristics of immunoblastic sarcoma. X380.

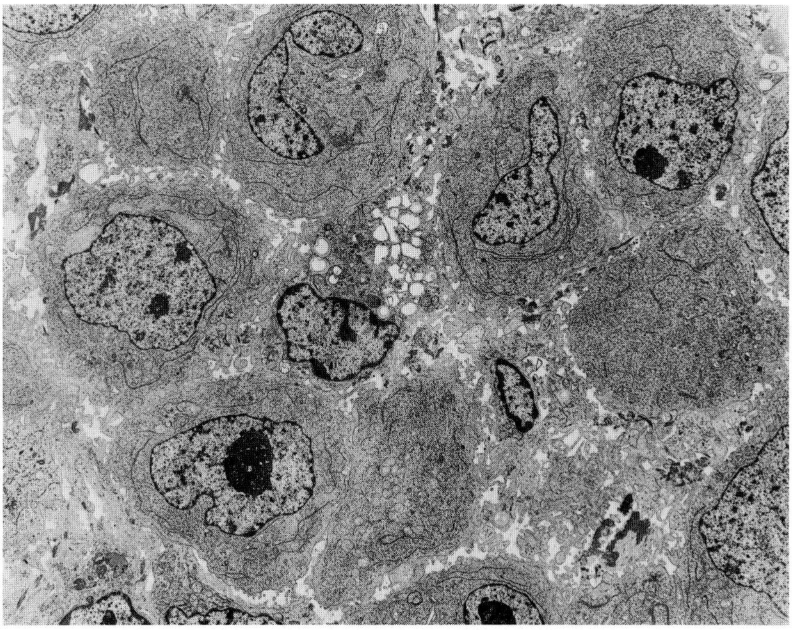

FIG. 7-22. Large cell (histiocytic) lymphoma, sinusoidal pattern. Ultrastructurally, these sinusoidal cells have the characteristics of transformed lymphocytes with plasmacytoid differentiation. The nucleus shows dispersed chromatin pattern with a prominent nucleolus. X1300.

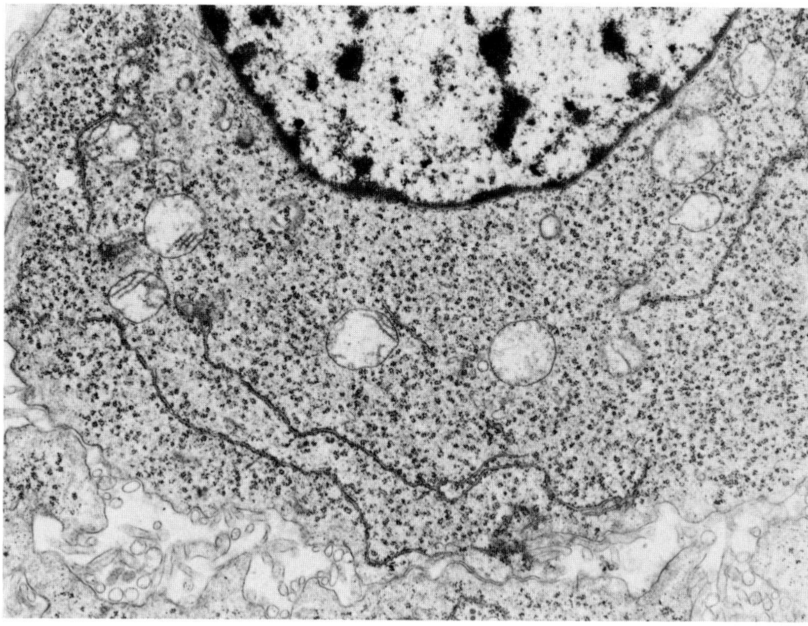

FIG. 7-23. Large cell (histiocytic) lymphoma, sinusoidal pattern. The cytoplasm shows many polyribosomes (indicating protein production), along with early plasmacytoid differentiation as evidenced by long strands of rough endoplasmic reticulum. The cytoplasmic border shows the characteristic undulating pattern. X9800.

In extranodal presentations of large cell lymphoma within soft tissue, the light microscopic appearance may be confused by fat necrosis. The resulting population of histiocytes can obscure the lymphoid nature of the neoplasm. In these cases ultrastructural studies may be helpful in suggesting or confirming the impression that the neoplastic cells are, in fact, transformed lymphocytes.

Immunoblastic sarcoma may contain extremely pleomorphic, often multinucleated cells that have undoubtly, in the past, led to a diagnosis of poorly differentiated carcinoma. Nevertheless, ultrastructural studies show the lack of cytoplasmic attachments, as well as the absence of any feature to suggest carcinoma. Prominent, large nuclei have the chromatin pattern of transformed lymphocytes with a variable nuclear outline and a variable degree of plasmacytoid differentiation within the cytoplasm.

Finally, in a number of patients with a known epithelial malignancy, EM may be helpful in establishing lymphoma as a second primary when metastasis has been suspected. For example, a patient with breast cancer may have an enlarged node sent for estrogen receptor assay. On several occasions, these lymph nodes and the patient's subsequent course have proven to be large cell lymphoma, a second primary, confirmed by ultrastructural studies.

Mycosis fungoides is one of a group of cutaneous T-cell lymphomas. Ultrastructurally, the typical mycosis cells have a similar morphology whether they are found in peripheral blood (Sézary cells) or in the skin lesions.[37] They contain extremely contorted nuclei with deep infoldings and peripheralized dense chromatin. They appear to have more cytoplasm and are larger than the cells of either poorly differentiated lymphocytic lymphoma (B-cell type) or T-cell lymphoblastic lymphoma, which may also show respectively cleaved or convoluted nuclei.

Myeloma and solitary plasma cell tumors can, on occasion, be extremely difficult to diagnose by light microscopy, particularly when the plasma cells are not well differentiated or when there is some cytologic distortion, possibly due to biopsy technique. Ultrastructural studies (Fig. 7-24) show eccentrically placed nuclei whose chromatin pattern may be that of well-differentiated plasma cells as evidenced by multiple peripheral clumps responsible for the "cartwheel pattern" appreciated by light microscopy, or they may show nuclear characteristics more common to immunoblasts, with dispersion of nuclear chromatin and prominent nucleoli. The cytoplasm is characterized by concentrically arranged long strands of rough endoplasmic reticulum. When the nuclei are large with dispersed chromatin and a large nucleolus, and when the cytoplasm is more mature, showing well-developed long parallel strands of rough endoplasmic reticulum, the term *plasma cell asynchrony* has been used by Bernier and Graham,[3] who conclude that the degree of asynchrony predicts the stage of the patient's disease.[17]

At present, we feel that there is a gray zone encompassing, on the one hand, plasmacytomas which are anaplastic and have nuclei more characteristic of immuno-

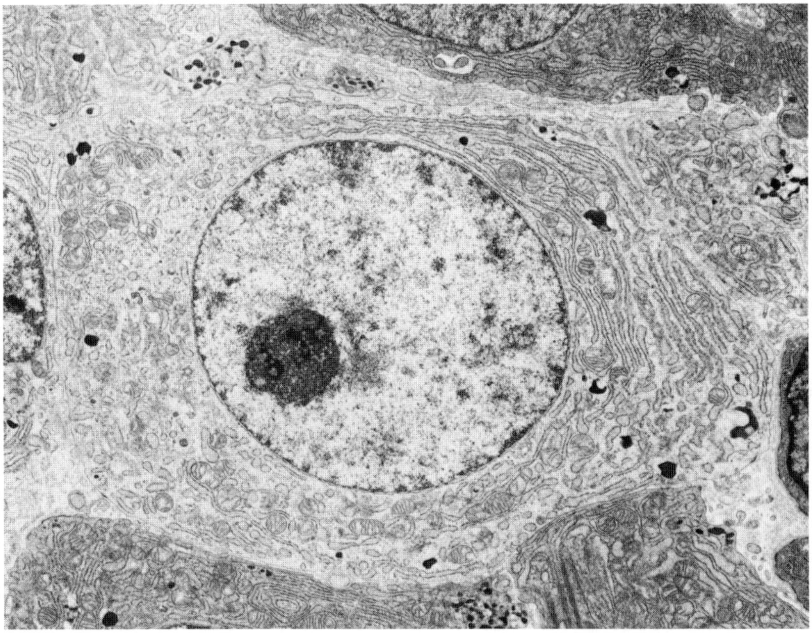

FIG. 7-24. Anaplastic plasmacytoma. This 75-year-old female had a soft tissue mass on her scalp, multiple lytic skeletal lesions, elevated IgA, and Bence Jones proteinuria. Nuclear-cytoplasmic asynchrony is demonstrated by nuclei whose chromatin pattern is dispersed and which contain a prominent, centrally placed nucleolus, while the cytoplasm shows a greater degree of maturity with concentrically arranged rough endoplasmic reticulum. X4600.

blasts, and, on the other hand, cases of a monomorphic proliferation of apparent immunoblasts, virtually all of which show plasmacytoid differentiation. In the case of a solitary lesion without a monoclonal immunoglobulin spike, the question is whether this represents multiple myeloma or immunoblastic sarcoma. It is uncertain whether the distinction can be made at an ultrastructural level.

Many atypical immunoblastic, lymphocytic, and histiocytic proliferations within lymph nodes are now being recognized. Within this rather broad category, immunoblastic lymphadenopathy has been most clearly defined histologically. Electron microscopic examination is not a great deal of help in distinguishing this entity from other reactive processes or abnormal proliferations because of its heterogeneous composition.

Similarly, in Hodgkin's disease and in the lymphomas fulfilling the requirements for a diagnosis of Lennert's lymphoma, because the morphology is a mixture of cells, the contribution of electron microscopy to diagnosis is limited. Many ultrastructural studies have concentrated on the morphology of the Reed-Sternberg cell and the similarity of L&H cells, of atypical mononuclear cells, and of lacunar cells to each other and to transformed lymphocytes, implying they are aberrant, transformed lymphocytes. At present, however, the findings from histochemical and immunopathologic studies seeking to determine the origin of the proliferating cells in Hodgkin's disease are contradictory.

HISTIOCYTOSES

Histiocytosis X, under which the clinical syndromes eosinophilic granuloma, Hand-Schüller-Christian disease, and Letterer-Siwe disease are included, is a proliferative disorder of histiocytes. Electron microscopy can be of great value in diagnosing atypical cases and in supporting the diagnosis in those that are more obvious. The characteristic histiocyte (Fig. 7-25) has an indented nucleus with ample cytoplasm containing sparse lysosomes, mitochondria, and short strands of rough endoplasmic reticulum. In addition, even at low power, one may appreciate a variable number of the characteristic Langerhan's granules (Fig. 7-26) with their trilaminar structure and, given a fortuitous plane of section, a bulbous dilatation at one end. Occasionally, enlarged lymph nodes may be the presenting or only manifestation of histiocytosis X.[5] In such involved lymph nodes, Langerhan's granules are present in the characteristic histiocytes, which by light microscopy are seen to be markedly distending the nodal sinuses along with a variable number of eosinophils.[5] Langerhan's granules are not absolutely specific for differentiated histiocytosis, as they have also been observed in reactive lymph nodes and, occasionally, in other neoplasms.[18] The morphology of the characteristic cell is similar, regardless of the degree of clinical involvement.

Malignant histiocytosis (histiocytic medullary reticulosis) is rare and somewhat overdiagnosed, in that it has been widely believed that any malignant sinusoidal proliferation within a lymph node that was not metastatic was malignant histiocytosis. These sinusoidal proliferations most frequently have the appearance of immunoblasts by light microscopy, an impression confirmed by electron microscopy, which indicates that they are actually an unusual "sinusoidal" pattern produced by the lymphoma. It is reported that Langerhan's granules have not been present in malig-

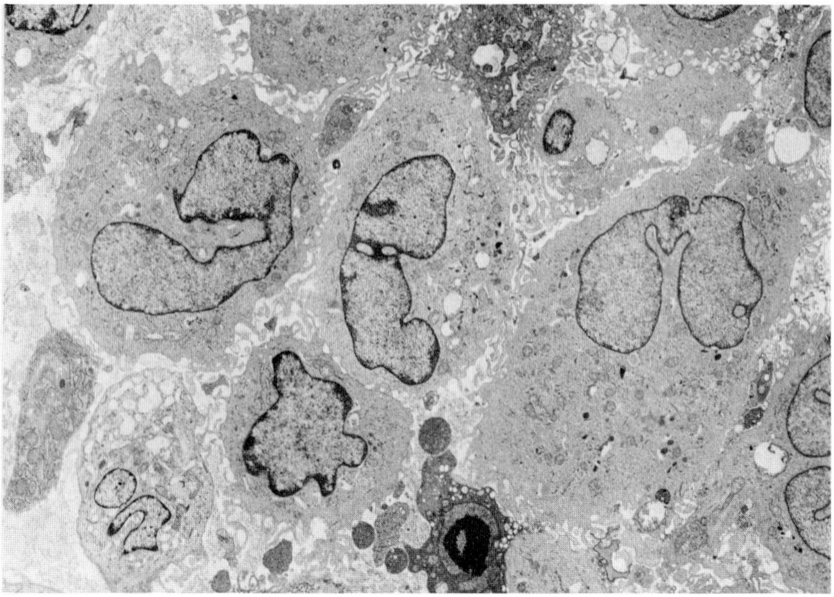

FIG. 7-25. Histiocytosis X. This adolescent female patient had a confirmed diagnosis of Hodgkin's disease with pulmonary recurrence. Initially, a bilateral lung infiltrate was thought to represent progression of disease (in spite of chemotherapy), and the patient underwent an open lung biopsy for documentation in anticipation of changing the chemotherapy. The characteristic histiocytes in histiocytosis show bean-shaped nuclei, ample cytoplasm with few lysosomal granules, and a variable number of characteristic Langerhans granules, some of which appear to form by invagination of the cytoplasmic membrane. X3600.

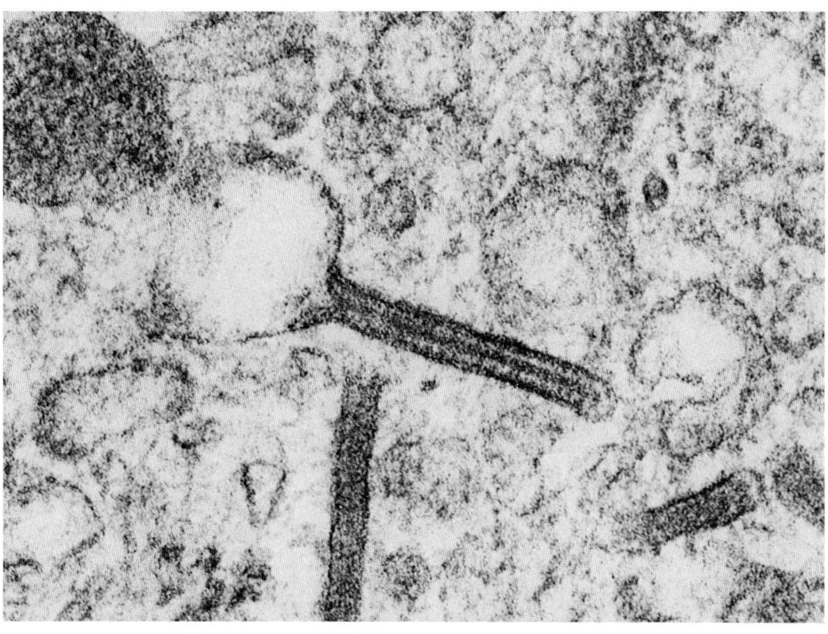

FIG. 7-26. Histiocytosis X. A high-power view of the characteristic Langerhans granules showing their trilaminar structure with a unipolar dilatation. X12,400.

nant histiocytosis and that the cytoplasm of the histiocytes contain many irregularly shaped lysosomal granules and a juxtanuclear filamentous zone.[34]

THYMOMA

The fine structure of the thymoma[36] mimics that of the normal thymus. Therefore, isolated electron micrographs will not distinguish neoplasia from normal or hyperplastic thymus. The light microscopic diagnosis of thymoma can be difficult, particularly in cases in which lymphocytes predominate, obscuring the epithelial cells, and in cases in which, for clinical reasons, only a small biopsy has been performed. Included in the differential diagnosis are mediastinal carcinoid, metastatic germ cell tumors, well-differentiated lymphocytic lymphoma, Hodgkin's disease, and large cell lymphoma.

The fine structure of thymomas[33,36] is characterized by distinctive epithelial cells showing many elongated cytoplasmic processes with prominent desmosomes between these processes (Figs. 7-27, 7-28, 7-29). Bundles of filaments may be present. There may be a prominent basal lamina surrounding groups of epithelial cells. In addition, the presence of gland-like structures has been reported[36] and should not necessarily discourage the diagnosis of thymoma in the presence of other distinctive characteristics. This finding in thymoma may represent an example of the tumor recapitulating the more primitive stage of thymic development, as the human fetal and infant thymus does contain gland-like spaces.[36]

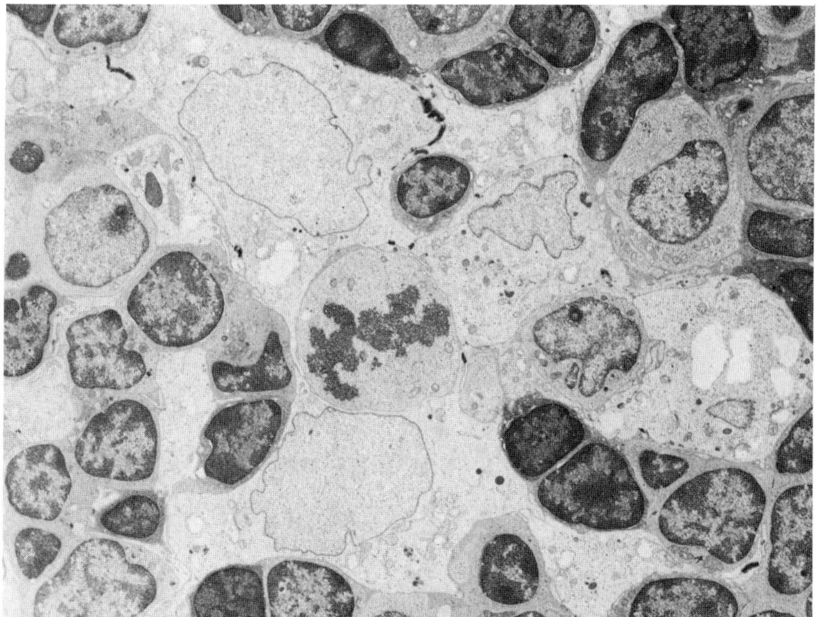

FIG. 7-27. Thymoma. This photograph demonstrates the distinction between the lymphocytes and epithelial cells. Even at low power, the latter show prominent and numerous desmosomes. X5800.

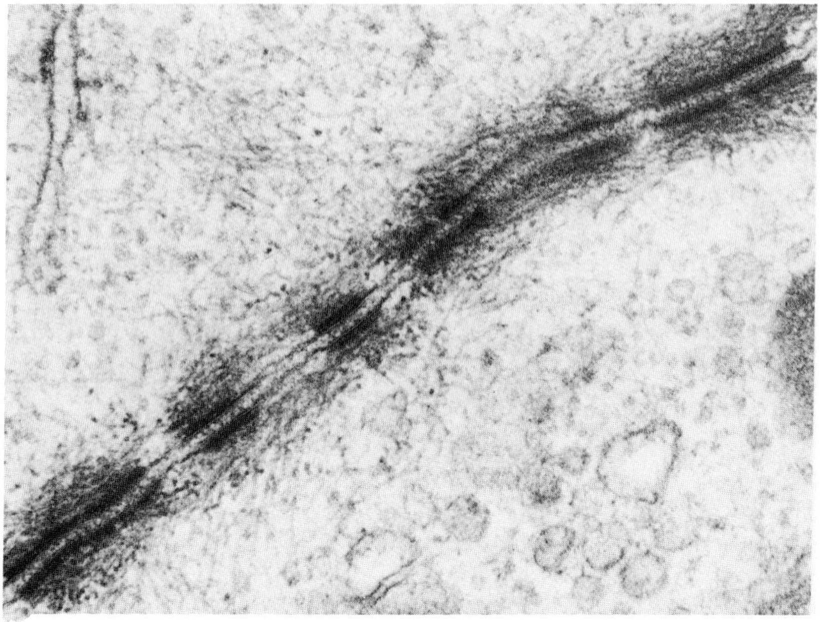

FIG. 7-28. Thymoma. A high-power view of the desmosomes uniting epithelial cells. X69,100.

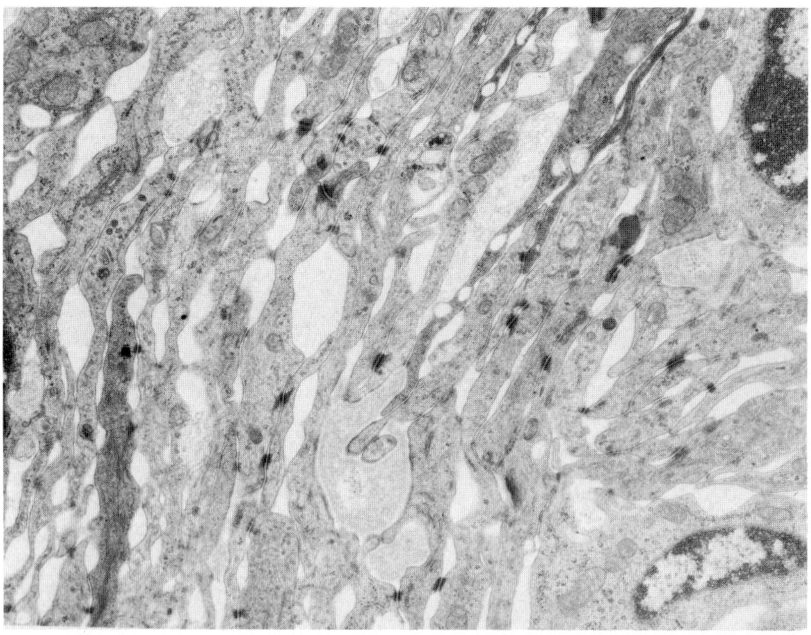

FIG. 7-29. Thymoma. The epithelial cells of thymoma frequently have long cytoplasmic extensions connected by well-formed desmosomes. X9600.

REFERENCES

1. **Barcos MP, Lukes RJ:** Malignant lymphoma of convoluted lymphocytes: A new entity of possible T-cell type. In Sink LF, Godden JO (eds): Conflicts in Childhood Cancer. New York, Alan R. Liss, 1975, pp 147–178
2. **Berard C, O'Conor LBT, Torloni H:** Histopathological Definition of Burkitt's Tumor. Geneva, World Health Organization, 1969
3. **Bernier GM, Graham RC Jr:** Plasma cell asynchrony in myeloma: Correlation of light and electron microscopy. Semin Hematol 13:239–245, 1976
4. **Bessis M:** Living Blood Cells and Their Ultrastructure. New York, Springer-Verlag, 1973
5. **Breton-Gorius J:** Apport de la microscopic electronique dans la classification cytologique des leucémies aiguës. Bull Cancer (Paris) 61:367–376, 1974
6. **Burke JS, Mackay B, Rappaport H:** Hairy cell leukemia (leukemic reticuloendotheliosis), pt 2. Ultrastructure of spleen. Cancer 37:2267–2274, 1976
7. **Byrnes RK, Golomb HM, Desser RK, et al.:** Acute monocytic leukemia. Am J Clin Pathol 65:471–482, 1976
8. **Carr I, Hancock LH, Ward AM:** Lymphoreticular Disease. London, Blackwell, 1977
9. **Daniel MT, Flandrin G:** Fine structure of abnormal cells in hairy cell (tricholeukocytic) leukemia, with special reference to their *in vitro* phagocytic capacity. Lab Invest 3:1–8, 1974
10. **Evans HL, Butler JJ, Youness EL:** Malignant lymphoma, small lymphocytic type: A clinicopathologic study of 84 cases with suggested criteria for intermediate lymphocytic lymphoma. Cancer 41:1440–1455, 1978
11. **Feldman JD:** Ultrastructure of immunologic processes. Adv Immunol 4:175–248, 1964
12. **Filippa DA, Lieberman PH, Erlandson RA, et al.:** A study of malignant lymphomas using light and ultramicroscopic, cytochemical and immunologic techniques. Am J Med 64:259–268, 1978
13. **Gillespie JJ:** The ultrastructural diagnosis of diffuse large cell ("histiocytic") lymphoma: Fine structural analysis of 30 cases. Am J Surg Pathol 2:9–20, 1978
14. **Glick AD, Horn RC:** Identification of promonocytes and monocytoid precursors in acute leukemia of adults: Ultrastructural and cytochemical observations. Br J Haematol 26:395, 1974
15. **Glick AD, Leech JH, Waldron JA, et al.:** Malignant lymphomas of follicular center cell origin: II. Ultrastructural and cytochemical studies. JNCI 54:23–36, 1975
16. **Glick AD:** Acute leukemia: Electron microscopic diagnosis. Semin Oncol 3:241–299, 1976
17. **Graham RC Jr, Bernier GM:** The bone marrow in multiple myeloma: Correlation of plasma cell ultrastructure and clinical state. Medicine (Baltimore) 54:225, 1975
18. **Hammer SP, Bockus D, Remington F, et al.:** Langerhans cells and serum precipitating antibodies against fungal antigens in bronchioloalveolar cell carcinoma: Possible association with pulmonary eosinophilic granuloma. Ultrastruct Pathol 1:19–37, 1980
19. **Henry K:** Electron microscopy in the non-Hodgkin's lymphomata. Br J Cancer 31[suppl 2]:73–93, 1975
20. **Imamura M, Sakamoto S, Hanazono H:** Malignant histiocytosis: A case of generalized histiocytosis with infiltration of Langerhans' granule-containing histiocytes. Cancer 28:467–475, 1971
21. **Ishikawa E, Yasuda H:** Electron microscopy in the study of malignant lymphoma. Gann Monograph on Cancer Research 15:293–297, 1971
22. **Kaiserling E:** Elektronenmikroskopishe Befunde bei non-Hodgkin's lymphomen. Haematol Bluttransfus 18:185–205, 1976
23. **Katayama I, Li CY, Yam LT:** Ultrastructural characteristics of the "hairy cells" of leukemic reticuloendotheliosis. Am J Pathol 67:361, 1972
24. **Kim H, Dorfman RF, Rappaport H:** Signet ring cell lymphoma: A rare morphologic and

functional expression of nodular (follicular) lymphoma. Am J Surg Pathol 2:119–132, 1978
25. **Levine JP, Dorfman RF:** Nodular lymphoma: An ultrastructural study of its relationship to germinal centers and a correlation of light and electron microscopic findings. Cancer 35:148–164, 1975
26. **Lukes RJ, Collins RD:** Lukes-Collins classification and its significance. Cancer Treat Rep 61:971–979, 1977
27. **McKenna RW, Parkin J, Brenning RD:** Morphologic and ultrastructural characteristics of T-cell acute lymphoblastic leukemia. Cancer 44:1290–1297, 1979
28. **Mollo F, Monga G, Coda R, Palestro G:** Ultrastructural features of human lymphomas. Acta Neuropathol (Berlin), suppl 6, 1975, pp 17–20
29. **Mori Y, Lennert K:** Electron Microscopic Atlas of Lymph Node Cytology and Pathology. New York, Springer-Verlag, 1969
30. **Murphy MJ, Hay JB, Norris B, Bessis MC:** Ultrastructural analysis of antibody synthesis in cells from lymph and lymph nodes. Am J Pathol 66:25–42, 1972
31. **Nathwani BN, Kim H, Rappaport H:** Malignant lymphoma, lymphoblastic. Cancer 38:964–983, 1976
32. **Osborne BM, Butler JJ, Mackay B:** Sinusoidal large cell "histiocytic" lymphoma. Cancer, 46:2484, 1980
33. **Pascoe HR, Miner MS:** An ultrastructural study of nine thymomas. Cancer 37:317–326, 1976
34. **Rausch CG, Herion JC, Carney CN, Weinstein P:** Malignant histiocytosis: A cytochemical and electron microscopic study of an unusual case. Cancer 44:2158–2164, 1979
35. **Robinowitz BN, Noguchi S, Roenigk HH:** Tumor cell characterization in mycosis fungoides. Cancer 37:1747–1753, 1976
36. **Rosai J, Levine GD:** Tumors of the thymus. In Atlas of Tumor Pathology. Washington, AFIP, 1976
37. **Rosas-Uribe A, Varikojis D, Molnar Z, Rappaport H:** Mycosis fungoides: An ultrastructural study. Cancer 34:636–645, 1974
38. **Schafer HE, Kruger GRF, Fischer R:** Morphological classification of malignant lymphomas: Ultrastructural, cytochemical and immunological results. Acta Neuropathol (Berlin), suppl 6, 1975, pp 21–29
39. **Schumacher HR, Szekely IE, et al.:** Acute leukemic cells: Quantitative and qualitative electron microscopy. Am J Pathol 73:27–40, 1973
40. **Strauchen JA, Young RC, DeVita VT, et al.:** Clinical relevance of the histopathological subclassification of diffuse "histiocytic" lymphoma. N Engl J Med 299:1382–1387, 1978
41. **Tanaka Y, Goodman JR:** Electron Microscopy of Human Blood Cells. New York, Harper & Row, 1972
42. **Tullig M, Breton-Gorius J:** Three types of Auer bodies in acute leukemia. Lab Invest 41:419–426, 1979
43. **Van den Tweel JC, Taylor CR, Parker JW, Lukes RJ:** Immunoglobulin inclusions in non-Hodgkin's lymphomas. Am J Clin Pathol 69:306–313, 1978
44. **Vernon S, Voet RL, Naeim F, Waisman J:** Nodular lymphoma with intracellular immunoglobulin. Cancer 44:1273–1279, 1979
45. **Williams WJ, Beutler E, Erslev AJ, Rundles RW:** Hematology (ed 2). New York, McGraw-Hill, 1977
46. **Williams JW, Dorfman RF:** Lymphadenopathy as the initial manifestation of histiocytosis X. Am J Surg Pathol 3:405–421, 1979
47. **Zucker-Franklin D:** Virus-like particles in the lymphocytes of a patient with chronic lymphocytic leukemia. Blood 21:509–572, 1963
48. **Zucker-Franklin D:** Cellular structure and function in normal and neoplastic lymphoid cells. Arch Intern Med 135:55–60, 1975

8

ELECTRON MICROSCOPY IN DIAGNOSTIC MICROBIOLOGY

RICHARD H. CONKLIN

Until recently, diagnostic microbiology laboratories have been reluctant to use the electron microscope in a routine manner. This has been in part due to lack of familiarity with the electron microscope and limited availability of an instrument. Today the electron microscope is used more commonly in surgical pathology and is therefore more accessible to the clinical microbiologist. The purpose of this chapter will be to examine some of the uses of the electron microscope in diagnostic microbiology.

BACTERIOLOGY AND MYCOLOGY

The electron microscope has very limited diagnostic use in bacteriology and mycology. The major application in diagnosis is the rare situation of examining the morphology of an unusual or new organism that cannot be fully classified by conventional biochemical and morphologic criteria. An example of this is shown in Fig. 8-1. This organism was found in a scrotal abscess. It had corkscrew motility and was extremely hard to grow. A sample of the organism was negatively stained and examined in the electron microscope. This answered two basic questions about the organism: (1) The absence of an axostile ruled it out as a spirochete; and (2) the presence of bipolar flagella were demonstrated and these were helpful in classifying the organism which was shown to be an anaerobic spirillium. This was an unusual isolate that had never been recorded from humans, though it has been found in animals.[1]

While this is a rare situation, use of the electron microscope should be considered with unusual isolates. Since there are increasing numbers of immunosuppressed patients surviving for longer periods of time, it is likely that the diagnostic laboratory will be required to identify more unusual or new organisms.[2,3]

VIRAL DIAGNOSIS

Diagnostic virology is the area in which the electron microscope has its greatest diagnostic utility. Recently, useful antiviral chemotherapeutic agents have been added to the armamentarium of physicians.[4,5,6] These agents are highly specific, demanding precise and rapid diagnosis of the viral etiologic agent.

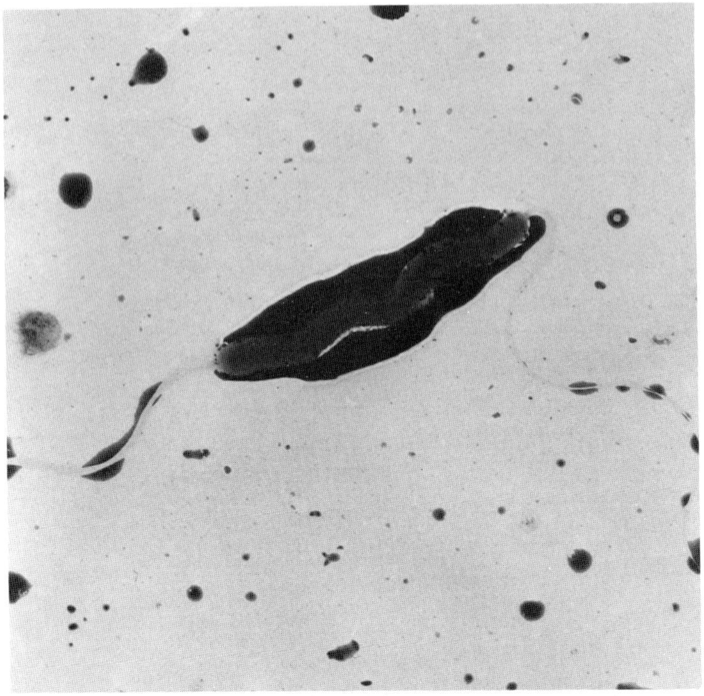

FIG. 8-1. Negatively stained bacterium showing bipolar flagella and the absence of an axostyle. ×19,000

One of the most important applications of the electron microscope in diagnostic virology is the direct examination of body fluids or tissue for virions. Direct examination has been used for the identification of:

Variola in vesicle fluid[7–10]
Varicella zoster in vesicle fluids[10, 11, 14, 15]
Herpes simplex in vesicle fluid[12, 13, 15] (Fig. 8-2)
Herpes zoster in CSF.[14]
Herpes simplex in CSF.[15, 12]
Herpes simplex in tears[15]
Adenovirus in tears[15]
Mumps virus in CSF[16]
Mumps virus in nasopharyngeal secretions[16]
Vaccinia virus in vesicles[9, 10]
Rotavirus in feces[17–19, 25]
Parvo-like virus in feces[17, 26]
Enteroviruses in feces[19]
Adenoviruses in feces[17, 28]
Coronaviruses in feces[27]
Astroviruses in feces[19, 36]

Cytomegalovirus in urine[15,20–22]
Cytomegalovirus in oropharyngeal secretions[20]
Wart virus in skin[23]
Orf-virus in vesicle fluid[24] (Figs. 8-3 and 8-4)
Hepatitis-B virus in sera[34]

The technique for examining specimens directly is rapid and simple. A drop of the specimen is placed on a Formvar carbon-coated copper grid. The excess fluid is removed with a filter paper. The grid is stained by placing a drop of 2% to 3% phosphotungstic acid at pH 7.0 on the grid. The excess stain is removed with a filter paper. The grid is dried and examined in the electron microscope. The entire procedure takes from 15 to 30 minutes to complete. A positive result is usually diagnostic; however, a negative result is not helpful. Since it takes between 10^6 and 10^9 virions/ml for detection by this method,[29,30] a negative result may only indicate that the number of viral particles present in a specimen is less than 10^6/ml.

Because of the large number of virions necessary for detection by direct electron microscopy, several techniques have been devised to concentrate the specimen. These are pseudoreplica, agar gel diffusion, differential centrifugation, airfuge centrifugation, specimen concentration by removing solute, and immunoelectron microscopy.

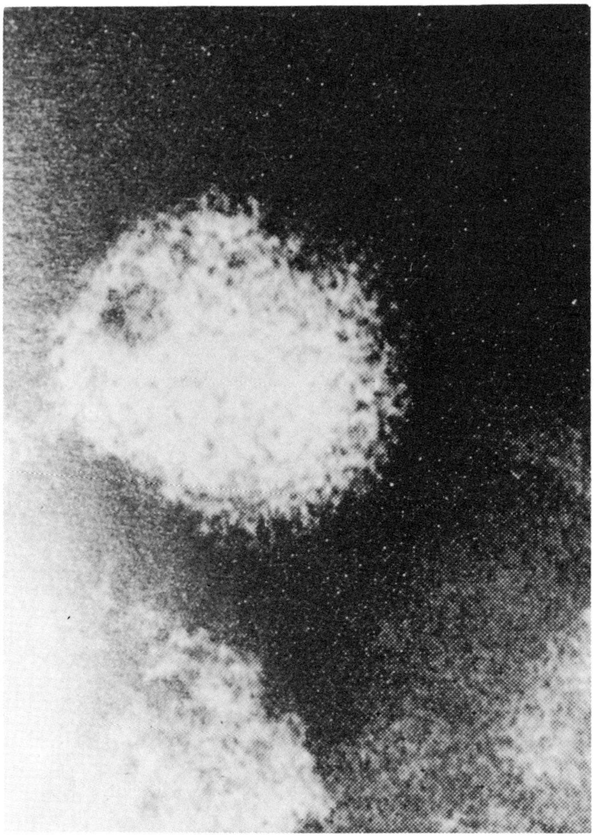

FIG. 8-2. Herpes virus from vesicle fluid in a patient with disseminated herpes simplex. ×266,600.

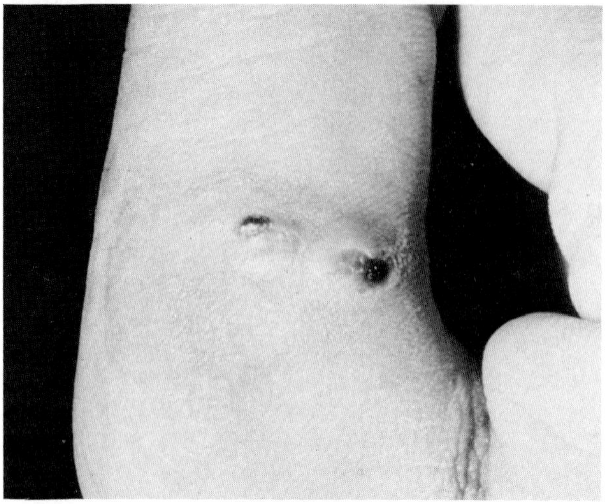

FIG. 8-3. Vesicular lesion of the hand on a 16-year-old farm laborer who had handled dead sheep, goats, and cattle. EM results from this fluid are shown in Fig. 8-4.

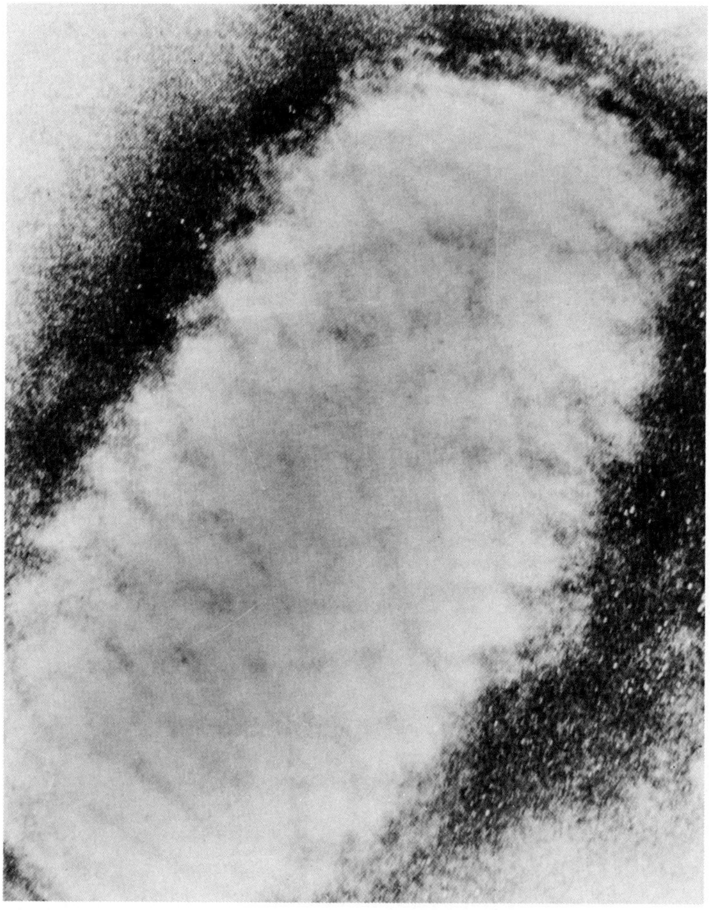

FIG. 8-4. Negatively stained orf virion in fluid from the vesicular lesion of the hand seen in Fig. 8-3. ×200,000.

Pseudoreplica

The pseudoreplica technique was originally introduced by Sharp[32] and was subsequently modified by Smith.[31,33] In this procedure (Figs. 8-5 and 8-6), the specimen is placed on an agarose (2%) block. The liquid is allowed to diffuse into the block, leaving the virus on the surface. The surface is then covered with 0.5% Formvar and allowed to dry. The block is then trimmed and the Formvar film is floated onto 0.5% uranyl acetate. A copper grid is placed on the Formvar film and picked up with a post. The resultant grid is stained with 2% phosphotungstic acid, dried, and examined.

This has been a very useful technique in detecting rotaviruses in stool[17] (Fig. 8-7) and cytomegaloviruses in urine.[15,20]

Agar Gel Diffusion

Agar gel diffusion is similar to the pseudoreplica technique. A sample of the specimen is placed on an agarose (2%) block. A Formvar carbon-coated grid is placed on the specimen drop, and the drop is allowed to diffuse into the agar block. The grid is then removed and negatively stained for EM examination. The pseudoreplica and agar gel diffusion give comparable results. A better preparation is obtained with specimens such as 10% fecal extracts if the specimen undergoes a low speed centrifugation to remove large debris prior to preparing it for EM examination.

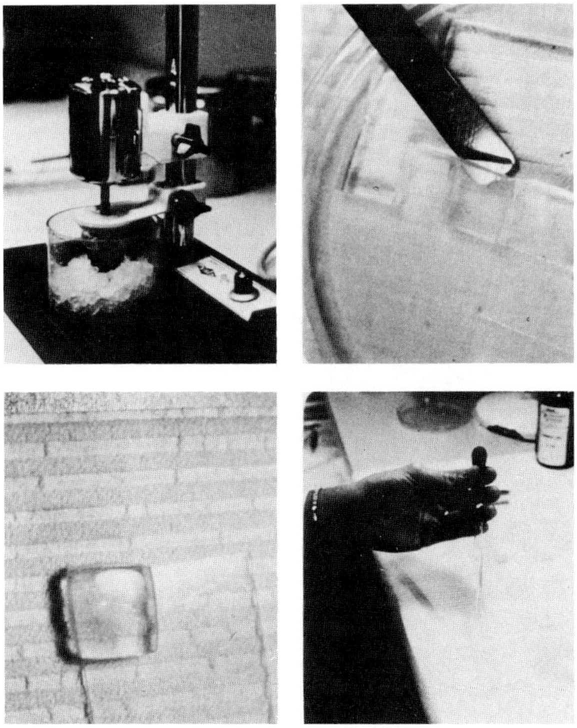

FIG. 8-5. Pseudoreplica technique. Upper left: homogenizing specimen. Upper right: cutting agarose block. Lower left: specimen drop on agarose block. Lower right: formvar applied to block.

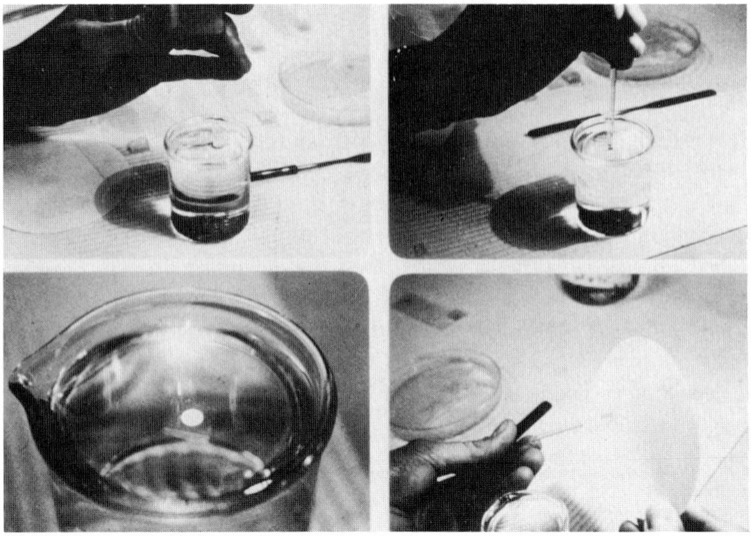

FIG. 8-6. Pseudoreplica technique. Upper left: floating formvar film off the agarose block. Lower left: carbon-coated grid on formvar film. Upper right: picking up grid with a post. Lower right: drying the grid.

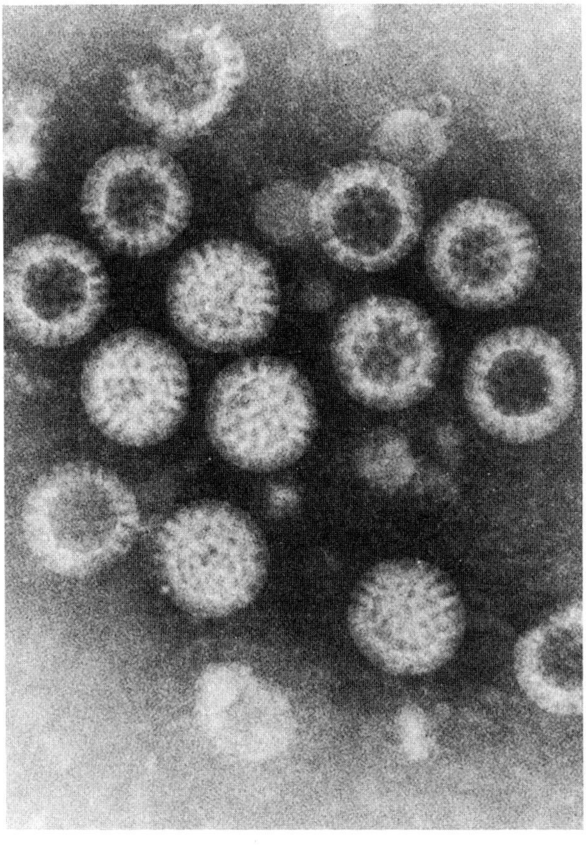

FIG. 8-7. Electron micrograph of rotavirus particles found in the stool of a 10-month-old child with diarrhea C stained with phosphotungstic acid. ×222,700.

Differential Centrifugation

In differential centrifugation, the specimen is centrifuged at low speed to remove large debris and then at high speed to pellet the virus. The virus pellet is then suspended in phosphate buffer or distilled water and prepared for direct EM examination. The results for rotavirus detection with the pseudoreplica technique and differential centrifugation are comparable.[17] Differential centrifugation takes several hours and involves expensive equipment, as opposed to the pseudoreplica technique which takes about one hour and minimal equipment. For the detection of stool rotaviruses, it is obvious that the pseudoreplica technique is optimal.

Airfuge Centrifugation

Airfuge centrifugation is a recently introduced technique that allows viruses to be directly sedimented on a grid by an air-pressure–driven centrifuge.[35] Suspensions with 10^4 particles/ml have been concentrated 100 times by this method. It has been used for rotaviruses in stool suspensions and adenoviruses in tissue culture supernatants.

Specimen Concentration by Removing Solute

Whenever there is a large volume of specimen with a low viral concentration, it is often helpful to concentrate the specimen by Amicon filtration or evaporation of solute. The resultant specimen may then be examined by direct EM.

Immunoelectron Microscopy

This method has been defined as the direct observation of antigen-antibody interaction by electron microscopy. The basis of this procedure was reported in 1941.[38] Since that time the technique has been standardized.[37,39,40,41] Immunoelectron microscopy is an excellent method for studying fastidious viral agents that either do not grow in culture or grow with great difficulty. In addition it is useful for demonstrating serologic evidence of viral infection, and for serotyping viral strains.[37] Immunoelectron microscopy has been extremely valuable in studying rubella,[42] the parvovirus-like agents of acute nonbacterial gastroenteritis,[43] and hepatitis A.[44]

Immunoelectron microscopy is not generally used as a routine diagnostic laboratory test at this time, but the rapidity, specificity, and simplicity of the test should make it increasingly useful in routine diagnostic virology.

Rapid Viral Diagnosis Using Tissue Specimens

Tissue examination with the electron microscope for rapid viral diagnosis is a little-used technique whose efficacy has not truly been evaluated.[30] The technique involves lysing the tissue either by freeze-thawing repeatedly or by homogenization. The resultant homogenate is centrifuged at low speed to remove debris, and the supernatant is processed in the same way that body fluids are prepared for EM examination. One report claims that the sensitivity for herpes detection in brain tissue is 90%.[15]

Future studies may show that EM tissue examination is a valuable diagnostic method. However, it should be compared to routine histopathologic and cytopathologic methods as well as to viral cultures to see if it indeed offers any advantage.

Routine Electron Microscopic Examination of Tissues for Viral Diagnosis

Electron microscopic examination of surgical pathology or autopsy tissues is often a useful adjunct procedure in viral diagnosis. In situations in which it is not possible to culture viral agents, or in the case of disease caused by agents that cannot be cultured, this is an extremely useful diagnostic modality.

Examination of brain biopsy specimens by electron microscopy is useful in corroborating evidence of herpes encephalitis (Fig. 8-8) from light microscopy, immunofluorescence, and cultures. In addition, viral agents of a slow viral illness, such as subacute sclerosing panencephalitis or progressive multifocal leukoencephalopathy (Fig. 8-9), may be detected in the tissue. Other viruses that may be involved in encephalitis, such as adenovirus,[45] Coxsackie virus, togavirus,[46] or mumps virus, may also be detected by EM examination.

An area in which EM has been valuable in diagnosis or in confirmation of viral culture results is viral pneumonias such as cytomegalovirus (Figs. 8-10 and 8-11), adenovirus,[48] and measles.[47] Electron microscopy is often a useful adjunct procedure in studying skin lesions (Fig. 8-12), liver biopsy specimens, myocardial biopsy speci-

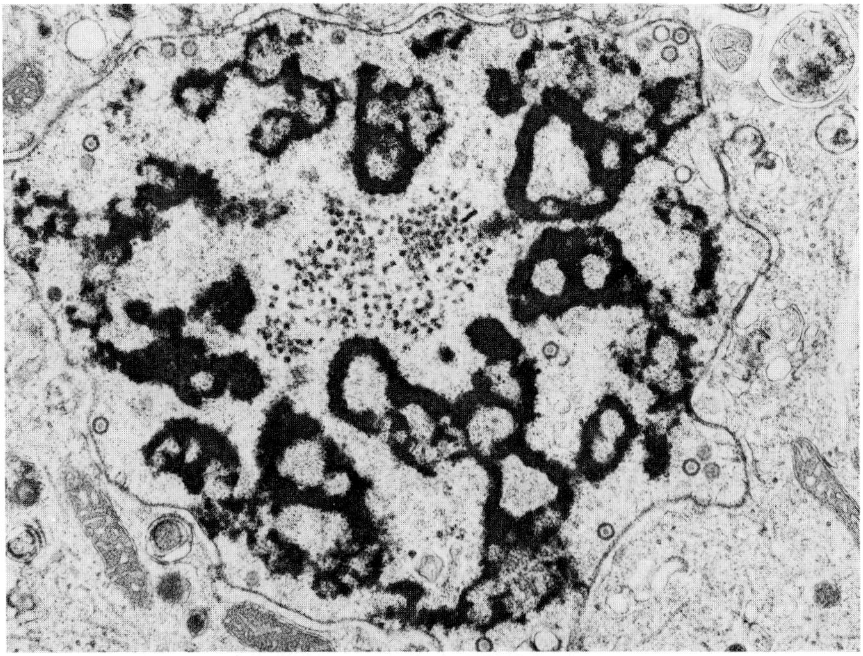

FIG. 8-8. Herpes virions in a brain biopsy specimen. ×19,000.

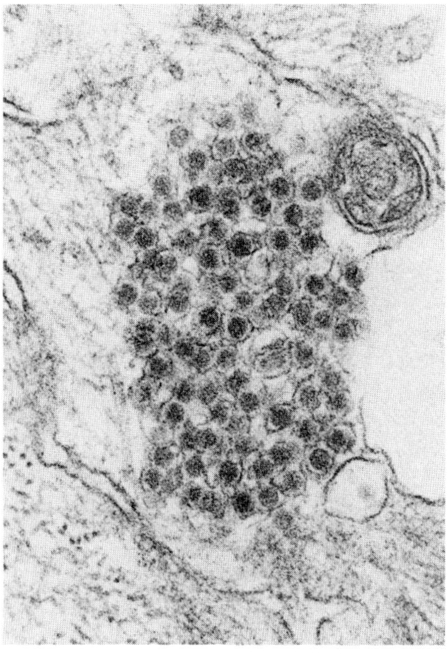

FIG. 8-9. Virions in an axon from a brain biopsy specimen obtained from a patient with progressive multifocal leukoencephalopathy. Virions were numerous in some astrocyte nuclei, but small groups were also found in the processes of neurons. ×60,000.

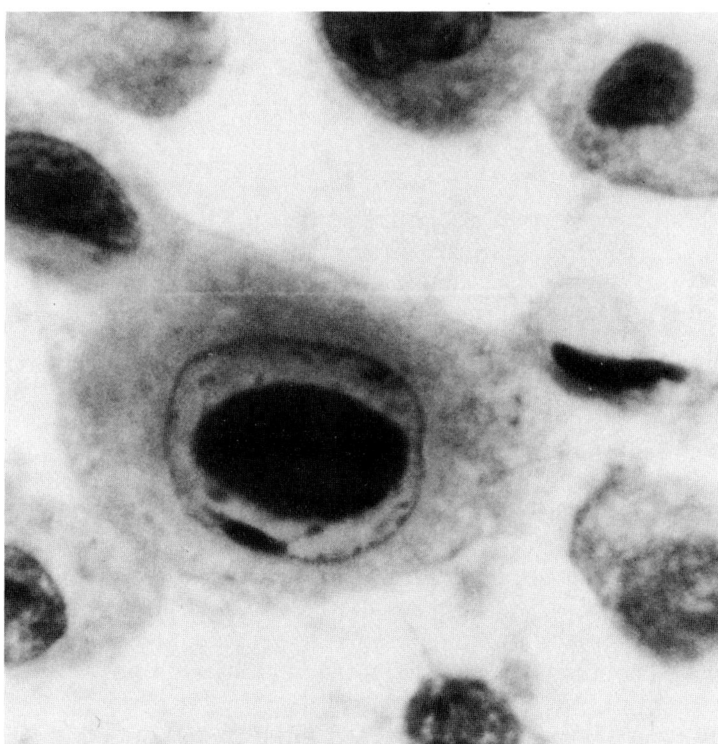

FIG. 8-10. Cytomegalovirus inclusion in a lung biopsy specimen of a renal transplant patient. Electron microscopy (Fig. 8-11) was confirmatory. ×900.

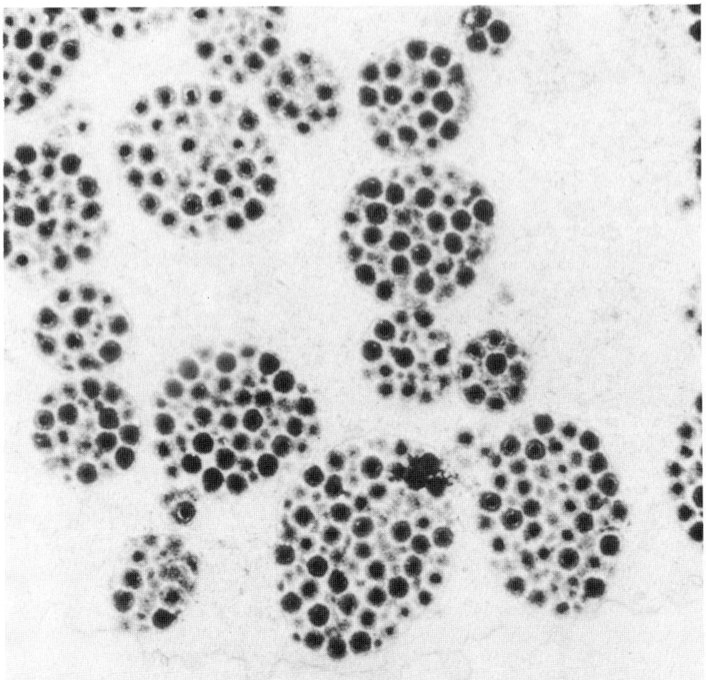

FIG. 8-11. Electron micrograph of cytomegalovirus in a lung biopsy specimen from a renal transplant patient. ×56,000.

mens, and other tissues. In this use of the EM, the virologist, the pathologist, and the electron microscopist must be a working team correlating data from a variety of sources.

It is important to be aware that while glutaraldehyde-fixed tissues are optimal for EM detection of viruses, tissues fixed in other fixatives, frozen, or embedded in paraffin or in H&E sections are also suitable for viral detection by electron microscopy.[19,48–51]

Electron Microscopy in Monitoring Cell Cultures

Electron microscopic examination of cell cultures or media helps the diagnostic virology laboratory to monitor primary cell cultures and cell lines for contaminating viruses and to detect cultured viruses that do not produce cytopathic effect.[30] The examination of cell cultures and fluid has been found to be helpful in distinguishing viruses that have similar cytopathic effect (e.g., reovirus and enterovirus) and decreases the number of antisera required for identification.[30]

Scanning Electron Microscopy in Diagnostic Virology

Although the scanning electron microscope has been used in a number of situations,[52,53] its value in diagnostic virology is unclear.

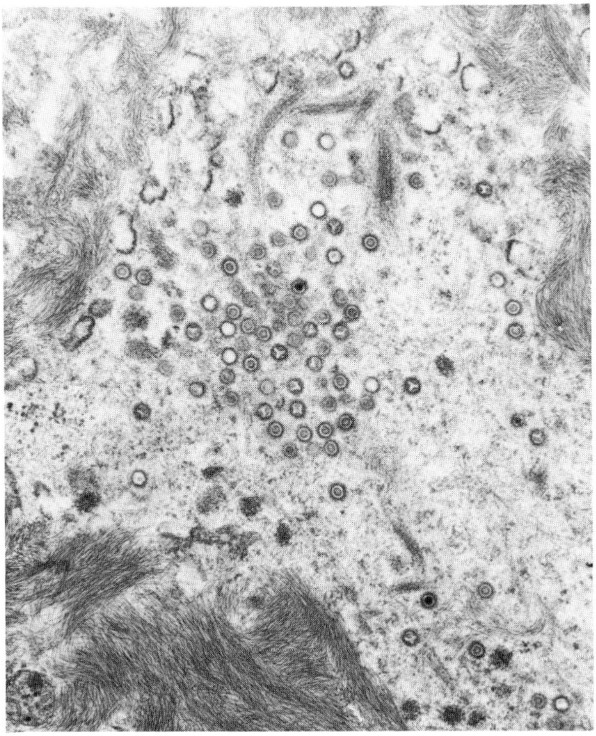

FIG. 8-12. Herpes virions in a skin biopsy specimen. ×20,900.

CONCLUSIONS

The electron microscope is a valuable adjunct tool in diagnostic virology. The disadvantages or shortcomings of electron microscopy are:

1. Lack of specificity
2. Relative insensitivity
3. Inability to process large numbers of specimens routinely

The advantages of the electron microscope in diagnostic virology are:

1. Detection of viruses that do not grow in cell cultures
2. Distinction between viruses that produce similar cytopathic effect in cell cultures
3. Quality control for contaminants in cell cultures
4. Not restricted to the specificity of immunologic rapid diagnostic procedures
5. When direct visualization of a virus in a specimen is positive, it is a very rapid diagnostic test
6. Detection of incomplete or noninfectious particles
7. Detection of virus particles in poorly transported specimens, fixed tissues, paraffin-embedded tissues, H&E sections

8. Immune EM with patient convalescent sera may identify viruses related to diseases of uncertain etiology
9. Viral detection in heavily contaminated specimens such as feces

It is clear that because the advantages are considerable, electron microscopy is destined to become a part of every routine viral diagnostic laboratory.

REFERENCES

1. **Davis CP, Cleven D, Brown J, Balish E:** Anaerobiospirillium: A new genus of spiral shaped bacteria. Int J Systemic Bacteriol 26:498–504, 1976
2. **Thomason BM, Ewing EP, Hecklin MD, et al.:** Tatlock bacterium (Pittsburgh pneumonia agent) presumptively identified in five cases of pneumonia. Ann Int Med 92:5–10, 1980
3. **Hibert GA, Moss CW, McDougal LK, et al.:** The Rickettsia-like organisms TATLOCK (1943) and HEBA (1959): Bacteria phenotypically similar to lut genetically distinct from legionella pneumophila and the WIGA bacterium. Ann Int Med 92:45–52, 1980
4. **O'Meara A, Deasy PF, Hillary IB, Bridgen WD:** Acyclovir for treatment of mucocutaneous herpes infection. Lancet 1:1196, 1979
5. **Kimberlin RH, Walker CA:** Antiviral compound effective against experimental scrapie. Lancet 2:591–592, 1979
6. **Buge A, Rancurel G, Metger J, et al.:** Isoprinosine in treatment of acute viral encephalitis. Lancet 2:691,1979
7. **Roogen CJ, Scott GD:** Smallpox diagnosis with special reference to the electron microscopy. Can J Public Health 39:467–477, 1948
8. **Cruickshank JG, Bedson HS, Watson DH:** Electron microscopy in the rapid diagnosis of smallpox. Lancet 2:527–530, 1966
9. **Nagler FPO, Rake G:** The use of the electron microscope in diagnosis of variola, vaccinia and varicella. J Bacteriol 55:45, 1948
10. **Nagington J:** Electron microscopy in differential diagnosis of poxvirus infections. Br Med J 2:1499–1500, 1964
11. **Almeida JD, Howatson AF, Williams MG:** Morphology of varicella (chickenpox) virus. Virology 16:353–355, 1962
12. **Ruiz-Palacios G, Pickering LK, Van Eys J, Conklin RH:** Disseminated herpes simplex with hepatoadrenal necrosis in a child with acute leukemia. J Pediat 91:757–759, 1977
13. **Smith KO, Melnick JL:** Recognition and quantitation of herpesvirus particles in human vesicular lesions. Science 137:543–544, 1962
14. **Evans AS, Melnick JL:** Electron microscope studies of the vesicle and spinal fluids from a case of herpes zoster. Proc Soc Exp Biol Med 71:283–286, 1949
15. **Lee FK, Macris MP, Nahmias AJ:** Electron microscopy for diagnosing viral infection. Lab Management 35–39, 1980
16. **Doane FW, Anderson N, Chatiyanouda K, et al.:** Rapid laboratory diagnosis of paramyxovirus infections by electron microscopy. Lancet 2:751–753, 1967
17. **Portnoy BL, Conklin RH, Menn M, et al.:** Reliable identification of reovirus-like agent in diarrheal stools. J Lab Clin Med 89:560–563, 1977
18. **Conklin RH, DuPont HL, Goldschmidt MC, Rodriguez T:** Occurrence of viral particles in diarrhea: Houston, Texas and Guatemala. N Engl J Med 292:644–645, 1975
19. **Flewett TH, Bryden AS, Davies H:** The viral flora of feces as seen by electron microscopy. J Clin Pathol 27:603–614, 1974

20. **Lee FK, Nahmias AJ, Stagno S:** Rapid diagnosis of cytomegalovirus infection in infants by electron microscopy. N Engl J Med 299:1266–1270, 1978
21. **Henry C, Hartsock RJ, Kirk Z, Behrer R:** Detection of viruria in cytomegalovirus infected infants by electron microscopy. Am J Clin Pathol 69:435–439, 1978
22. **Montplaisir S, Belloncik S, Leduc NP, et al.:** Electron microscopy in the rapid diagnosis of cytomegalovirus: Ultrastructural observation and comparison of methods of diagnosis. J Infect Dis 125:533–538, 1972
23. **Williams MG, Howatson AF, Almeida JD:** Morphological characterization of the virus of the human common wart (Verruca vulgaris). Nature 189:895–897, 1961
24. **Conklin RH:** A case report of Orf. Hospital Practice, in press, 1981
25. **DuPont HL, Portnoy BL, Conklin RH:** Viral agents and diarrheal illnesses. Annu Rev Med 28:167–177, 1977
26. **Appleton H, Buckley M, Thom BT, et al.:** Virus-like particles in winter vomiting disease. Lancet 1:409–411, 1977
27. **Caul EO, Paver WK, Clarke SKR:** Coronavirus particles in feces from patients with gastroenteritis. Lancet 1:1192, 1975
28. **Flewett TH, Bryden AS, Davies H, et al.:** Relationship between viruses from acute gastroenteritis of children and newborn calves. Lancet 2:61–63, 1974
29. **Flewett TH, Boxall E:** The hunt for viruses in infections of the alimentary system. Clin Gastroenterol 5:359–385, 1976
30. **Doane FW, Anderson N, Zbitnew A, Rhodes AJ:** Application of electron microscopy to the diagnosis of virus infections. Can Med Assoc J 100:1043–1049, 1969
31. **Smith KO:** Preparation of viruses for electron microscopy. In Busch H (ed): Methods in Cancer Research. New York, Academic Press, 1967, vol 1, pp 562–567
32. **Sharp DG:** Sedimentation counting of particles via electron microscopy. In Proc Intern Congr Electron Microscopy, 4th Congr. Berlin, 1958, pp 542–548
33. **Smith KO, Melnick JL:** A method for staining virus particles and identifying their nucleic acid type in the electron microscope. Virology 17:480–490, 1962
34. **Bayer M, Blumberg S, Werner B:** Particles associated with Australian antigen in the sera of patients with leukemia, Down's syndrome and hepatitis. Nature 218:1057, 1968
35. **Gelderblom H, Reupke K:** Rapid viral diagnosis using the airfuge. Abst Fourth Int Cong Virol, p 630, 1978
36. **Madeley CR, Cosgrove BP:** Viruses in infantile gastroenteritis. Lancet 2:124, 1975
37. **Kapikian AZ, Dienstag JL, Purcell RH:** Immune electron microscopy as a method for the detection, identification, and characterization of agents not cultivable in an in vitro system. In Rose NR, Friedman H (eds): Manual of Clinical Immunology. Washington, D.C., American Society for Microbiology, 1976
38. **Anderson TF, Stanley WM:** Study by means of the electron microscope of reactions between tobacco mosaic virus and antiserum. J Biol Chem 139:339–344, 1941
39. **Almeida JD, Waterson AP:** The morphology of virus-antibody interaction. Adv Virus Res 15:307–338, 1969
40. **Doane FW:** Identification of viruses by immunoelectron microscopy. In Kurstak E, Morrisset R (eds): Viral Immunodiagnosis. New York, Academic Press, 1974, pp 237–255
41. **Kapikian AZ, Feinstone SM, Purcell RH, et al.:** Detection and identification by immune electron microscopy of fastidious agents associated with respiratory illness, acute nonbacterial gastroenteritis and hepatitis A. Perspect Virol 9:9–45, 1975
42. **Best JM, Banatvala JE, Almeida JD, Waterson AP:** Morphological characteristics of rubella virus. Lancet 2:237–239, 1967
43. **Kapikian AZ, Wyatt RG, Polin R, et al.:** Visualization by immune microscopy of a 27 nm particle associated with acute infectious non-bacterial gastroenteritis. J Virol 10:1075–1081, 1972

44. **Feinstone SM, Kapikian AZ, Purcell RH:** Hepatitis A: Detection by immune electron microscopy of a virus-like antigen associated with acute illness. Science 182:1026–1028, 1973
45. **Chou SJ, Burrell R, Harley J, et al.:** Subacute focal adenovirus encephalitis, abstracted. J Neuropathol Exp Neurol 31:173, 1972
46. **Bastian FO, Wende RO, Singer DB, Zeller RS:** Eastern equine encephalomyelitis: Histopathologic and ultrastructural changes with isolation of the virus in a human case. Am J Clin Pathol 64:10–14, 1975
47. **Breitfield V, Hashida Y, Sherman FE, et al.:** Fatal measles infection in children with leukemia. Lab Invest 28:279, 1973
48. **Pinkerton H, Carroll S:** Fatal adenovirus pneumonia in infants: correlation of histologic and electron microscopic observations. Am J Pathol 65:543–546, 1971
49. **Zimmerman LE, Font RL, Ts'o M, Fine BS:** Application of electron microscopy to histopathologic diagnoses. Trans Am Acad Ophthalmol Otolaryngol 76:101–107, 1972
50. **Blank H, Davis C, Collins C:** Electron microscopy for the diagnosis of cutaneous viral infections. Br J Dermatol 83:69–81, 1970
51. **Yunis EJ, Atchison RW, Michaels RH, DeCicco FA:** Adenovirus and ileocaecal intussusception. Lab Invest 33:347–351, 1975
52. **Aranzi C, Fenters J, Tolkacz J:** Examination of virus infected culture cells by scanning electron microscopy. Appl Microbiol 20:633–637, 1970
53. **Enlander D, Drew L, Hoo R, et al.:** The cytopathic effect of herpes simplex virus on Hep-z cells as shown by scanning electron microscopy. J Gen Virol 25:313–316, 1974

9

ELECTRON MICROSCOPY IN CYTODIAGNOSIS

NELSON G. ORDÓÑEZ

Cytologic specimens for ultrastructural study are generally in the form of fluids. With the exception of cerebrospinal fluid, there is usually an abundant volume containing a large number of cells for study. Several simple techniques of filtration and centrifugation are available for cells that are suspended in fluids (see Chapter 2). Other procedures and modifications of the standard techniques have been developed for processing small specimens.[1,2] Differential flotation techniques using solutions of different densities, such as bovine serum, albumin, and Ficoll-Hypaque, have been successfully applied to mixtures of benign and malignant cells to separate them into more homogeneous fractions.[2–4] These techniques are particularly helpful when the ratio of malignant to normal cells is so low that the former are unlikely to be detected in routine preparations. Techniques are also available for the retrieval of specific cells from Papanicolaou-stained smears on glass slides for electron microscopy. These procedures have proved useful for the identification of viral particles, permitting a diagnosis of viral infection before the results of cultures become available.[5–7]

The advantages of electron microscopy over light microscopy are increased magnification and greater resolution. Cells obtained from serous fluids usually are of a variety of types. Light microscopic cytologic diagnosis is mainly based on the nuclear-cytoplasmic ratio, nuclear shape, arrangement of the nuclear chromatin, and the size and location of the nucleolus. Limited diagnostic information is obtained from the cytoplasm and cell membrane. Although the light microscopic examination of cytologic preparations stained with Papanicolaou, Giemsa, and hematoxylin and eosin stains in most instances permits a distinction between benign and malignant cells, it does not consistently determine the precise type or origin of malignant tumors. Furthermore, some benign cells may be atypical and present a diagnostic problem.

Under the electron microscope, on the other hand, the cytoplasm appears as a highly complex structure, rich in organelles and a variety of inclusions. The ultrastructure of the cytoplasm of cancer cells varies with the site of origin, histologic type, and degree of differentiation of the tumor. In some cases where cells are derived from an effusion of a tumor having characteristic cytoplasmic inclusions (such as a melanoma), a specific tumor diagnosis can be made. Electron microscopy can also be useful in the identification of tumors that have specialized cell organelles, such as endocrine tumors and some sarcomas. With the electron microscope, specific types of cell junctions can be identified. For instance, well-developed desmosomes are often present in epithelial tumors, whereas no cell junctions are seen in lymphomas.

A few short microvilli can regularly be seen distributed over the surface of different types of benign cells, but with only a few exceptions (such as in oat cell carcinoma), the presence of numerous regular microvilli is a diagnostic feature of carcinoma (or mesothelioma) cells, regardless of the anatomic origin and histologic type.[8] In some instances, the base of the microvillus is broader than the tip, but branching and fusion are not infrequent.[9,10] Although these structures can be recognized by transmission electron microscopy, they are better visualized using the scanning electron microscope, which, in addition to providing a three-dimensional view, permits the examination of much of the cell surface (Figs. 9-1 and 9-2).[8,9,11]

As mentioned above, an application of electron microscopy in the diagnosis of effusions lies in distinguishing atypical cells from cancer cells. The two types of benign cells commonly confused with cancer are macrophages (histiocytes) and mesothelial cells. By electron microscopy, the macrophage has an irregular outline showing elongated cytoplasmic projections. The nucleus is indented and generally eccentrically located with peripheral margination of dense chromatin. The cytoplasm contains numerous lysosomes that reflect its phagocytic activity. A small Golgi apparatus, lipid droplets, mitochondria, and short profiles of rough endoplasmic reticulum are also present (Fig. 9-3). Some histiocytes may have multiple small cytoplasmic vacuoles that, on occasion, fuse to form a single vacuole that gives the cells a signet

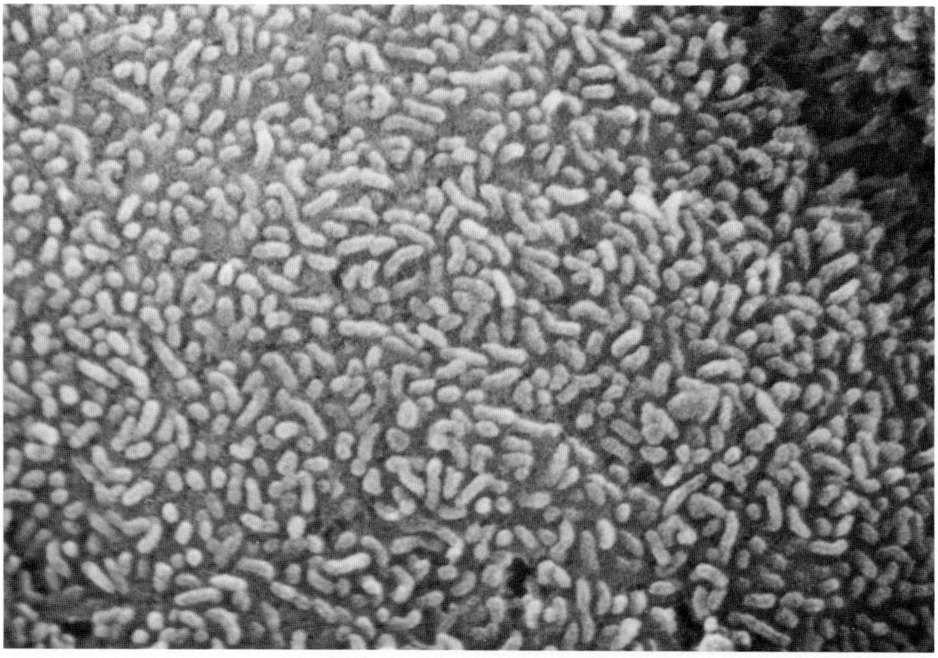

FIG. 9-1. Scanning electron micrograph of breast carcinoma cells showing dense irregular microvilli. X17,000.

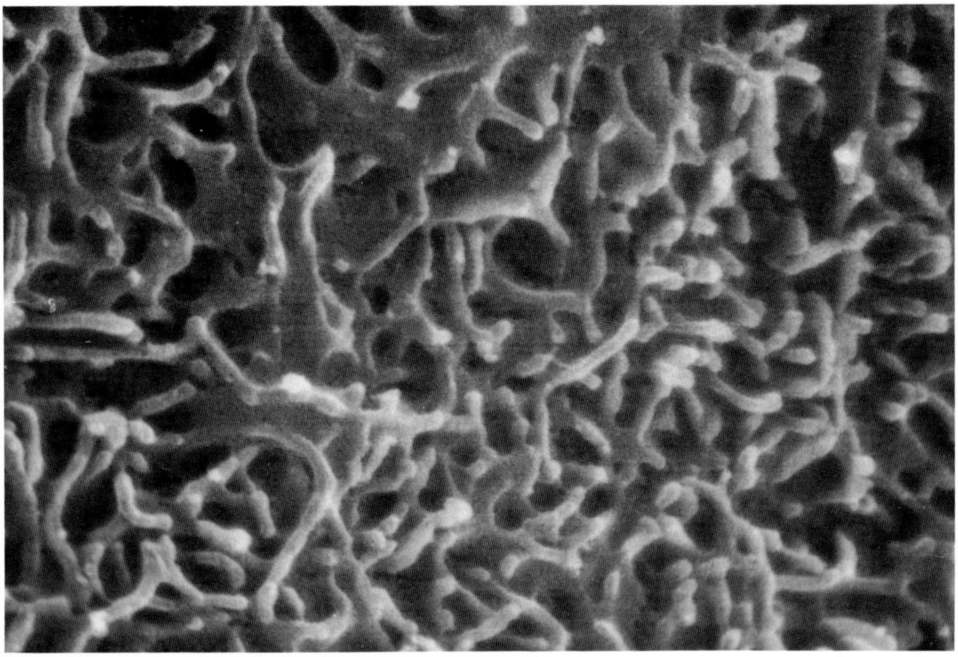

FIG. 9-2. Scanning electron micrograph of an ovarian adenocarcinoma cell demonstrating its markedly irregular surface. X19,500.

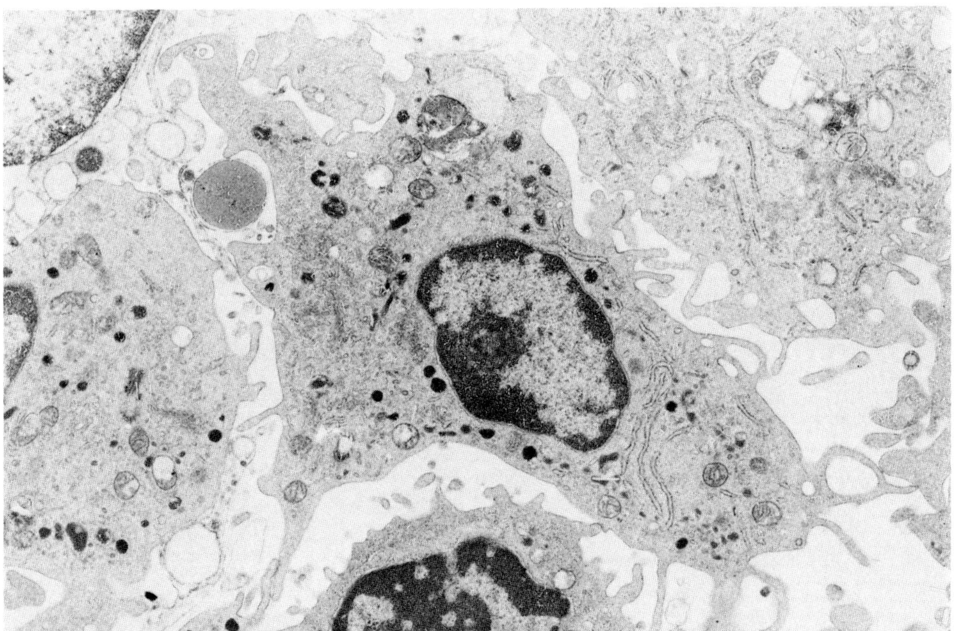

FIG. 9-3. Macrophage from pleural fluid. The nucleus is eccentric with marginated dense chromatin and many cytoplasmic processes. X5900.

ring-like appearance. These cells may form small groups and simulate adenocarcinoma. By scanning electron microscopy, the macrophages appear covered with irregular ridges or ruffles of varying size, shape, and width (Fig. 9-4).

MESOTHELIAL CELLS

The mesothelial cells lining serous body cavities are flattened. However, in effusions they become round or oval in shape and have a central round nucleus with a homogeneously distributed chromatin pattern and a small nucleolus. The cytoplasm is abundant. The most characteristic cytoplasmic feature is the accumulation of concentrically arranged, fine filaments, usually restricted to the inner half of the cytoplasm. Elongated mitochondria, many ribosomes, abundant rough endoplasmic reticulum, and a prominent Golgi apparatus are mainly located in the outer part of the cytoplasm. Sparse, short microvilli may cover the cell surface (Figs. 9-5 and 9-6). In some reactive and atypical mesothelial cells, there is an accumulation of intracytoplasmic glycogen. The nucleolus can be prominent and the cytoplasmic outline slightly irregular, with blunt, thick, short cytoplasmic processes (Fig. 9-7).[11,12] Mesothelial cells are sometimes arranged in groups joined by cell junctions, and are thus easily confused with adenocarcinoma. Adenocarcinoma cells, however, may have an irregular nuclear outline and nuclear chromatin pattern, numerous cytoplasmic vac-

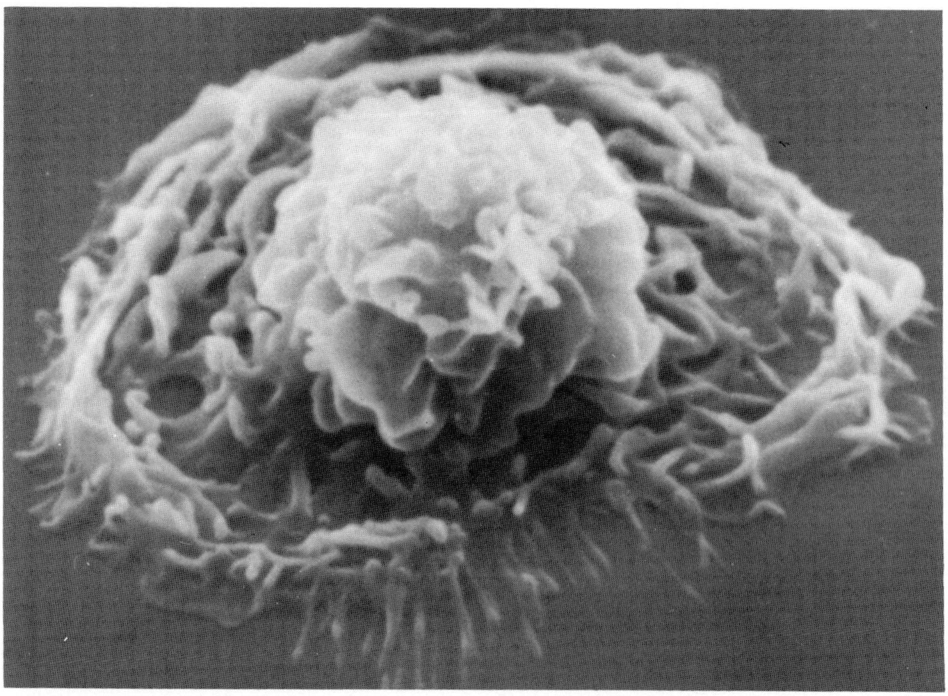

FIG. 9-4. Scanning electron micrograph of a macrophage demonstrating its irregular surface with cytoplasmic ridges of different widths. X8,300. (Courtesy of Sherrie Heston and Shashi Mehta.)

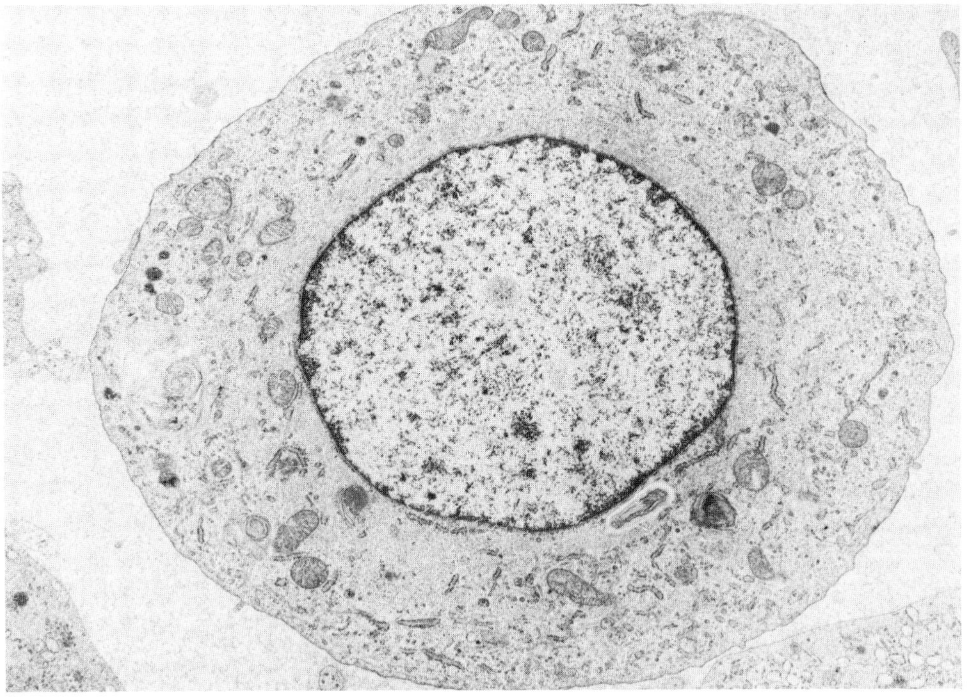

FIG. 9-5. A mesothelial cell with central nucleus and evenly distributed chromatin. Note the few short cytoplasmic processes. X6500.

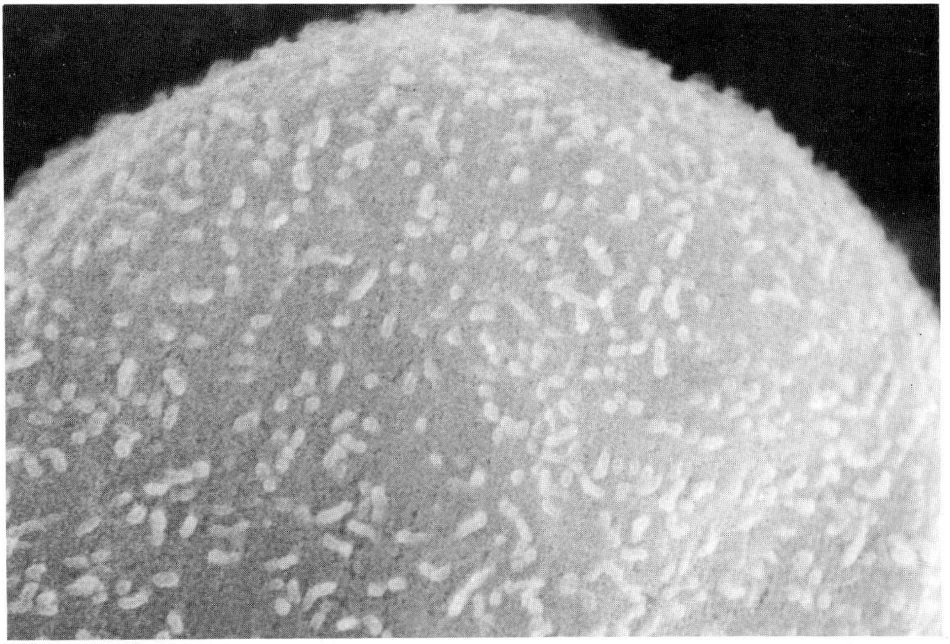

FIG. 9-6. Scanning electron micrograph of a mesothelial cell. Its surface is covered by short microvilli. X19,500.

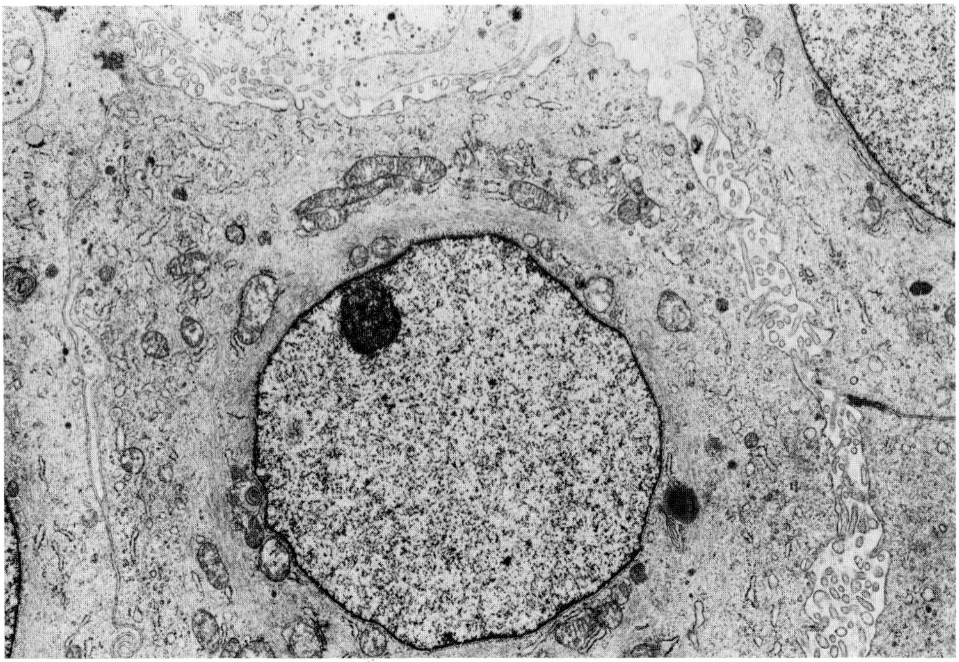

FIG. 9-7. Reactive mesothelial cells showing perinuclear fibrils, prominent nucleolus, numerous cytoplasmic processes, and cell junctions. X6900.

uoles containing lipid and sometimes mucin, many uniform microvilli covering the entire or a portion of the cell surface, and, on occasion, an intracytoplasmic lumen (Fig. 9-8). Electron microscopic features may thus permit a distinction between malignant cells and atypical mesothelial cells.

NEEDLE BIOPSIES

In recent years, the use of fine needle aspiration biopsy of masses has expanded considerably in the United States (see Chapter 2). This procedure has been demonstrated to be a rapid and convenient way of obtaining tumor cells for both cytologic smears and electron microscopic studies. The specimens obtained in this way are generally well preserved, despite some mechanical disruption of cells (Figs. 9-9 and 9-10). Needle biopsies of tumors involving bone usually require decalcification, and the cells are more prone to artifactual distortion (Fig. 9-11). Different techniques for the handling and processing of these specimens have been successfully used to minimize cellular artifactual changes.[13] Since one of the advantages of electron microscopy is the need for only a small amount of tissue, fine needle aspiration has proved particularly well suited for providing material for ultrastructural study, and is a powerful diagnostic tool when combined light and electron microscopic studies are performed.

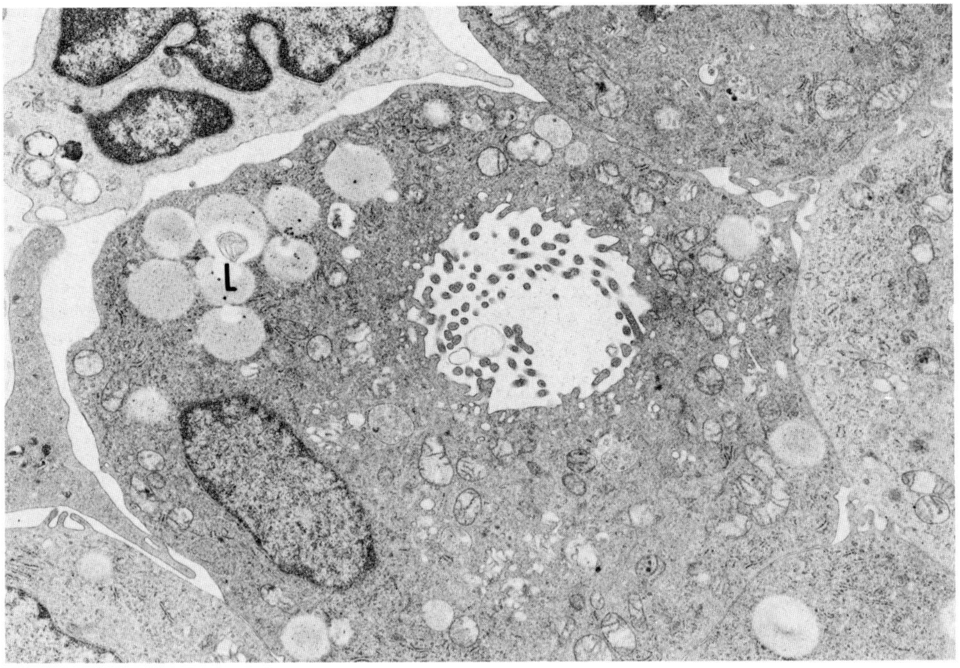

FIG. 9-8. Metastatic breast carcinoma in pleural fluid. The cell in the center contains an intracytoplasmic space lined by microvilli as well as numerous lipid droplets (L). Portion of a macrophage is seen in the left upper corner. X7600.

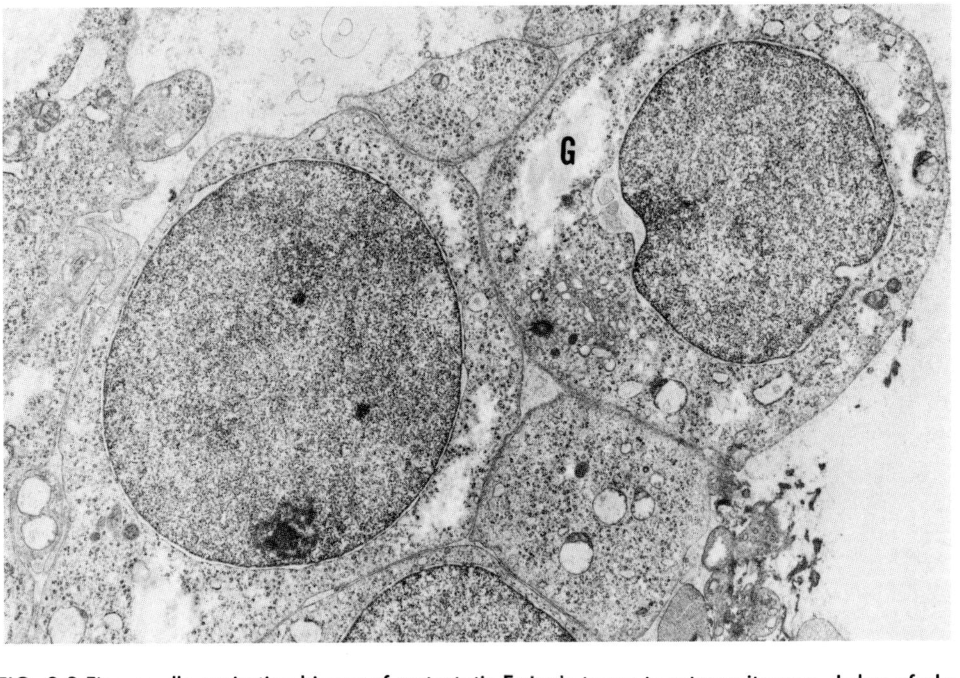

FIG. 9-9 Fine needle aspiration biopsy of metastatic Ewing's tumor to retroperitoneum. Lakes of glycogen (G) appear as clear zones in the cells. X7700.

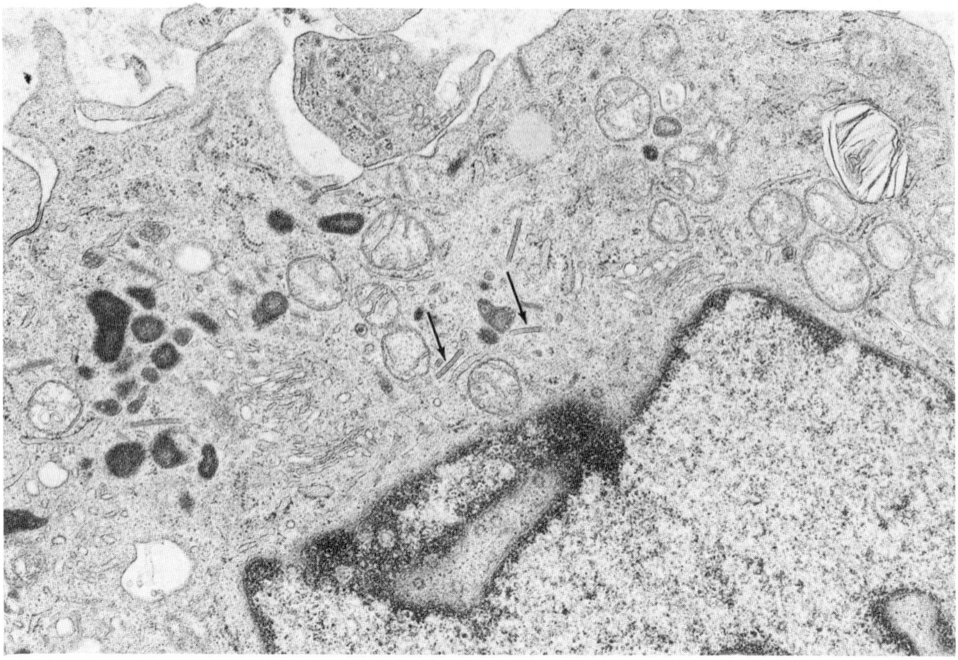

FIG. 9-10. Histiocytosis X cell showing Langerhans granules (arrows). X14,000.

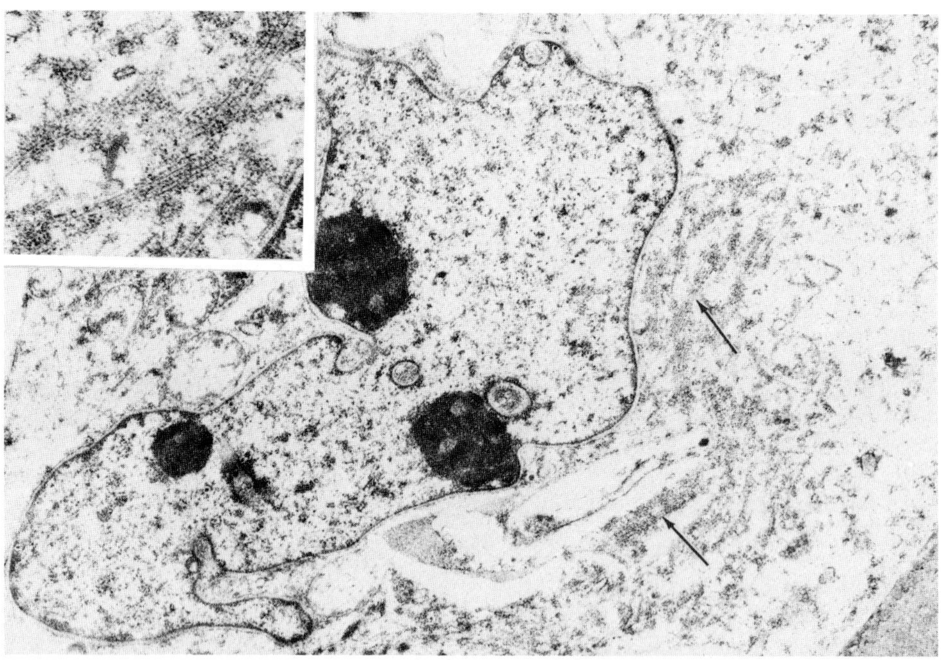

FIG. 9-11. Needle aspiration biopsy of a metastatic embryonal rhabdomyosarcoma from bone marrow. In spite of the moderate artifactual distortion, myofilaments are visible in the cytoplasm (arrows). X4,100. Insert: Note the myofilaments. X29,000.

REFERENCES

1. **Kolmell H:** A method for concentrating cerebrospinal fluid cells. Acta Cytol (Baltimore) 21:154, 1977
2. **Elequin FT, Muggia FM, Ghossein N, Schreiber K:** A quick method for concentrating and processing cancer cells from serous fluids and fine needle aspirates. Acta Cytol (Baltimore) 21:596, 1977
3. **Mavligit GM, Gutterman JU, Hersh EM:** Separation of viable from non-viable tumor cells using Ficoll-Hypaque density solution. Immunol Commun 2:463, 1973
4. **Katz RL, Lukeman JM:** The comparative diagnostic accuracy of cancer-cell detection with Ficoll-Hypaque gradiant separation and standard centrifugation techniques. Am J Clin Pathol, 74:18, 1980.
5. **Coleman DV, Russell WJ, Hodgson J, et al.:** Human papovavirus in Papanicolaou smears of urinary sediment detected by transmission electron microscopy. J Clin Pathol 30:1015, 1977
6. **Coleman DV:** Cytological diagnosis of virus infected cells in Papanicolaou smears and its application in clinical practice. Am J Clin Pathol 32:1075, 1979
7. **Hills E, Laverty CR:** Electron microscopic detection of polyoma virus particles in selected koilocytotic cells in routine cervical smears. Acta Cytol (Baltimore) 23:53, 1979
8. **Domagala W, Woyke S:** Transmission and scanning electron microscopic studies of cell effusions. Acta Cytol (Baltimore) 19:214, 1975
9. **Domagala W, Koss LG:** Configuration of surfaces of human cancer cells in effusions: A scanning electron microscopic study of microvilli. Virchows Archiv [Cell Pathol] 26:27, 1977
10. **Legrand M, Pariente R:** Electron microscopy in the cytological examination of metastatic pleural effusions. Thorax 31:443, 1976
11. **Gondos B, Lai CE, King EB:** Distinction between atypical mesothelial cells and malignant cells by scanning electron microscopy. Acta Cytol (Baltimore) 23:321, 1979
12. **Murad TQ:** Electron microscopic study of cells in pleural and peritoneal effusions. Acta Cytol (Baltimore) 12:401, 1973
13. **Mackay B, Osborne BM:** The contribution of electron microscopy to the diagnosis of tumors. Pathobiol Annu 8:359, 1978

10

TUMOR DIAGNOSIS

BRUCE MACKAY

Transmission electron microscopy is being used with increasing frequency to aid in the identification of tumors that cannot be classified by routine light microscopy. Its application is, at the present time, restricted by the limited information available on ultrastructural diagnostic criteria for many tumors. Accumulation of the necessary data is complicated by the breadth of the ultrastructural spectrum that most neoplasms display and the degree of overlap in their fine structure. Single case reports are of limited value in delineating the range of morphology of a particular tumor; one need only cite the light microscopy of melanoma as a comparable example. In addition, published reports describing the ultrastructure of a tumor depend for their validity on a correct diagnosis by light microscopy. The process of gathering information or tumor ultrastructure, determining the range of morphology for a particular tumor type, and distilling criteria that are reliable and reproducible for use in diagnosis will take some time and require the contributions of many pathologists who are willing to study, analyze, and report groups of cases. In the meantime, those using the electron microscope for tumor diagnosis must constantly be mindful of the range of variation and the relative lack of specificity of many of the ultrastructural features of tumor cells.

Used with appropriate caution, the transmission electron microscope is a powerful diagnostic tool that can lead to the identification of many neoplasms that by light microscopy would either remain unclassified or be incorrectly labelled. It does have its limitations. Sometimes it is possible to decide from electron microscopy that a tumor is benign or malignant, but other than select findings, such as the absence of a basal lamina in an infiltrating carcinoma, the criteria are essentially those employed by the cytologist or pathologist at the light microscopy level, where the assessment is based on many more cells and is probably more accurate. A nonneoplastic cell, such as a macrophage or reactive fibroblast, can present a bizarre and alarming ultrastructural appearance, while in contrast there are many aggressive neoplasms whose cells have bland nuclei, small nucleoli, and a generally innocuous histology. The electron microscopist should therefore be wary of concluding that a particular tumor is benign or malignant solely on the basis of the ultrastructural observations.

It is difficult to provide a precise figure that would indicate the relative usefulness of electron microscopy in tumor diagnosis, but it can be said that ultrastructural study provides at least some information about tumors that cannot be confidently identified by light microscopy. This information is not always specific, since some tumors, especially when poorly differentiated, do not have sufficiently distinctive features. But in this situation, a diagnosis of primitive neoplasm can be rendered without the uncertainty that would persist from light microscopy alone. Much of the time, the contribution of the electron microscope is to narrow a differential diagnosis.

In many laboratories, it is the practice for the pathologist to leave the preparation, viewing, and photography of diagnostic specimens to a competent technologist who provides electron micrographs from which the pathologist attempts to derive the necessary diagnostic information. While many technologists are knowledgeable in the field of histology, a pathologist will usually be better able to appreciate deviations from normal tissue structure and to assess their significance in differential diagnosis. Actual viewing of the thick and thin sections by a pathologist is therefore strongly recommended.

The quality of results from a diagnostic study is directly proportional to the quality of the material that is provided for the study. One advantage of electron microscopy is the small amount of tissue necessary. However, this tissue must be representative, and should be correctly procured, fixed, and processed to yield optimal return. Interpretation based on artifactually distorted, degenerating, or poorly preserved cells is, to say the least, hazardous. It may be worth subjecting a patient to an additional biopsy procedure to obtain acceptable tissue rather than attempting to interpret inadequate material. Efforts directed toward the routine procurement of properly handled, well-preserved tumor tissue specimens will reap dividends in the quality of the electron microscopic preparations they produce, and in the information that can be derived from them.

APPROACH TO THE STUDY OF A TUMOR SPECIMEN

There is no hard and fast rule for the amount of tissue that should be taken from a tumor specimen for electron microscopic processing. Little is gained by taking large quantities from a single area of a neoplasm though sampling of a large soft tissue tumor from several sites for both light and electron microscopy can be useful when variations in histology are present. Our routine practice is to embed five cubes of tissue from a tumor, each about 1 mm, and to place the remaining tissue, generally not exceeding 10 additional blocks, in a flat embedding dish from which it can be retrieved on the few occasions that this proves necessary. When it can be anticipated that the specimen may not be representative, it is advantageous to embed the entire tissue in capsules. The delay occasioned by having to reembed pieces of additional tissue can then be avoided. It is possible to cut out tissue cubes from the surplus tissue block and mount them with rapid epoxy on tooled ultramicrotome slugs, but this too is inconvenient and time-consuming.

The first step in the diagnostic study is the examination of the semithin sections. Our practice is to prepare them on two blocks from each case, cutting more blocks only when the initial sections are, for any reason, not acceptable. Study of the semithin sections provides reassurance that the block selected for thin sections is representative, and that the tissue it contains is adequately preserved. It also gives information on the composition of the specimen. When a block is only partially representative, the significant area must be retained as the block is trimmed (see Chapter 2). This information can be conveyed to the technician verbally, by diagram, or by simultaneous viewing of the semithin section. A double-head light microscope is useful in the EM laboratory, since pathologist and technician can examine the sections together and discuss each case, agreeing on any precautions that need to be observed during trimming or sectioning. The technician can note blocks that contain areas of fat or hard tissue and anticipate problems in sectioning. A separate diamond knife, or area of knife edge, can be reserved for cutting blocks that contain particularly hard areas. It is a good policy to encourage technicians to examine the semithin sections. Their interest is stimulated, their sense of participation heightened, and their capability extended, so that they can perform the review of the semithin sections on occasions when the pathologist is not available, thereby expediting the diagnostic process.

Before studying the thin sections with the electron microscope, the pathologist should be aware of the composition of the particular block from which the thin sections were prepared. Is it purely tumor, or is there an admixture of other components, such as inflammatory cells, connective tissue and fibroblasts, or vessels? While this may be obvious by light microscopy, it can often be difficult to identify individual neoplastic cells in the electron microscope when they are admixed with cells of other types. Inflammatory cells can distort the appearance of hematopoietic or lymphopoietic neoplasms; plump endothelial cells can be confused with those of an epithelial tumor; and reactive fibroblasts can simulate sarcoma cells.

When the thin section is placed in the electron microscope, it is useful to first scan the grid at low magnification, mentally recalling the thin section in order to obtain some orientation. When the tissue is only partially representative, or if the quality of the tissue or section varies, suitable areas can be selected for more detailed study. In some instances it is convenient to examine the grid square by square, as when the tissue is a uniform population of similar appearing cells and features necessary to establish the diagnosis are sparse; e.g., rare cell junctions, occasional small acini in an otherwise solid carcinoma, rare dendritic processes, or premelanosomes in a predominantly amelanotic melanoma.

If a pathologist examines the thin section, diagnostic information can often be obtained during the course of viewing, and a verbal report rendered immediately, followed later by the written report. Only a few representative photographs need then be taken for documentation. In some difficult cases it is necessary to defer the diagnostic decision until micrographs have been prepared and studied. An automatic printer in the darkroom will greatly expedite their processing. The mechanism for relaying the diagnosis to the clinician responsible for the management of the patient, and for including it in the patient's medical record, should be standardized for efficiency. Our policy is that specimens that are submitted for electron microscopy only, without companion tissue for light microscopy, are reported by the pathologist responsible for the electron microscopy diagnostic service. If tissue for light microscopy is also provided, the results of the ultrastructural study are sent to the pathologist responsible for the case, who can then incorporate the findings in his report of the light microscopy, submit a supplemental report, or, at his discretion, take no further action. Where the clinician is aware that tissue has been provided for electron microscopy, a report should be submitted for the patient's medical record even when the findings are not contributory. The responsibility for determining whether a charge should be levied for the ultrastructural diagnostic study rests with the pathologist performing the electron microscopy.

Assessment of a tumor by electron microscopy requires a combination of knowledge of cell fine structure, the histologic appearance of tumors, and tumor ultrastructure. Until more extensive data are available on the range of morphology and degree of variation and overlap among human neoplasms at the ultrastructural level, experience will continue to be a valuable asset. A degree of restraint is necessary, since it is relatively easy to overinterpret ultrastructural findings. It has already been stressed that poorly preserved material or artifactually distorted tissue should be evaluated with caution or preferably not at all. The pathologist should make full use of the available information on any problem case, including the light microscopic histology and the clinical data.

Because of our currently limited knowledge of the range of fine structure of the cells of human neoplasms, and because of the considerable duplication of features among tumors of different histologic types, a detailed, precise delineation of the ultrastructure of each human tumor with a listing of diagnostic features is not possible at the present time. Similarly, a comprehensive review of the available information would of necessity be lengthy, require profuse illustration, and would be inappropriate for a short introductory book. An alternative approach will therefore be followed, in which basic features of tumor cells are considered and their value in tumor diagnosis discussed.

STROMA

Most of the significant information concerning the stroma of a tumor can be obtained by light microscopy, aided, if necessary, by such special staining procedures as the trichrome, reticulin, and alcian blue procedures. Occasionally, however, it is useful to examine the extent or composition of the intercellular tissues. The presence and quantity of collagen in a myxoid sarcoma are apparent at the ultrastructural level (Fig. 10-1), and slender sheets of collagen between cells can be identified in some tumors whose cells appear closely apposed by light microscopy. The eosinophilic tissue between groups of nuclei in a Schwann cell tumor can be seen to be predominantly cytoplasmic extensions or mixtures of cytoplasm and collagen, and in some neurogenic tumors, segment long spacing collagen is prominent in the stroma. When amyloid is present in small quantities in the stroma of an endocrine tumor, it can be identified ultrastructurally. In most instances, however, the stroma of a tumor can be adequately evaluated by light microscopy, and it is unusual to make observations of the intercellular tissues by electron microscopy that suggest or modify a diagnosis. Nevertheless, examination of the stroma should routinely be part of the study of a problem case.

CELL SURFACE

It may be difficult to detect the boundaries of tumor cells by light microscopy, but they are readily evident with the electron microscope, allowing precise assessment of the size, and, more important, the shape of the cells and the contours of their membranes.

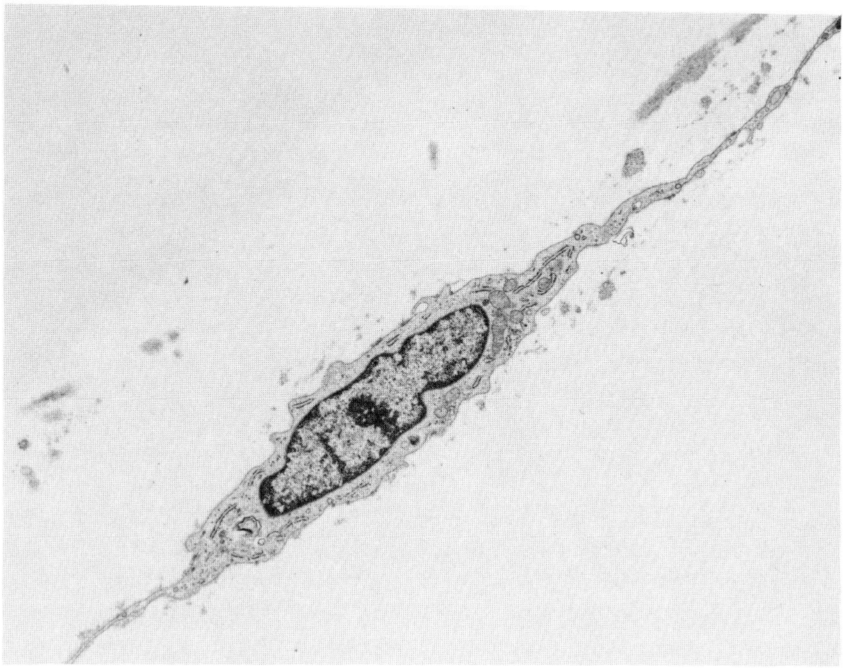

FIG. 10-1. Sparse aggregates of collagen in the stroma of a myxoid sarcoma. X4400.

Cell Profile

It is frequently the case that the more anaplastic a tumor, the more irregular the shape of its cells. Specialized structures on the cell surface, such as microvilli in an adenocarcinoma, become fewer and more variable in size and shape as the tumor dedifferentiates. However, the cells of many rapidly proliferating tumors are uniform with smooth nuclear and cytoplasmic profiles. Ewing's tumor, whether arising in bone or soft tissues, is typically composed of cells with round nuclei and smooth plasma membranes. However, the inconsistency of tumors is illustrated by cases of Ewing's tumor with irregular cell and nuclear profiles.[1,2]

Cell Apposition

Cells that appear to be compactly grouped and intimately apposed by light microscopy may be found on electron microscopy to be separated by bands of intervening collagen fibrils. When adjacent cell membranes are in contact, they may closely parallel one another, an appearance common in epithelial neoplasms. In adenocarcinomas, adjacent plasma membranes may display infoldings (Fig. 10-2) that result from interposed microvilli being sandwiched between the cells, or from true interdigitation. In a tumor of cells that are not normally cohesive, the plasma membranes are more likely to be separated by gaps that vary in width. Some large cell lymphomas can be distinguished from poorly differentiated carcinomas partly because of the relationships of their cell membranes.

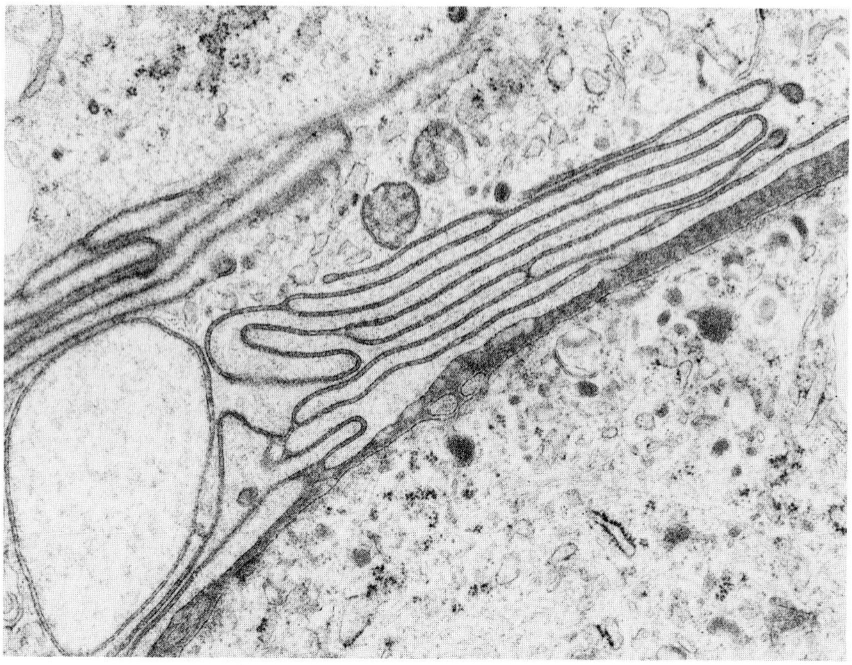

FIG. 10-2. Interdigitations of the cell membranes of adenocarcinoma cells. X19,000.

Cell Junctions

Closely allied to apposition of tumor cells is the presence of cell junctions. The most common and distinctive is the desmosome, seen to good advantage in the better differentiated squamous carcinomas where it is typically associated with bundles of tonofilaments. As a tumor whose cells are united by desmosomes dedifferentiates, the tonofilaments become short and sparse and ultimately disappear, and the intermediate line that separates the cell membranes at the desmosome is also lost. The resulting attachment site is merely thickened cell membranes. Cell junctions also decrease in size and number with loss of differentiation. The components of the junctional complex are present in adenocarcinomas in which acinar formation is occurring, but in a poorly differentiated adenocarcinoma they become reduced to scattered focal densities similar to those seen in squamous tumors.

Cell junctions are not confined to epithelial neoplasms, but it is extremely questionable whether they can occur between cells of hematopoietic or lymphopoietic proliferative disorders such as the malignant lymphomas. They can form between macrophages,[3] and we have seen a rare large cell lymphoma where small thickenings of adjacent cell membranes simulated cell junctions. Attachment specializations in these tumors, if indeed they occur, must be extremely rare and poorly developed, and a tumor with frequent cell junctions is probably not a leukemia or lymphoma. Junctions are frequently found between cells of mesenchymal tumors, though they are almost always fewer and less well developed than those of epithelial tumors. The plaque-like attachments of smooth muscle cells are present in benign smooth muscle tumors and are prominent in some leiomyosarcomas. Their presence does not relate to the behavior of the tumor, as they are often developed in metastases (Fig. 10-3), and can be scanty in primary tumors. In the epithelioid type of transformation that occurs in some gastrointestinal tract smooth muscle sarcomas, the junctions are lost, together with other smooth muscle characteristics including myofilaments, as the cells progressively assume the epithelioid appearance. Small attachment plaques, generally mere thickenings of adjacent cell membranes, occur in fibrosarcomas and malignant fibrous histiocytomas. In Schwann cell sarcomas, densities may connect the plasma membranes of cytoplasmic extensions.

Long desmosomes, often several in a series, unite the epithelial cells of a thymoma, providing a striking contrast with the lymphocytic component of the tumor. Tonofilaments are usually only prominent in scattered cells.

Basal Lamina

A well-defined basal lamina is present around groups of cells of well-differentiated carcinomas, but with dedifferentiation it fragments and is ultimately lost. Its presence, therefore, indicates a high grade carcinoma. It is rarely seen in a tumor that is infiltrating connective tissue as individual cells or small groups of cells. In mesenchymal tumors, a basal lamina forms around neoplastic Schwann cells in schwannomas and neurofibromas, and in the better differentiated Schwann cell sarcomas.

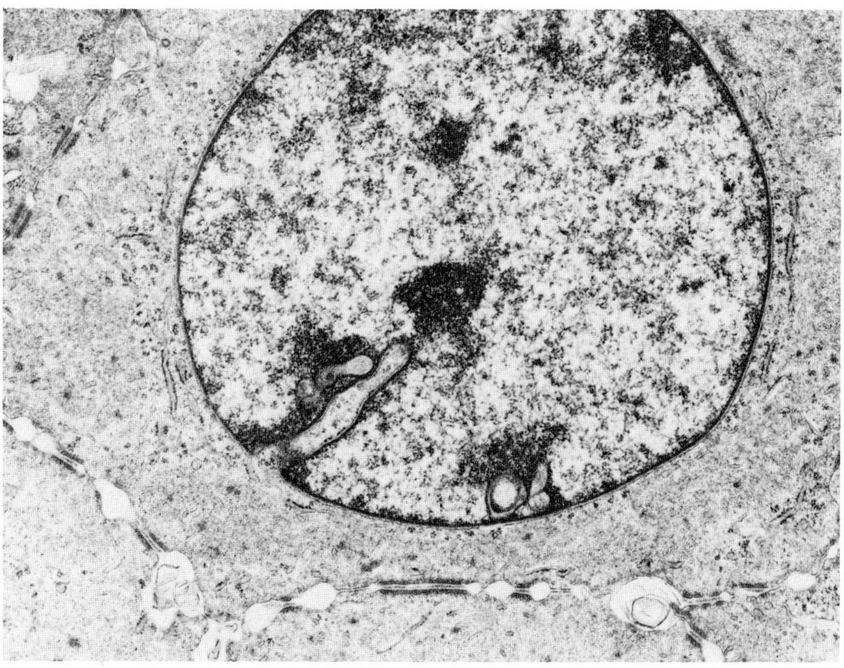

FIG. 10-3. Attachment plaques between cells of a metastatic leiomyosarcoma. X10,500.

Once the latter become solid aggregates of cells with closely apposed membranes, a basal lamina can no longer be seen. Reduplication of the basal lamina occurs in many nonneoplastic conditions, where it represents a nonspecific response to injury. In tumors this appearance is not common and is best seen in some salivary gland neoplasms. The eosinophilic zone that surrounds groups of cells in an adenoid cystic carcinoma is formed by replication of the basal lamina (Fig. 10-4), often to an extensive degree.

Cell Projections

Cell projections take various forms, ranging from short microvilli, through the longer and more elaborately structured cilia, to dendritic processes and the long cytoplasmic extensions of neurogenic tumors.

Microvilli. It is necessary to distinguish between true microvilli, which are uniform finger-like projections of the apical cell membrane exemplified by those of the absorptive cells of the small intestine (see Chapter 4), and microvillus-like protrusions of the cell surface, which are more irregular in size, shape, number, and occurrence. The latter may be observed on a wide range of neoplastic cells.

True microvilli are found on the apical surface of carcinoma cells of acinar or duct cell origin that are forming lumens, and the cells are generally united by junctional complexes. With dedifferentiation, the number and regularity of the microvilli di-

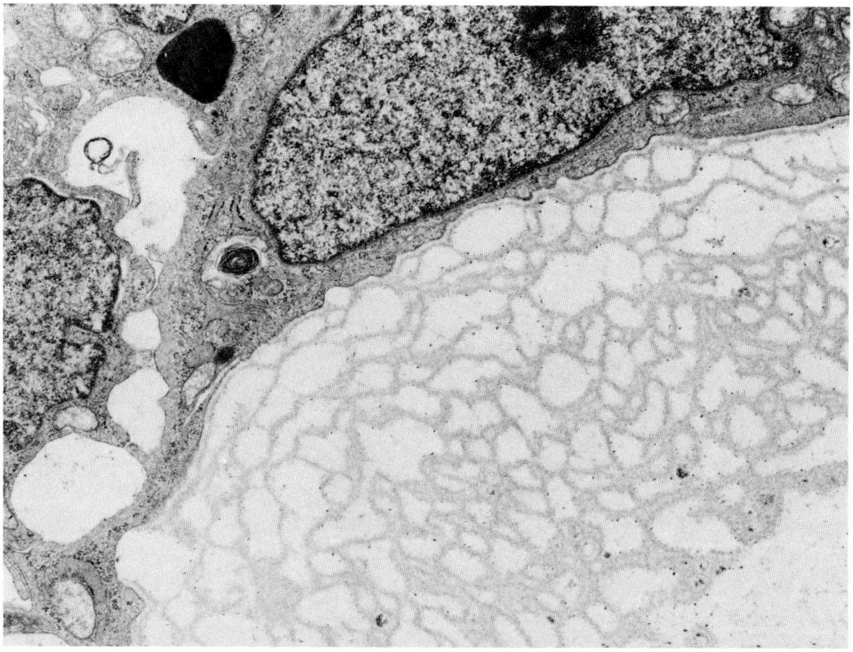

FIG. 10-4. Replication of the basal lamina in an adenoid cystic carcinoma. X10,700

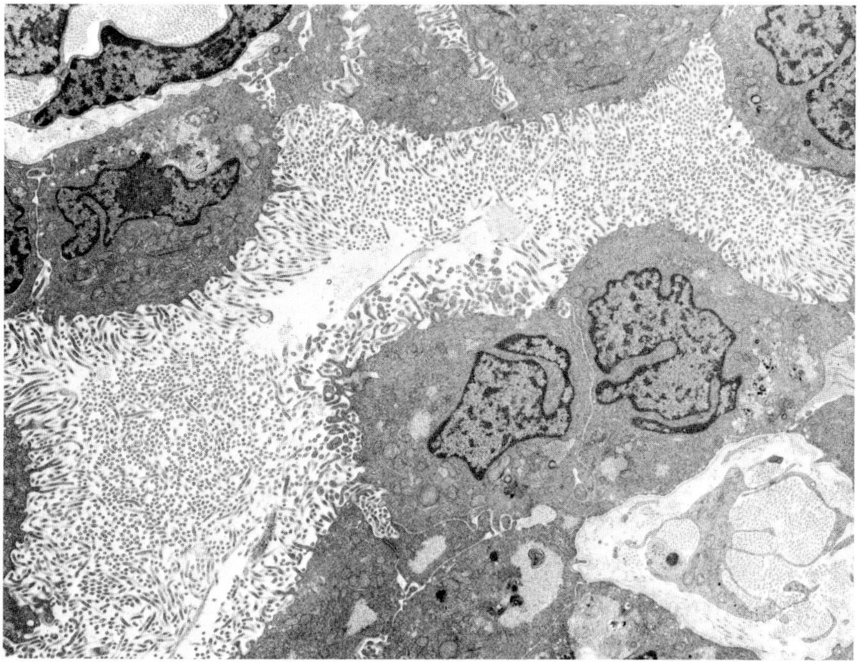

FIG. 10-5. In this well-differentiated mesothelioma, many long slender microvilli project from the apical surfaces of the cells. X4200.

minish. True microvilli are particularly well developed in mesotheliomas (Fig. 10-5), where they are long, slender, sinuous, and branching. In dedifferentiating mesotheliomas, they tend to clothe many of the free surfaces of the cells as they lose cohesion, but become progressively shorter and fewer to the point where distinction from an adenocarcinoma is difficult or impossible. In some adenocarcinomas, microvilli may be present on the lateral margins of the cells as well as on the surface facing the lumen, and in a liver cell carcinoma, microvilli may project into a slender pericapillary space that resembles the space of Disse.

Irregular, slender projections of the peripheral cytoplasm can readily simulate true microvilli and create the erroneous impression that a tumor is an adenocarcinoma. As a squamous cell carcinoma dedifferentiates, desmosomes are lost and the cells separate. Residual projections of the cell surface are often numerous but lack the organization of true microvilli. Similar projections may be seen in some melanomas and lymphopoietic tumors, including large cell lymphomas and hairy cell leukemia (leukemic reticuloendotheliosis), where they are characteristically profuse. The projections on hairy cells are best seen in blood or bone marrow preparations, but they can retract if the cells are centrifuged before they are fixed. When hairy cells pack the sinusoids of the splenic red pulp, their projections are pressed between adjacent cells. The protrusions on hairy cells occasionally are sparse or absent ("bald" hairy cells), and the diagnosis should not be excluded solely for this reason.

Microvilli of adenocarcinomas arising from the alimentary canal and its embryologic derivatives, the biliary and pancreatic duct systems, may contain microfilament cores. More cases must be studied to determine how frequently this occurs and whether it is reliable in differential diagnosis.

Cilia. The system of microtubules that forms the axoneme of a cilium is produced by the centriole. It is a common finding to observe a centriole in a tumor cell sprouting microtubules either within the cytoplasm or projecting from the surface covered by cell membrane. Aberrant formation of a cilium is of no diagnostic value, since it can be seen in an occasional cell in a wide variety of different tumors.

True cilia are relatively uncommon on neoplastic cells. They occur in some adenocarcinomas of the female genital tract, particularly the uterine tube, and are useful in the differential diagnosis of central nervous system tumors, since they intermingle with the microvilli of ependymoma cells (Fig. 10-6).

Cell Processes. Cell processes are extensions of the cytoplasm. They are often broad and long and may contain cytoplasmic components similar to those of the cell body. Usually only one or, at most, a few are formed by one cell. The formation of cell processes is well developed on neurogenic cells, and is seen in neoplasms of the central and peripheral nervous systems. Astrocytomas have processes that contain mitochondria, ribosomes, microtubules, and microfilaments, whereas a neoplastic oligo-

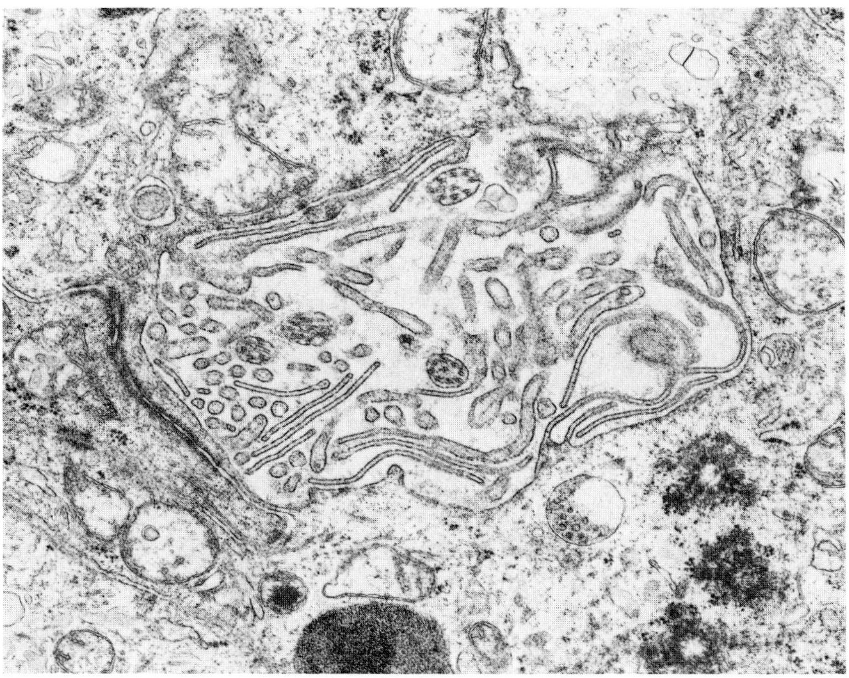

FIG. 10-6. A few transversely sectioned cilia can be distinguished from the microvilli on these ependymoma cells by their component microtubules. X36,000.

dendrocyte is generally round. The zone of tissue devoid of nuclei that surrounds capillaries in ependymomas is produced by microfilament-containing extensions of the tumor cells that weave a dense layer, limited close to the vessel wall by an undulating basal lamina. Meningioma cells typically have attenuated cytoplasmic extensions that curve, undulate, branch, and enclose small spaces, and when loosely arranged create a filigree pattern (Fig. 10-7).

Cell processes are a feature of benign Schwann cell tumors (Fig. 10-8). The processes branch, and those of neighboring cells may be in contact. A basal lamina separates them from the stroma. They contain a few mitochondria, sometimes a few cisternae of endoplasmic reticulum, groups of ribosomes, and longitudinally aligned microfilaments with varying numbers of microtubules. In Schwann cell sarcomas, processes are often well developed and may be the most distinctive feature of the cells at the ultrastructural level. In the more cellular Schwann cell sarcomas, processes and cell bodies become apposed, the basal laminae disappear, and microtubules are sparse or absent. In myxoid Schwann cell sarcomas, partitions between compartments of stroma are made up of many attenuated cytoplasmic extensions. It can be difficult in the two-dimensional electron microscope image to determine if a process is tubular or a thin sheet of cytoplasm.

In the differential diagnosis of small round cell tumors, neuroblastomas can be identified by the presence of dendritic cell processes that contain, in varying numbers, longitudinally oriented microtubules and small (200 mm) dense-core, membrane-bound granules (Fig. 10-9). There does not appear to be any close relationship

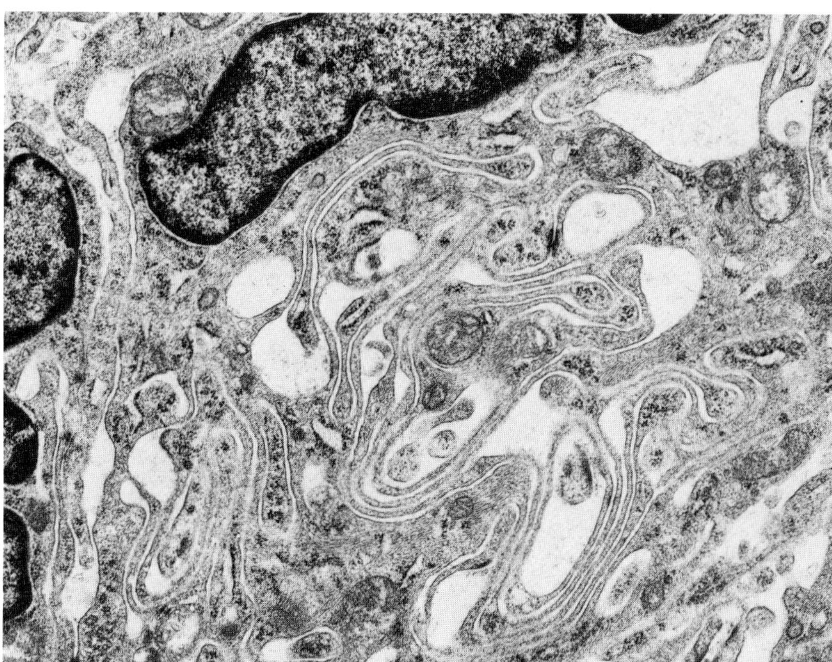

FIG. 10-7. Cytoplasmic extensions on meningioma cells. X4000.

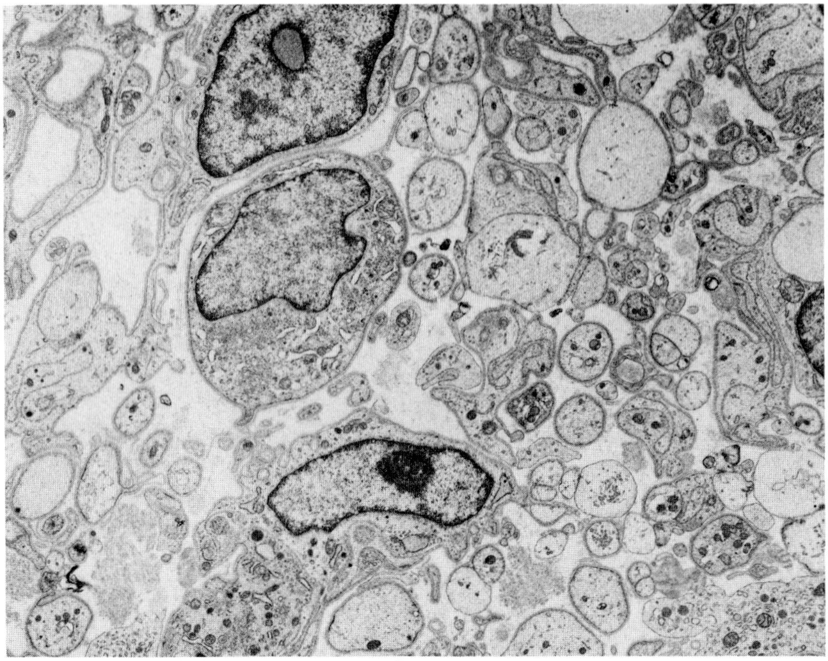

FIG. 10-8. Many transversely sectioned processes of the neoplastic Schwann cells lying alongside the cell bodies in a schwannoma. X4200.

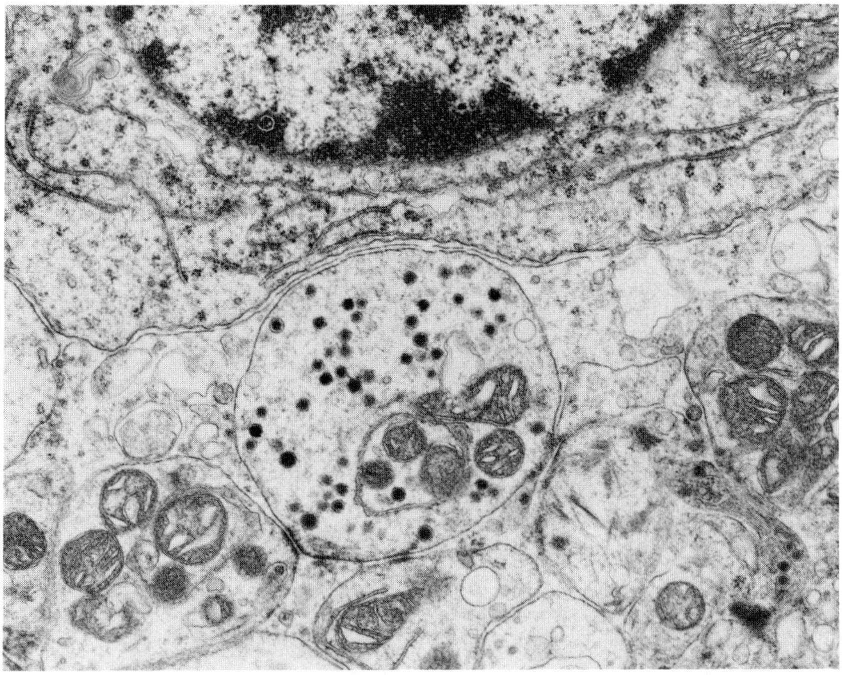

FIG. 10-9. A dendritic neuroblastoma cell process containing uniformly small granules. X15,300.

between the number of processes and the degree of differentiation of a neuroblastoma. However, as it matures, the processes elongate, and they dominate the histology in the mature ganglioneuroma, forming bundles between the groups of ganglion cell bodies.

NUCLEUS

The shape of the nucleus is of particular value in the subclassification of hemopoietic neoplasms. Although the nuclear contour is seen to good advantage with the electron microscope (Fig. 10-10), a better appreciation of the nuclear shape is obtained from the semithin sections of the plastic-embedded tissue prepared for light microscopy, and careful examination of these sections is recommended. An impression of the number of cleaved and uncleaved cells in a large cell lymphoma can more reliably be obtained by light microscopy, since a thin section does not always indicate the presence of a nuclear cleft. The nuclear profile exhibits a wide range of appearances in human neoplasms and is relatively limited in its diagnostic value. Nuclear shape does not consistently relate to tumor behavior. Marked irregularities in nuclear contour may be seen in some aggressively infiltrating carcinomas, but many rapidly proliferating neoplasms have cells whose nuclei are bland in appearance.

Nuclear pockets or blebs have been described in cells of granulocytic leukemia and equated with polyploidy.[4] They are even more common in some large cell lym-

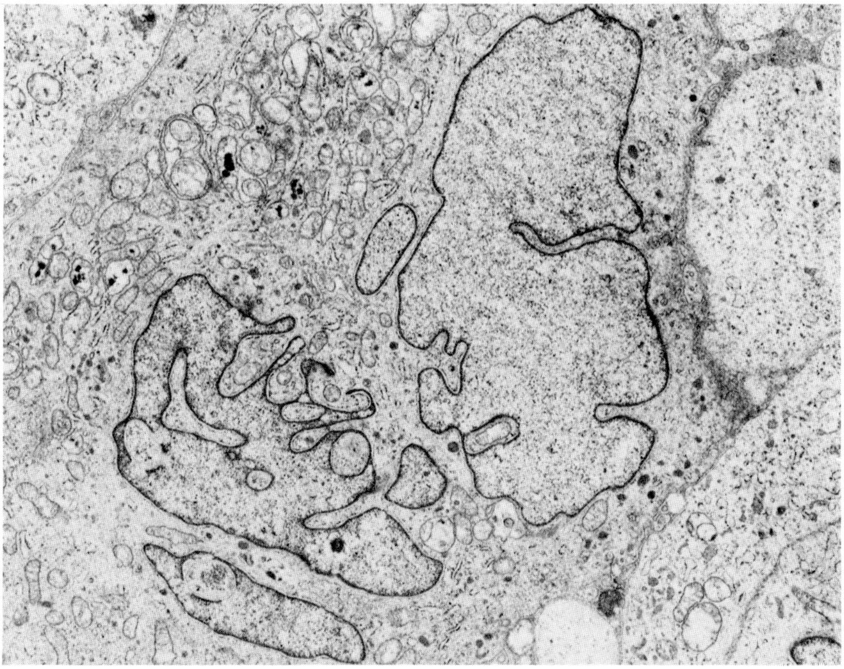

FIG. 10-10. Irregular nuclear profile in an osteosarcoma cell. X5800.

phomas, and may also be seen in other lymphomas and infrequently in other tumors. Proliferation and festooning of the nuclear envelope produces the so-called annulate lamellae, which can occur in a wide range of cells and are consequently not of diagnostic value.[5]

While examining the nuclei of a tumor, the chromatin pattern should be noted. It is necessary that it be assessed in cells that are well preserved, since chromatin clumping occurs when cells undergo pyknosis. Figs. 10-11 and 10-12 contrast the chromatin of nonstimulated and transformed lymphocytes. In the latter, the chromatin is fine and dispersed, typical of its appearance in large cell lymphomas.

In many rapidly growing tumors, the nucleoli are particularly large. Some melanomas and anaplastic carcinomas show this dramatically, but perhaps the largest nucleoli are seen in the so-called clear cell sarcoma, which is probably a melanoma arising within the soft tissues.[6] Inclusions are occasionally described within tumor cell nuclei, but they are not of diagnostic value.

CYTOPLASM

Most of the diagnostic information obtained by the ultrastructural examination of tumor cells is derived from the appearance of the cytoplasm. Its more important components will be considered.

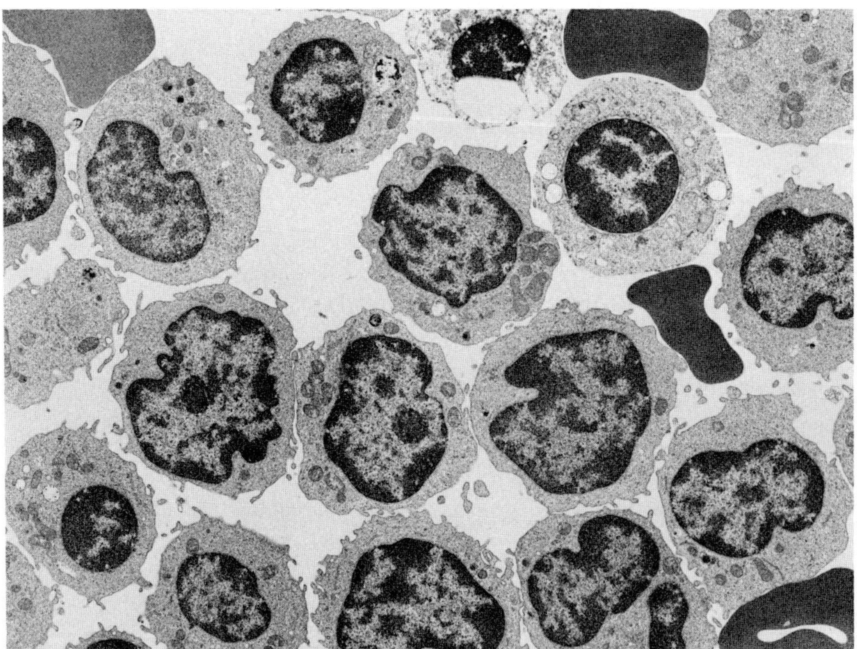

FIG. 10-11. Nonstimulated human T lymphocytes. X4000.

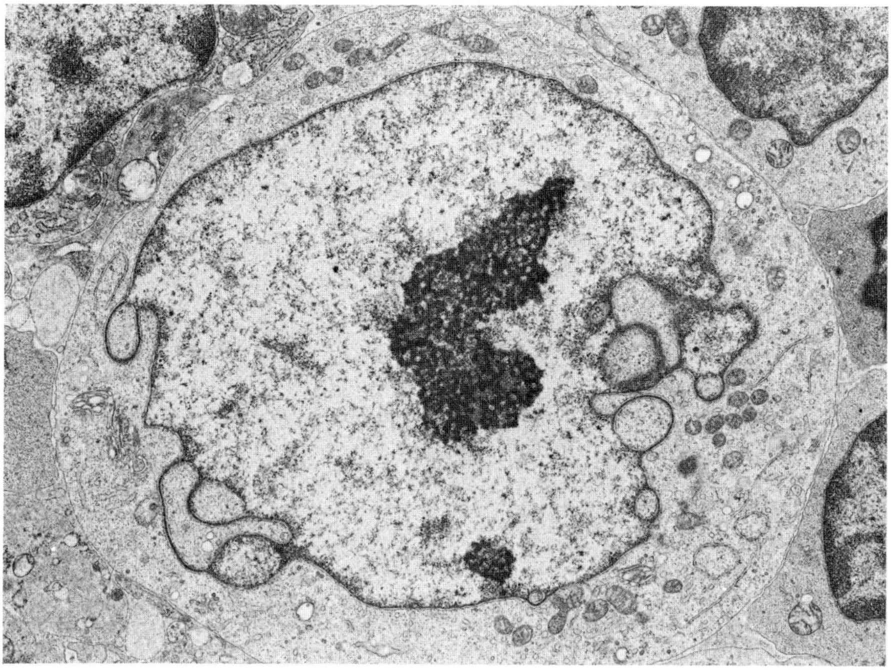

FIG. 10-12. The nucleus of a transformed lymphocyte has fine dispersed chromatin and a prominent nucleolus. X5900.

Density

A frequent observation in descriptions of tumors is variations in cytoplasmic density among the cells. Because of this, it may be inferred that more than one cell type is present. This is generally not the case. Changes in the electron density of the cytoplasm can result from a number of causes. The more organelles a cell contains, the denser its cytoplasm will appear at low magnifications. The degree of hydration also influences the density of the cytoplasm; e.g., accumulating fluid in a degenerating cell will produce electron-lucent zones (Fig. 10-13). So-called clear cells, by light microscopy, are not always clear at the ultrastructural level, since the appearance may reflect the presence of glycogen, which can form granular (see Chapter 4) or electron-lucent (Fig. 10-14) aggregates, or lipid, which appears as well-defined droplets.

Cytoplasmic Spaces

Cells of a gland-forming tumor may contain intracytoplasmic acini (Fig. 10-15). They are more common in breast carcinomas,[7] but can occur in other tumors. The possibility of a transverse section through a surface invagination must be excluded. Artifactual clefts in the cytoplasm produced by inadequate fixation or processing are not membrane bound and do not possess microvilli. Cytoplasmic spaces may be dis-

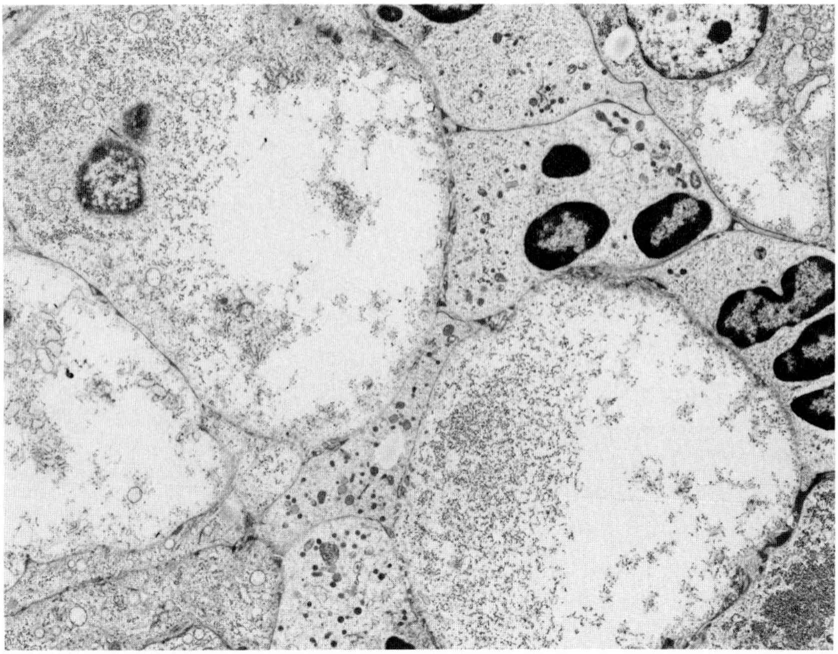

FIG. 10-13. Clear zones in the cytoplasm of cells of a bronchogenic adenocarcinoma are a degenerative change. Note the contiguous neutrophils. X3700.

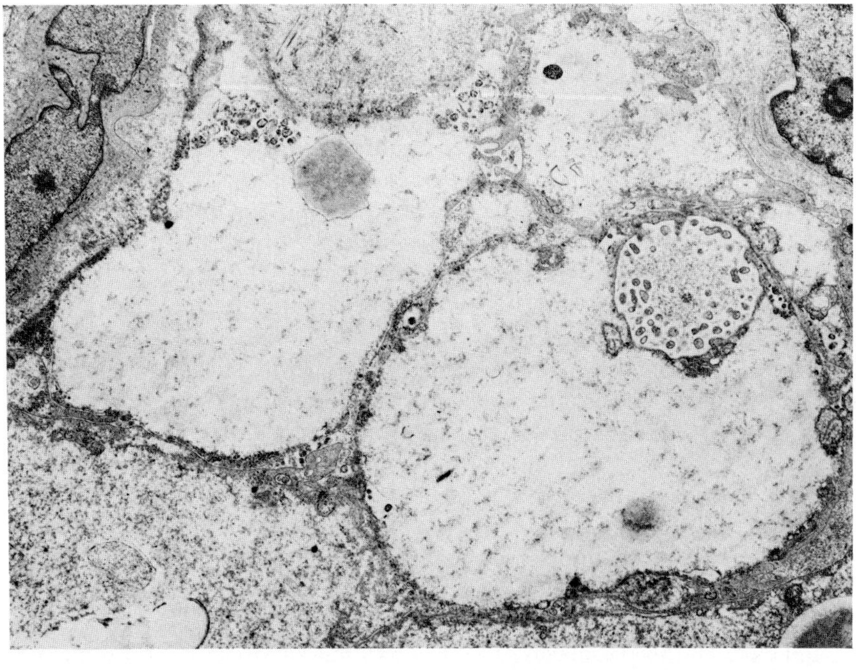

FIG. 10-14. Electron-lucent zones in the cytoplasm of these tumor cells indicate the presence of lakes of glycogen. X7400.

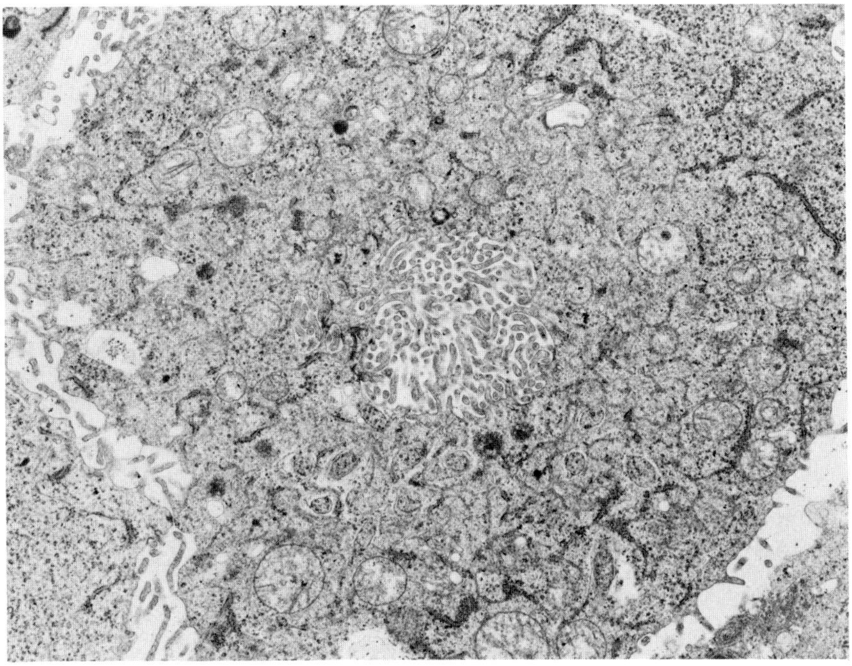

FIG. 10-15. A small intracytoplasmic acinus within a breast carcinoma cell. X7200.

tended cisternae of the rough endoplasmic reticulum, and lakes of mucin in an adenocarcinoma can appear as clear zones. The rare signet ring cell lymphoma owes its appearance to the accumulation of cytoplasmic immunoglobulin, formed in small vesicles which coalesce to produce a large vacuole that pushes the nucleus to the side of the cell. Small vesicles are common in many tumors, but pinocytotic vesicles aggregating beneath the cell membrane are best seen in smooth muscle tumors. They are particularly numerous in the glomus tumor (Fig. 10-16).

Mitochondria

There is a wide variation in the number of mitochondria within cells of various tumors. In many poorly differentiated neoplasms they are sparse, but at the other extreme they may literally pack the cytoplasm. The latter appearance is seen in the so-called oncocytic neoplasms, which include Hürthle cell tumors of the thyroid, some salivary neoplasms, and occasionally a bronchogenic, renal, or endocrine tumor such as a carcinoid[8] or a pituitary adenoma.[9] Part of the cytoplasm of a mitochondrion-rich parotid tumor is illustrated in Figure 10-17.

Steroid-forming cells may have mitochondria with tubular cristae. They can occur in tumors of these cells (Fig. 10-18), usually mixed with mitochondria that have the usual shelf-like cristae. We have found them in cells of 13 of 22 adrenocortical carcinomas. They may be prominent in endocrine tumors of the gonads, but, again, are not a consistent finding. Occasional mitochondria with what appear to be tubular cristae

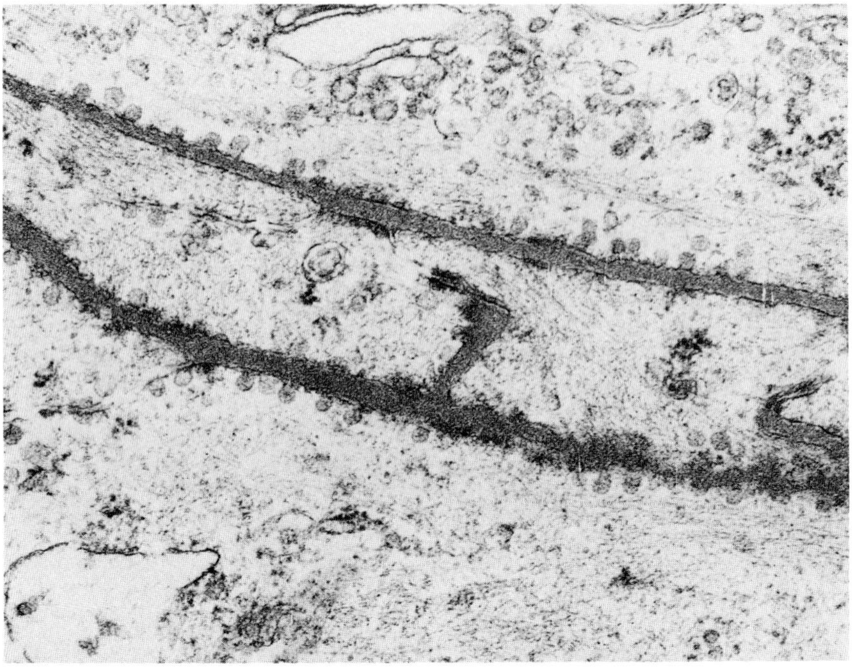

FIG. 10-16. Pinocytotic vesicles subjacent to the cell membranes in a malignant glomus tumor. X37,000.

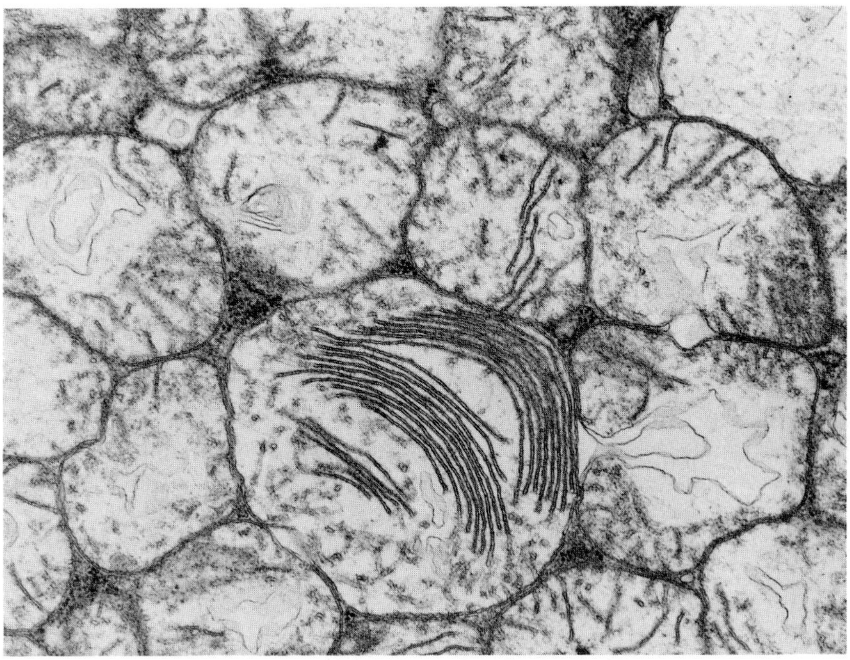

FIG. 10-17. Closely packed mitochondria from a Warthin's tumor cell. X22,000.

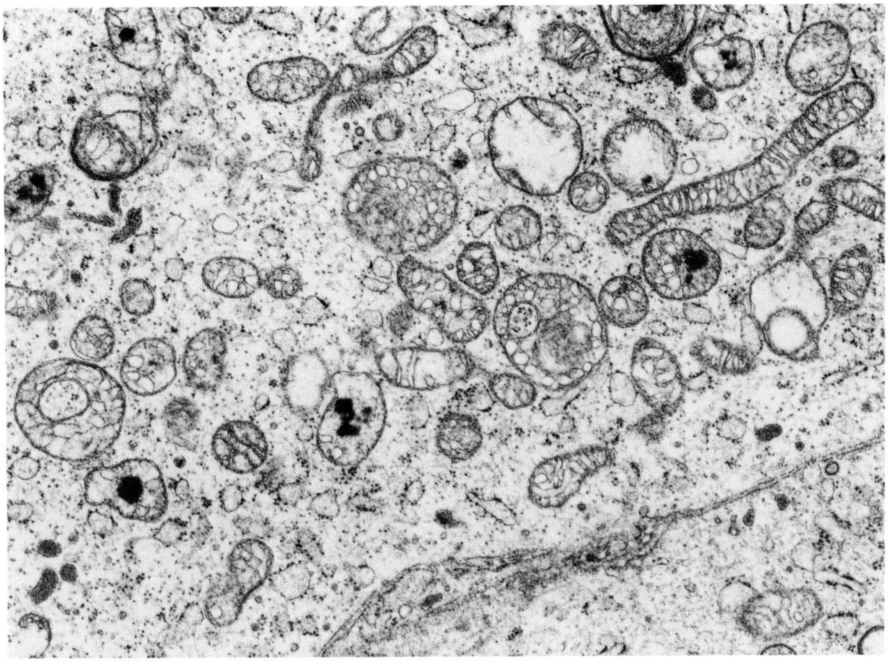

FIG. 10-18. In this adrenocortical carcinoma cell, many of the mitochondrial cristae are tubular, and some contain large, dense, matrical granules. X15,000.

should not be considered proof of a tumor of steroid-forming cells in the absence of other criteria. They may be present in degenerating cells, and can be encountered in some melanomas and in small numbers in many other neoplasms.

Small matrical granules are normally present in mitochondria, but larger dense bodies within the mitochondrial matrix have been seen in adrenocortical cells and their tumors, both benign and malignant, and in animal tissue.[10,11] They are spherical, sharply defined, and uniformly electron-dense, in contrast to the irregular dense zones that appear in hypoxic cells within the mitochondrial matrix. Crystalline inclusions have been noted in liver cell[12] and adrenocortical cell[13] mitochondria. We have seen them in 1 of 40 liver cell carcinomas, but they were not observed among 22 cases of adrenocortical carcinomas.

Endoplasmic Reticulum

Considerable quantities of rough endoplasmic reticulum are seen in cells that are forming protein secretory products. The plasma cell, and exocrine gland cells, such as those of the pancreas, are examples. Comparable quantities of rough endoplasmic reticulum can be found in tumors of these cells (Fig. 10-19). Distended cisternae are characteristic of fibrosarcomas but may also be seen in other mesenchymal tumors, including malignant fibrous histiocytoma and liposarcoma. Accumulation of elec-

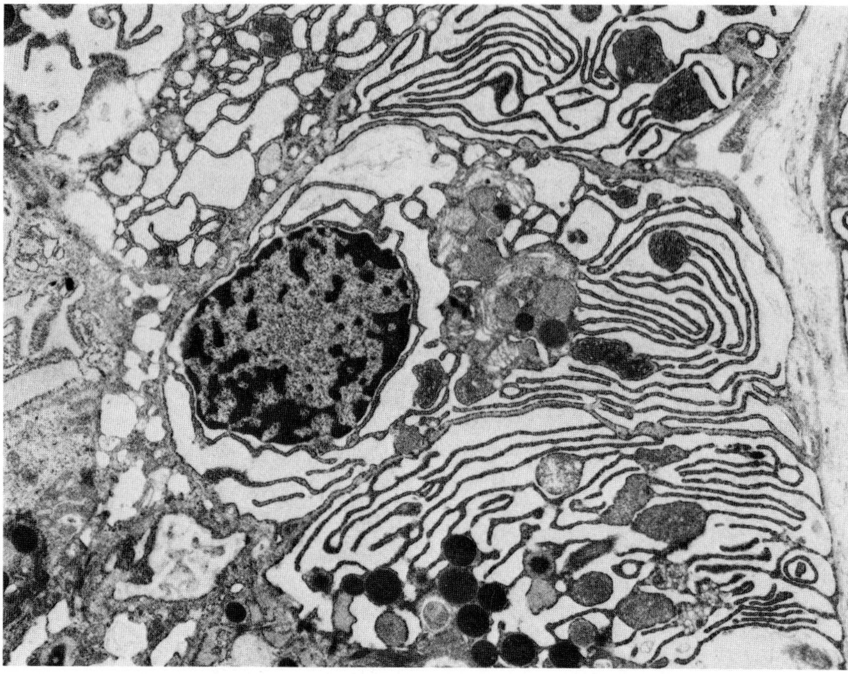

FIG. 10-19. Acinar cell carcinoma of the pancreas, metastatic to liver. The tumor cells contain large, dense granules and abundant, distended cisternae of rough endoplasmic reticulum. X3600.

tron-dense material (Fig. 10-20), which may be focally condensed, can be observed in some malignant fibrous histiocytomas and more frequently in liposarcomas. It is possible that this accumulation represents disordered lipid synthesis. Scanty cisternae with many unattached ribosomes (Fig. 10-21) are features of transformed lymphocytes and are seen in many large cell lymphomas, although in some of these tumors the cells accumulate moderate numbers of mitochondria, and a few lipid droplets are a not uncommon finding.

The presence of microtubules within cisternae of the endoplasmic reticulum has been reported in various nonneoplastic cell types,[14] but it is not common in human tumors. They have been found in melanomas (Fig. 10-22),[15] and in our experience they are seen in at least 10% of metastatic melanomas, but their number varies greatly and in some tumors they are only present in occasional cells. Nevertheless, these tubules are a particularly unusual and distinctive feature and the observation may suggest the diagnosis in a tumor of undetermined type. Unfortunately, they are not specific, Myxoid sarcomas may contain arrays of intracisternal microtubules.[15,16] The significance of the tubules is not known, and it is not clear whether their presence in a sarcoma is an indication of a particular cell type. In tumor cells, the tubules are usually straight and parallel, roughly 30 nm in diameter, with a fine helical periodicity and, occasionally, a dense rather than a hollow core (Fig. 10-22). The tubuloreticular inclusions seen in some lymphocytes are often bundles of curving or undulating microtubules.[14]

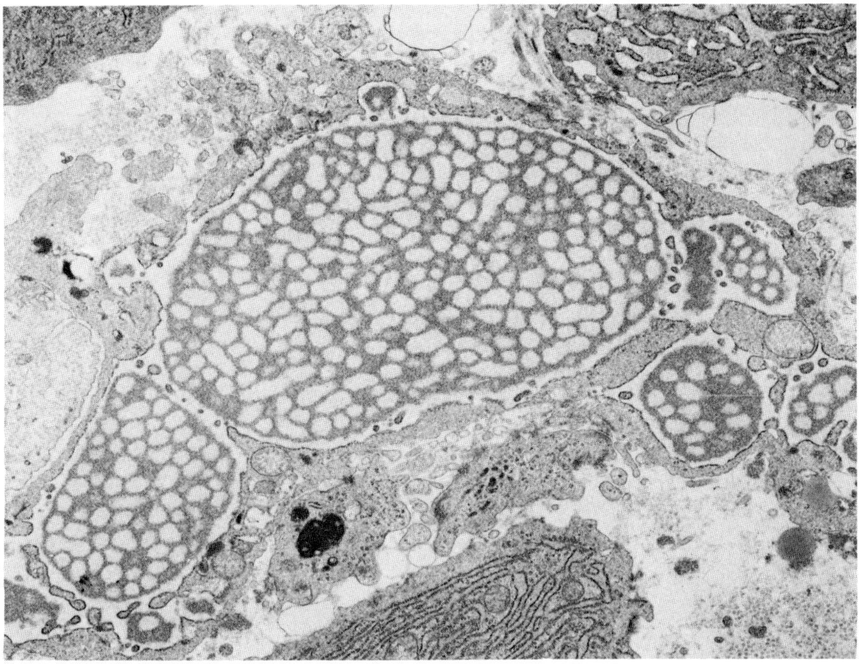

FIG. 10-20. Cribriform condensations of electron-dense material in cisternae of a liposarcoma cell. X7000.

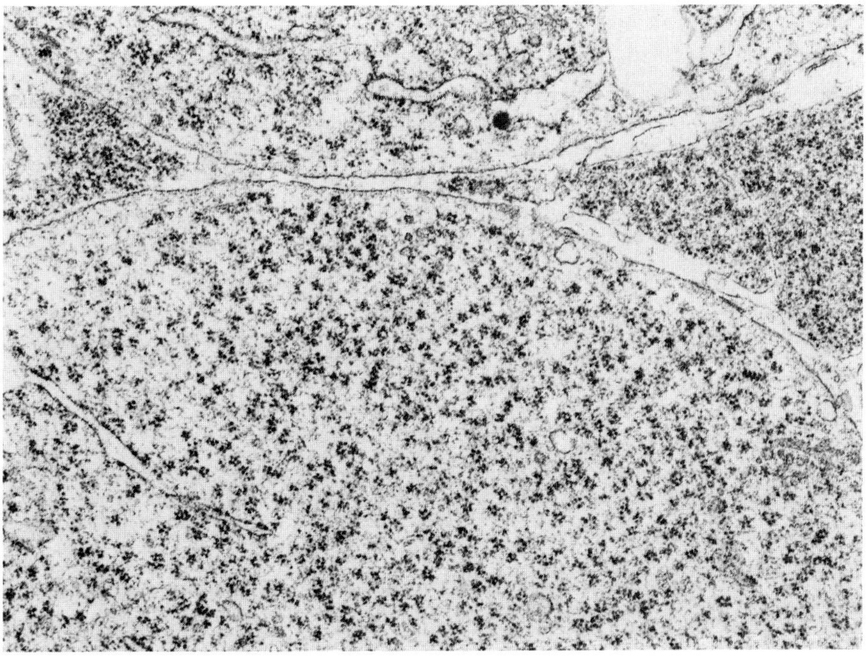

FIG. 10-21. Numerous free ribosomes in the cytoplasm of cells of a large cell lymphoma. X20,000.

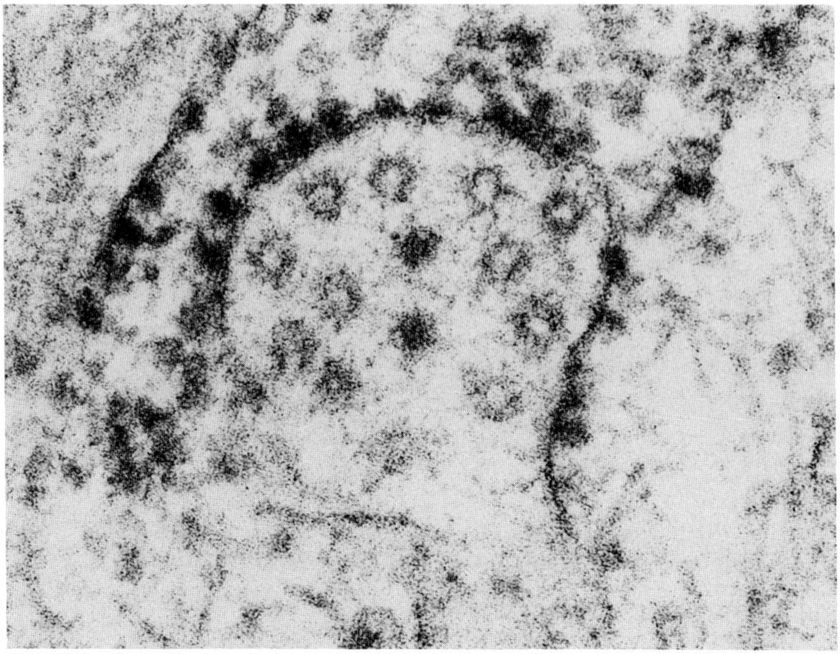

FIG. 10-22. Dense cores can be seen within two of the microtubules in a cisterna of a melanoma cell. X88,000.

Most cells contain at least small quantities of smooth endoplasmic reticulum, but plentiful amounts are characteristic of steroid-forming cells and may be found in adrenocortical adenomas and carcinomas. They are by no means a universal finding in these tumors. We found them in considerable quantities in 3 adrenocortical carcinomas and in moderate amounts in 9 others in a study of 22 cases. There was no correlation between the presence of extensive zones of smooth reticulum and the presence or absence of functional activity by the tumors. Smooth endoplasmic reticulum can also be plentiful in interstitial cell tumors of the testis, in ovarian hilus cell tumors, and in some liver cell carcinomas, as well as in cells of the enigmatic alveolar soft part sarcoma.[17]

Lysosomes

In the Chediak-Higashi syndrome, large lysosomal granules are present in the proliferating cells.[18] Extensive accumulations of lysosomes form in the lozenge-shaped cells of the granular cell tumor, formerly termed a myoblastoma (Fig. 10-23). Since any degenerating cell may possess moderate numbers of lysosomes, their presence is in general of limited diagnostic value, and it is more important to be aware that they may be plentiful and that they can simulate secretory granules. In a thyroid tumor, they can give rise to the mistaken impression that a follicular carcinoma is of C cell origin (Fig. 10-24). Primary lysosomes and secretory granules can look remarkably similar, and the cautionary advice below should therefore be heeded.

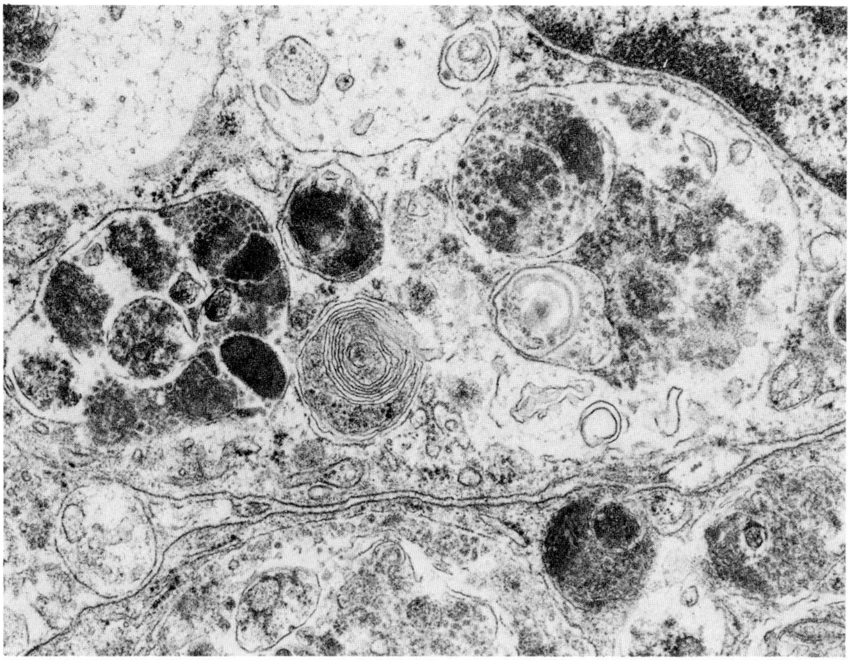

FIG. 10-23. Lysosomal aggregates in a granular cell tumor of the breast. X28,000.

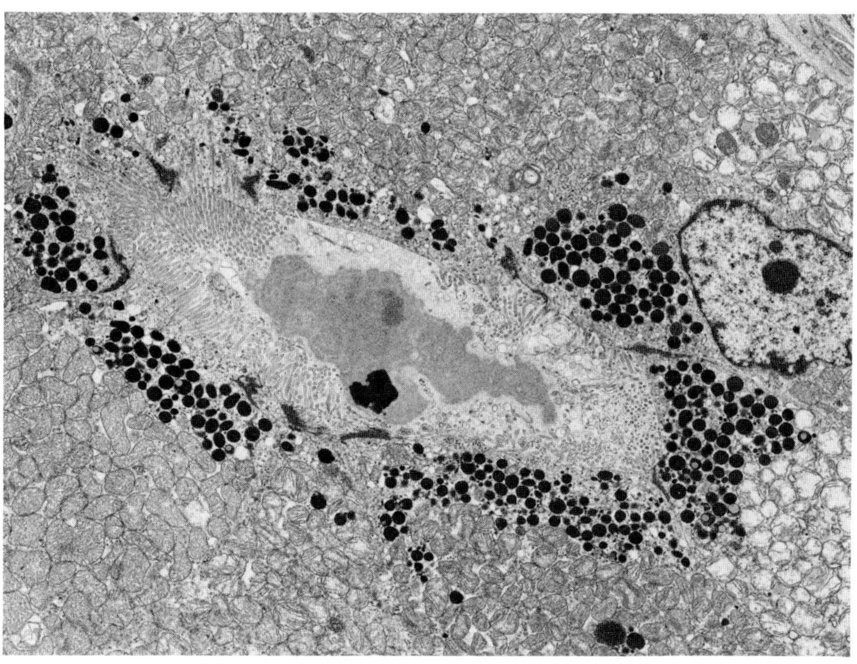

FIG. 10-24. The accumulation of mitochondria in the cytoplasm of these Hürthle cells has pushed the lysosomes to the apex, where they simulate secretory granules. X3700.

Secretory Granules

Secretory granules are formed by accumulations within the cytoplasm of protein secretions, which in most instances are spherical. The electron-dense contents may be homogeneous or vary in density, and the limiting membrane can be intimately apposed to the contents or separated by a clear halo.

Dense bodies in the cytoplasm should only be accepted as true secretory granules when they are numerous within most cells of a tumor and fall within a relatively narrow range of diameters. If these criteria are adhered to, confusion with primary lysosomes should be avoided.

It is convenient to subdivide secretory granules into three groups based on the average granule diameter. The large (usually over 700 nm) granules of exocrine cells are often numerous in cells of salivary gland tumors, notably acinic cell carcinomas, and in acinar cell carcinomas of the pancreas (Fig. 10-19).[19] Granules of intermediate caliber, mostly between 250 and 600 nm, are found in the polypeptide-forming endocrine cells (APUD cells) and their tumors (Fig. 10-25). As a rule it is not possible to determine the functional behavior of a tumor of this type from the appearance of the granules, with the exception of the crystalline cores in an insulinoma (Fig. 10-26) and the granules of many pheochromocytomas in which the limiting membrane is ovoid and the dense core eccentrically positioned (Fig. 10-27). Correlation with the

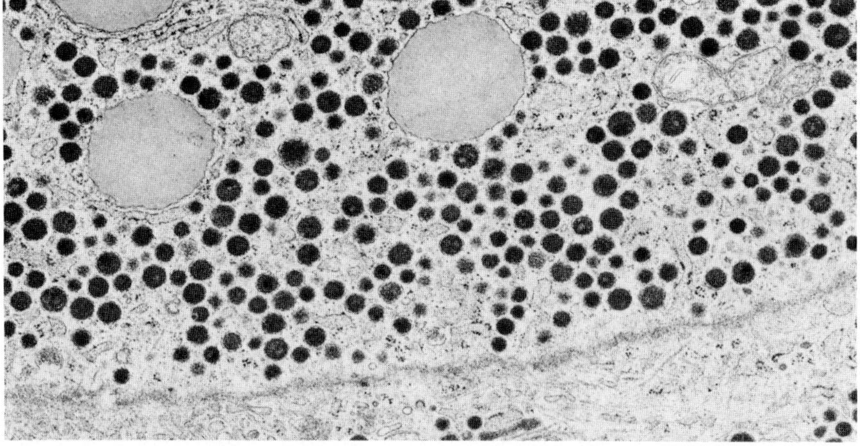

FIG. 10-25. Islet cell carcinoma, metastatic to the liver, in a patient with the Zollinger-Ellison syndrome. The cytoplasm contains many secretory granules and a few lipid droplets. X13,000.

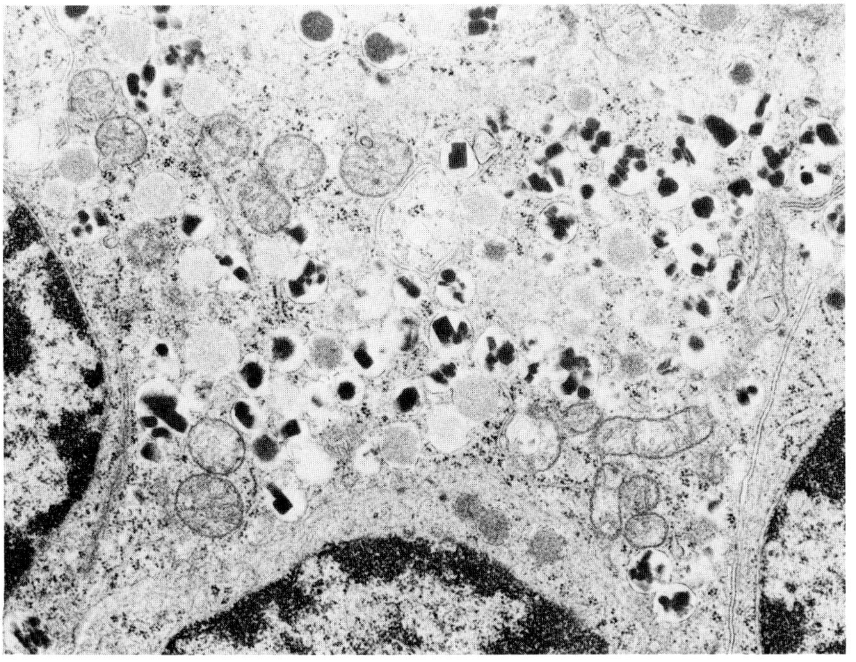

FIG. 10-26. Crystalline dense cores in granules of an insulinoma. X6000.

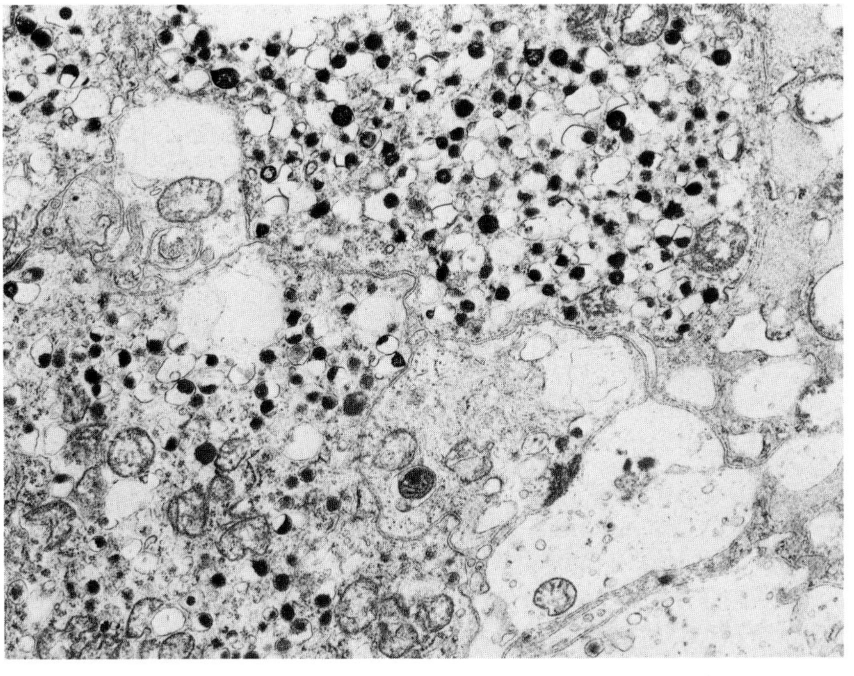

FIG. 10-27. Eccentric dense cores within ovoid limiting membranes are characteristic of a pheochromocytoma. X10,000.

light microscopy and clinical information is usually necessary to determine the cell type, and immunofluoresence and immunoperoxidase preparations are useful. A cell of a dedifferentiating endocrine tumor, or one that is being stimulated to secrete rapidly, will generally contain fewer and smaller granules than a comparable cell under resting conditions in normal tissue.

Small granules less than 200 nm in diameter are seen in thyrotropic cell pituitary adenomas,[20] where they form a single row subjacent to the plasma membrane. The cells are smaller and more angular than those of other pituitary adenomas. Misplaced exocytosis may be seen in prolactin-forming tumors.[21] The advent of transphenoidal hypophysectomy has resulted in a deterioration in the quality of many pathologic specimens of pituitary disorders, and fragments of normal anterior lobe may be mixed with pieces of an adenoma. When the ultrastructural study shows a marked variation in granule caliber from one cell to the next, and the sinusoidal pattern is maintained, it is likely that the tissue is nonneoplastic anterior pituitary.

Small granules are also found in small cell carcinomas of the lung (Fig. 10-28). They are uniform, often sparse, and even absent in some tumors, and failure to recognize them in a tumor that otherwise fulfills the criteria for a small cell bronchogenic carcinoma should not be viewed as contraindicating the diagnosis. Granules of similar caliber concentrate in the dendritic processes of neuroblastomas and the cutaneous Merkel cell tumor.[22,23]

Glycogen

Small quantities of glycogen that are not detectable by light microscopy can be recognized with the electron microscope as dense granules or electron-lucent zones, depending on the technical procedures used in preparing the tissue. It appears that many tumors can contain glycogen at least occasionally. Its value in differential diagnosis is somewhat limited, even in the small round cell tumors where the presence of glycogen does not exclude neuroblastoma,[24] while Ewing's tumor, which typically does possess glycogen, may not contain any when the patient has been treated.[25]

Lipid

Most liposarcoma cells contain cytoplasmic lipid droplets (Fig. 10-29) that tend to coalesce, so that in one cell there are several different lipid droplet sizes, or a single large vacuole that compresses the nucleus and organelles in a rim of peripheral cytoplasm. In some myxoid and pleomorphic tumors, there is scanty lipid. The accumulation of electron-dense material within cisternae has been described above (Fig. 10-20). Lipid droplets are common in the cells of malignant fibrous histiocytomas, but they are usually small and uniform and are only numerous in occasional cells. A few lipid droplets can also be found in fibrosarcoma cells. Lipid droplets with lakes of glycogen should suggest a renal cell carcinoma, and in the sarcomatoid variant may be the only indication of the diagnosis. In lymphopoietic tumors, lipid is often present in small quantities in large cell and undifferentiated lymphomas. Lipid tends to accumulate in cells under hypoxic conditions, and therefore should be disregarded in degenerating cells or in viable cells adjacent to an area of necrosis.

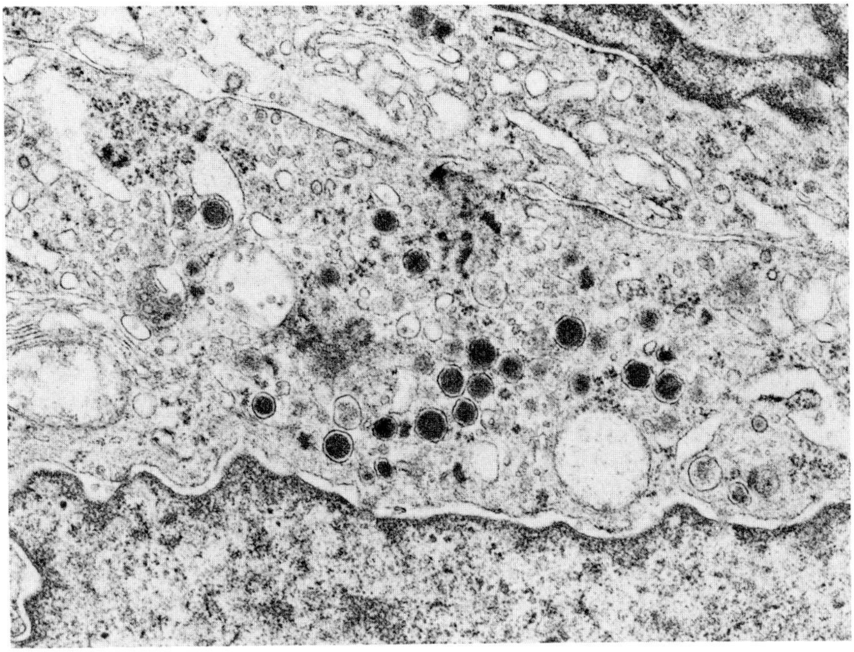

FIG. 10-28. Small dense-core granules in a bronchogenic small cell undifferentiated carcinoma. X20,000.

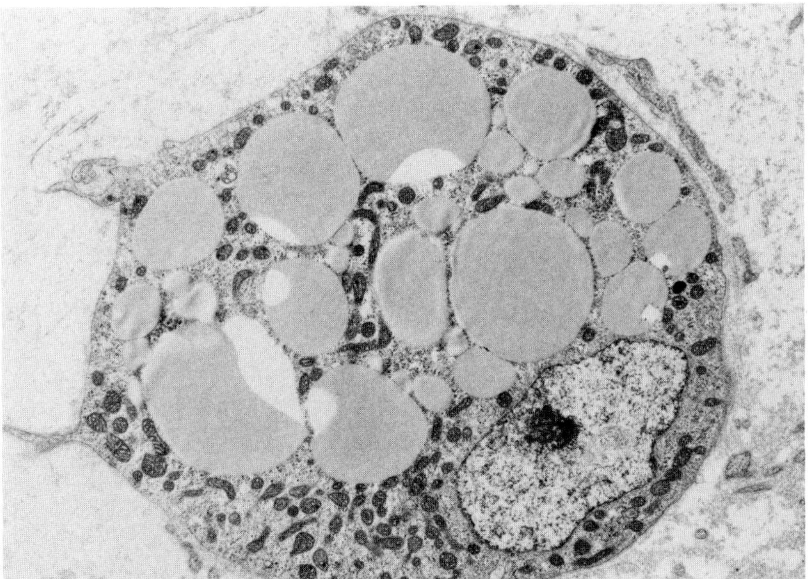

FIG. 10-29. Many lipid droplets of varying size in a liposarcoma cell. X5000.

Mucin

Dense cores at the center of aggregates of cytoplasmic mucin can simulate secretory granules, but they are not membrane limited or particularly uniform in size and shape. Confusion is nevertheless possible at low magnifications, as in a lung neoplasm where a carcinoid tumor may be erroneously suspected. Coexistence of mucin with endocrine-type granules occurs in goblet cell carcinoid tumors.[26]

Filaments

Nonspecific actin filaments are present in many different types of cell, and they may form aggregates in tumor cells. In some soft tissue tumors, bands or irregular bundles of filaments may be found, but they are not known to be of diagnostic value. In some epithelioid sarcomas, the filaments run concentrically around the nucleus (Fig. 10-30), but this is not a consistent finding. In the case illustrated, they were sparse in a subsequent metastasis.

Of more significance is the identification of specific myofilaments of either smooth or skeletal muscle. Smooth muscle myofilaments can be identified by the presence of interspersed densities within the actin filaments, and are found in benign smooth muscle tumors and many of the better differentiated leiomyosarcomas (Fig. 10-3). Paradoxically, smooth muscle tumors of the gastrointestinal tract may not contain

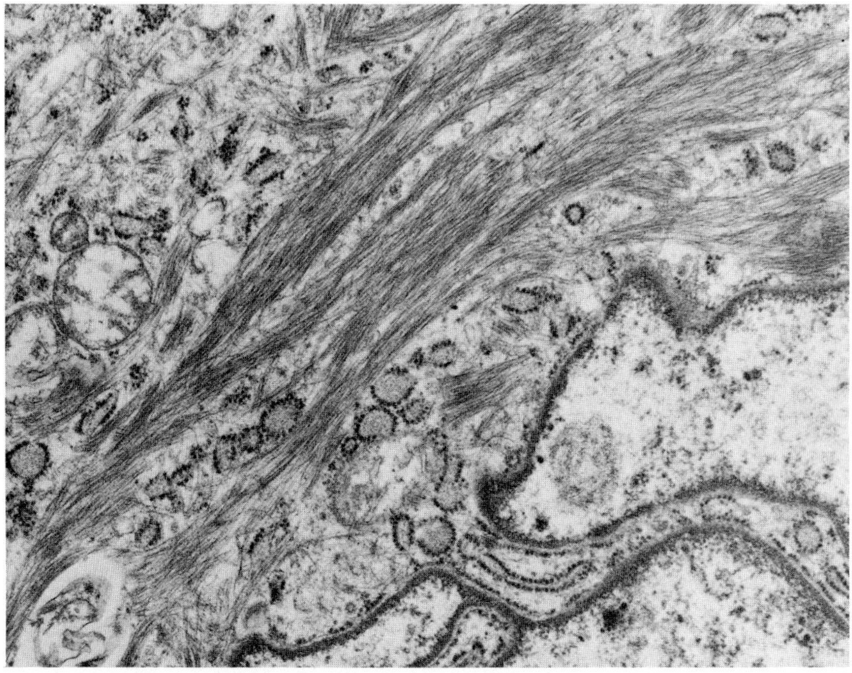

FIG. 10-30. Epithelioid sarcoma of the perivaginal tissues. Microfilaments form loose bundles in the cytoplasm alongside the nucleus. X24,000.

cytoplasmic filaments. This is particularly the case with the so-called epithelioid tumors, in which a transformation to more mesenchymal features with loss of all smooth muscle characteristics occurs. In a recent study of 20 malignant smooth muscle tumors arising in the stomach or intestines, we could find myofilaments in only 5 cases. Bundles of smooth muscle myofilaments occupy the peripheral cytoplasm of myofibroblasts (Fig. 10-31), and anchoring filaments extend from the outer aspect of the cell to attach to adjacent collagen, providing a fulcrum for contraction. Myofibroblasts are common in nonneoplastic, benign, and the better-differentiated malignant fibroblastic tumors, including some malignant fibrous histiocytomas.

The presence of skeletal muscle myofilaments is the ultrastructural hallmark of a rhabdomyosarcoma (Fig. 10-32), and in their absence the diagnosis is not possible ultrastructurally, although it may be suspected. They vary greatly in number and in their degree of organization. They are seen to best advantage in the botryoid rhabdomyosarcomas, in which the actin and myosin filaments may form long segments of myofibrils, so that cross-striations are obvious by light microscopy. Rhabdomyosarcoma cells may, however, form copious myofilaments that fail to organize into fragments of myofibrils with bands that can be detected at the light microscopy level. In alveolar and embryonal rhabdomyosarcomas, many of the small round cells do not form myofilaments, or may only produce them in small numbers that are difficult to find without a detailed search. It may be useful in this situation to scan the tumor at low magnification, searching for cells that are larger, have more cytoplasm, or contain zones of cytoplasm devoid of organelles, and then to examine these cells in more

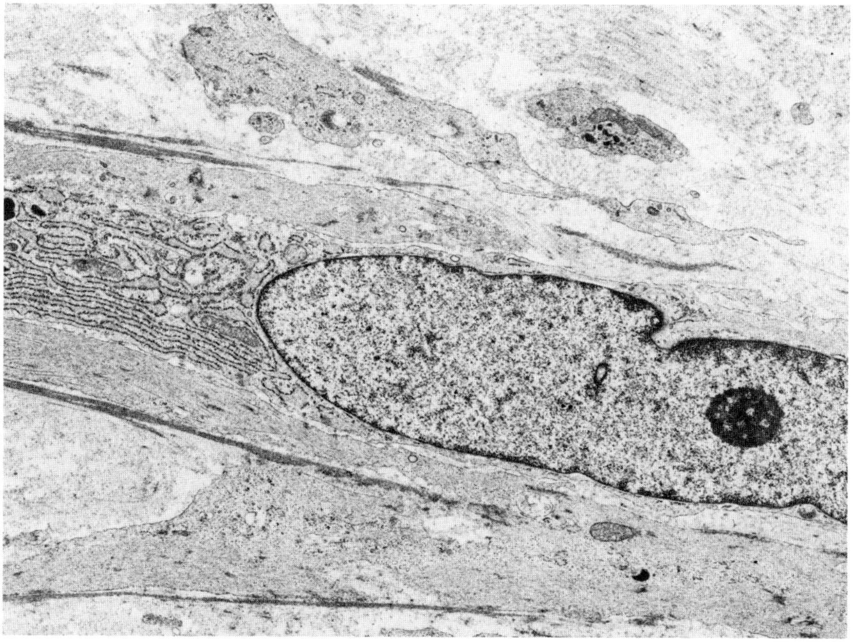

FIG. 10-31. Myofibroblasts from a well-differentiated fibrosarcoma. The dense lines alongside the cells are anchoring filaments. X6000.

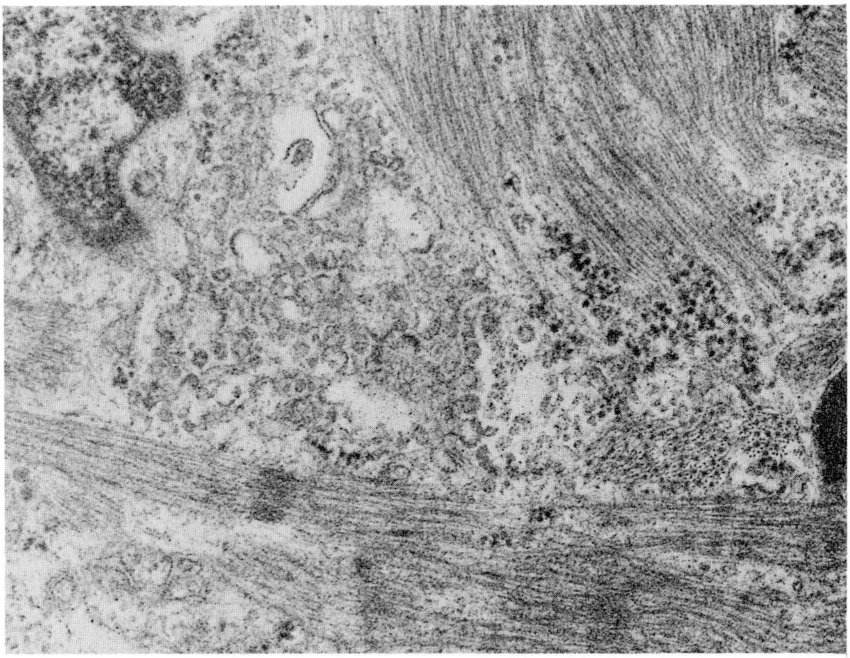

FIG. 10-32. Skeletal muscle myofilaments forming short fragments of myofibrils in a rhabdomyosarcoma. X35,000.

detail. Many rhabdomyosarcoma cell nuclei have irregular profiles. Pleomorphic rhabdomyosarcomas are rare, but on ultrastructural study an occasional sarcoma with large cells and abundant cytoplasm will be found to be replete with myofilaments.

Microtubules

The unusual finding of intracisternal microtubules in some melanomas and myxoid sarcomas has been described. Longitudinally oriented microtubules can be seen in the cytoplasmic extensions of neoplastic Schwann cells (Fig. 10-33), but although they are typically present in the benign tumors, they may be sparse or absent in the sarcomas.They are also seen in astrocytomas, and can be numerous in dendritic processes of neuroblastoma cells, even when the small secretory granules are difficult to find. Microtubules can be encountered in the occasional cell of a primitive tumor that has a high rate of mitotic activity.

Crystals

The crystalloids of Reinke seen in Leydig cells may not be found in tumors of these cells. Some neoplastic plasma cells contain elaborate cytoplasmic crystals. The granules with crystalline cores in B cell tumors of the pancreatic islets are usually ad-

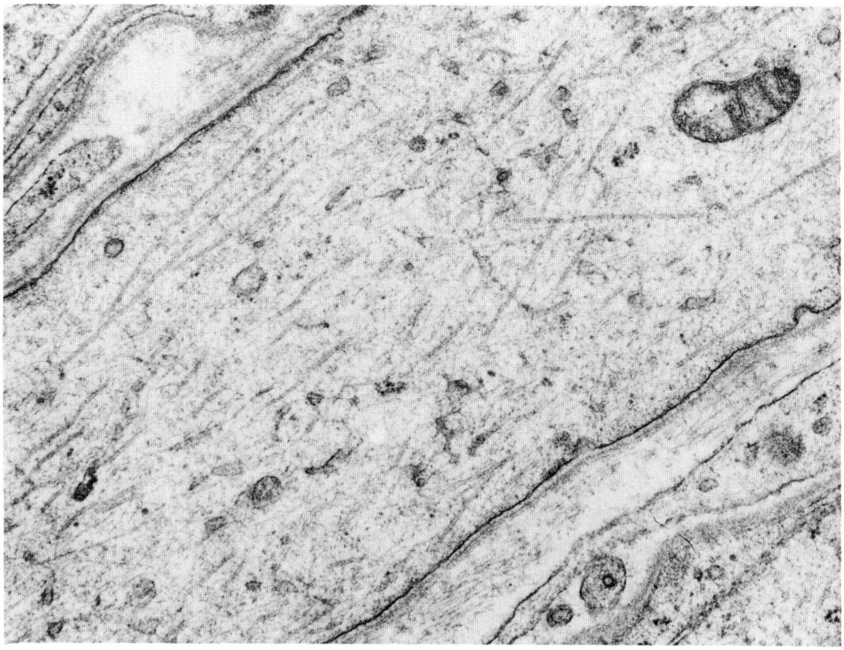

FIG. 10-33. A segment of a cytoplasmic extension of a schwannoma cell showing the microtubules and limiting basal lamina. X22,000.

mixed with spherical granules, and the latter may predominate. Probably the best known example of crystal formation in tumor cells occurs in the alveolar soft part sarcoma, a soft tissue neoplasm of undetermined histogenesis. The crystalline deposits form within membrane-bound sacs in the cytoplasm and mature to rhomboid structures (Fig. 10-34), around which the limiting membrane can still be detected.

Other Dense Bodies

Melanosomes may on occasion be small and round, and at low power suggest secretory granules. As has already been noted, the distinction between a primary melanosome and a secretory granule may be impossible. In melanomas, however, there is more variation in size, shape, and composition among the cytoplasmic densities than is seen in cells that contain true secretory granules, and examination of several cells will usually disclose at least some evidence of features consistent with melanosomes, such as multiple small granular densities, chain-like aggregates of these bodies coiling within the limiting membrane, or bar-shaped inclusions with features of premelanosomes (Figs. 10-35 and 10-36). The distinctive periodicity of the premelanosome, and its bounding membrane, are well seen in Fig. 10-36. Its identification requires viewing at a relatively high magnification, and in a tumor where these bodies are sparse, the process of locating them can be inordinately time-consuming. It can be facilitated by scanning at a lower power, and by focusing on cells that contain large

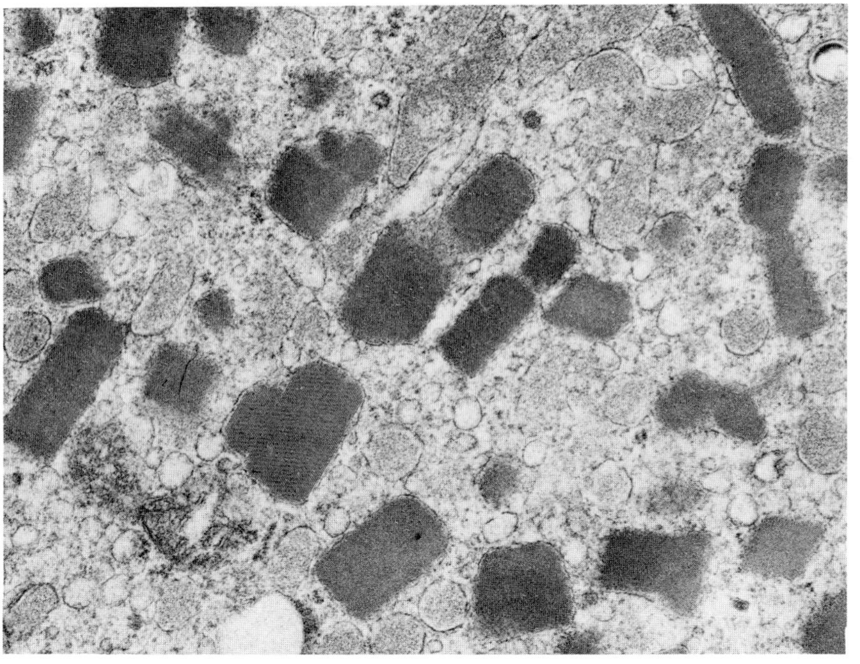

FIG. 10-34. Rhomboid crystals in an alveolar soft part sarcoma. X26,000.

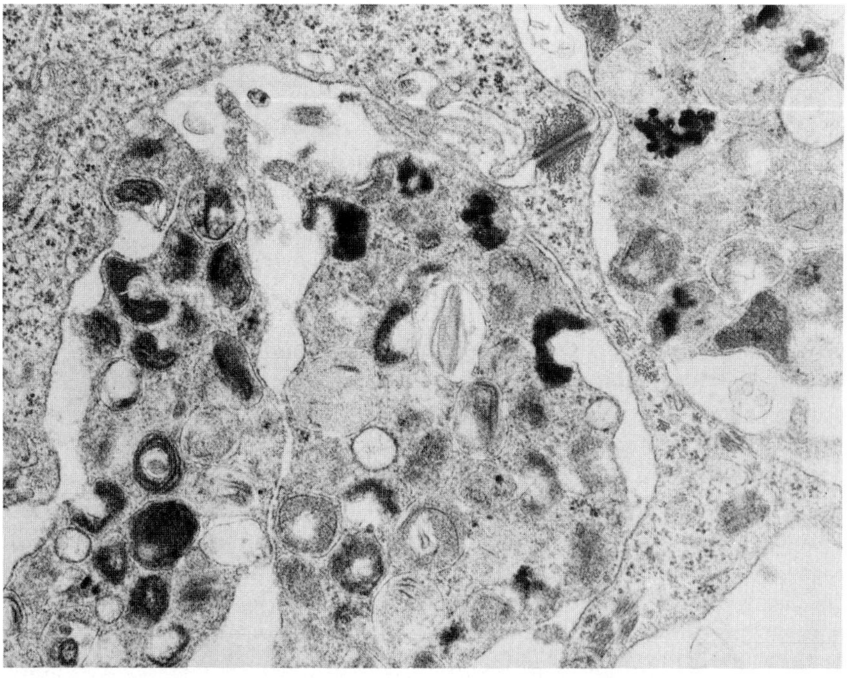

FIG. 10-35. Processes of a melanoma cell from a primary cutaneous melanoma insinuating themselves between squamous cells. X22,000.

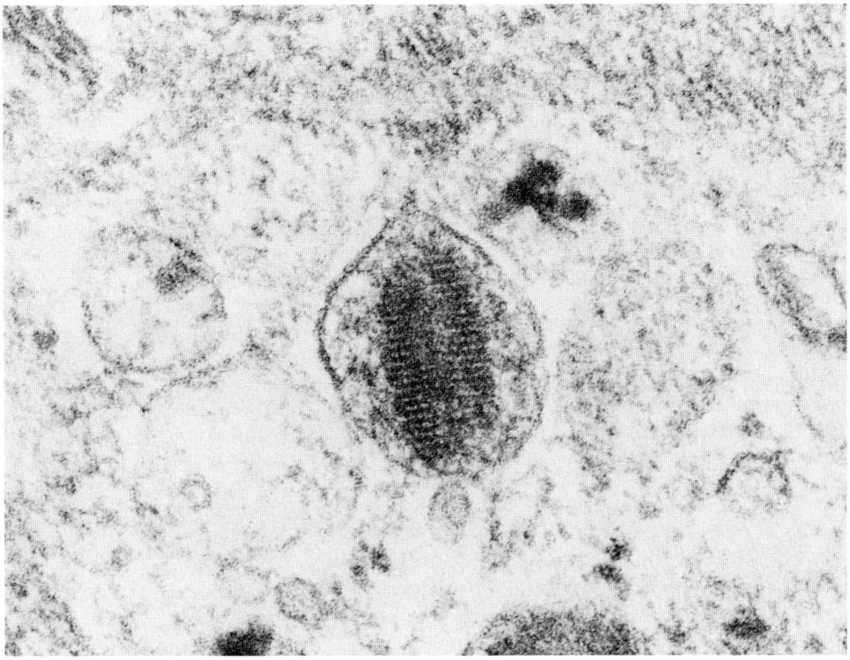

FIG. 10-36. A premelanosome from a cutaneous melanoma. X12,000.

numbers of cytoplasmic dense bodies, particularly densities that contain an elongated core.

The Weibel-Palade bodies that are common in capillaries in many locations may infrequently appear in the cells of endothelial tumors, but they are not seen in many angiosarcomas. Large, round dense bodies may be present in liver cell carcinomas,[27,28] and may be antitrypsin or alpha fetoprotein aggregates. In the case illustrated (Fig. 10-37), antitrypsin staining was strongly positive within the cytoplasm and particularly around the large bodies, but their interior did not appear to stain. Some densities in tumor cells result from phagocytosis, as in Kaposi's sarcoma, in which the spindle-shaped tumor cells, which are apparently vasoformative, may engorge extravasated red blood cells from the stroma.

CONCLUSION

The main themes of this chapter are the recapitulation of normal morphology in cells of well-differentiated tumors, the progressive loss of specific features as a tumor dedifferentiates, and the degree of overlap in the ultrastructure of many tumors of different types. The importance of correlating findings from electron microscopy with those from routine light microscopy and with the available clinical information in any problem case is reiterated. A more extensive bibliography on the contribution of electron microscopy to tumor diagnosis can be found in References 29 and 30.

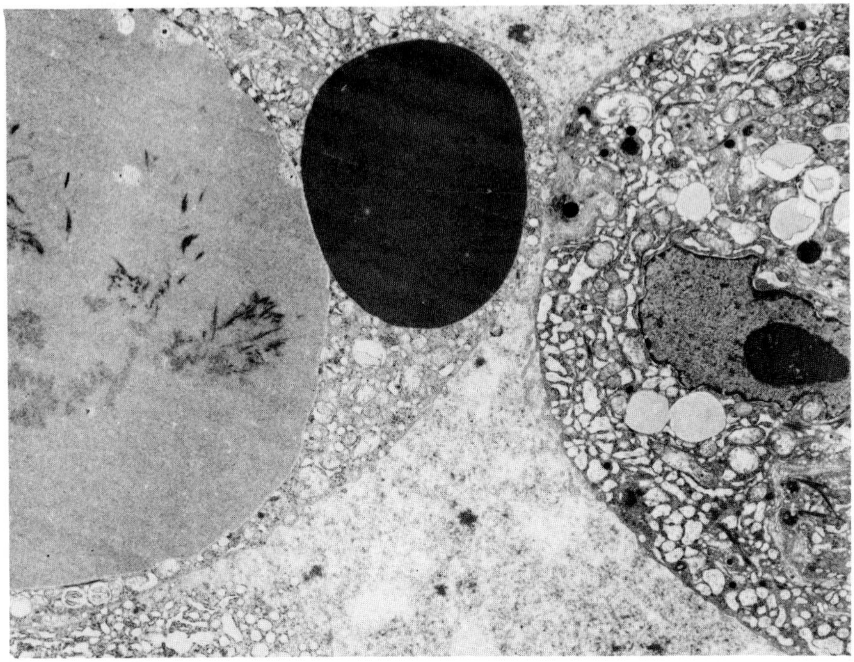

FIG. 10-37. Large cytoplasmic densities in a liver cell carcinoma. X5300.

REFERENCES

1. **Llombart-Bosch A, Blache R, Peydro-Olaya A:** Ultrastructural study of 28 cases of Ewing's sarcoma: Typical and atypical forms. Cancer 41:1362, 1978
2. **Nascimento AG, Cooper KL, Unni KK, et al.:** A clinicopathologic study of 20 cases of large-cell (atypical) Ewing's sarcoma of bone. Am J Surg Pathol 4:29, 1980
3. **Mackay B, Ferrans VJ, Henry K, et al.:** Cell junctions between cardiac macrophages: A case for the panel. Ultrastruct Pathol 1:67, 1980
4. **Ahearn MJ, Trujillo JM, Cork A, et al.:** The association of nuclear blebs with aneuploidy in human acute leukemia. Cancer Res 34:2887, 1974
5. **Wischnitzer S:** The annulate lamellae. Int Rev Cytol 27:65, 1970
6. **Chung EB, Enzinger FM:** Clear cell sarcoma of tendons and aponeuroses: Further observations. Lab Invest 38:8, 1978
7. **Battifora H:** Intracytoplasmic lumina in breast carcinoma. Arch Pathol 99:614, 1975
8. **Sajjad SM, Mackay B, Lukeman JM:** Oncocytic carcinoid tumor of the lung. Ultrastruct Pathol 1:171, 1980
9. **Landolt AM, Oswald UW:** Histology and ultrastructure of an oncocytic adenoma of the human pituitary. Cancer 31:1099, 1973
10. **Macadam RF:** Fine structure of a functional adrenal cortical adenoma. Cancer 26:1300, 1970
11. **Kai O, Fujioka T, Yasuda M:** Intramitochondrial bodies in bovine adrenocortical cells. Cell Tissue Res 185:69, 1977

12. **Soares JO, Moura Nunes JF:** Intramitochondrial paracrystalline inclusions in human hepatocytes. Am J Clin Pathol 71:497, 1979
13. **Magalhaes MM, Magalhaes MC:** Inclusions intramitochondriales a structure cristalline dans la cortico-surrenale du rat. J Microscopie 7:549, 1968
14. **Grimley PM, Schaff Z:** Significance of tubulo-reticular inclusions in the pathobiology of human diseases. Pathobiol Annu 6:221, 1976
15. **Mackay B, Ayala AG:** Intracisternal tubules in human melanoma cells. Ultrastruct Pathol 1:1, 1980
16. **Vernick SH, Kay S, Escobar M, et al.:** Intracisternal tubules in myxoid chondrosarcoma. Arch Pathol Lab Med 101:556, 1977
17. **Unni KK, Soule EH:** Alveolar soft part sarcoma: An electron microscopic study. Mayo Clin Proc 50:591, 1975
18. **White JG, Clawson CC:** Chediak-Higashi syndrome: Variable cytochemical reactivity of giant inclusions in polymorphonuclear leukocytes. Ultrastruct Pathol 1:223, 1980
19. **Osborne BM, Culber SJ, Cangir A, Mackay B:** Acinar cell carcinoma of the pancreas in a 9-year-old child: Case report with electron microscopic observations. South Med J 70:370, 1977
20. **Samaan NA, Osborne BM, Mackay B, et al.:** Endocrine and electron microscopic studies of pituitary adenomas secondary to hypothyroidism. J Clin Endocrinol Metab 45:903, 1977
21. **Bloodworth JMB, Kovacs K, Horvath E:** Light and electron microscopy of pituitary tumors: In Linfoot JA (ed): Recent Advances in the Diagnosis and Treatment of Pituitary Tumors. New York, Raven Press, 1979, p 141
22. **Mackay B, Masse SR, King OY, Butler JJ:** Diagnosis of neuroblastoma by electron microscopy of bone marrow aspirates. Pediatrics 56:1045, 1975
23. **Silva EG, Mackay B:** Neuroendocrine (Merkel cell) carcinoma of the skin: An ultrastructural study of 9 cases. Ultrastruct Pathol, 2:1, 1980
24. **Triche TJ, Ross WE:** Glycogen-containing neuroblastoma with clinical and histopathologic features of Ewing's sarcoma. Cancer 41:1425, 1978
25. **Ayala AG, Mackay B:** Ewing's sarcoma: An ultrastructural study. In Management of Primary Bone and Soft Tissue Tumors. Chicago, Year Book Medical Publishers, 1977, p 179
26. **Warner TFCS, Seo IS:** Goblet cell carcinoid of appendix. Cancer 44:1700, 1979
27. **Cohen C:** Intracytoplasmic hyaline globules in hepatocellular carcinomas. Cancer 37:1754, 1976
28. **Palmer PE, Wolfe HJ:** Alpha-1 antitrypsin deposition in primary hepatic carcinomas. Arch Pathol Lab Med 100:232, 1976
29. **Mackay B, Osborne BM:** The contribution of electron microscopy to the diagnosis of tumors. Pathobiol Annu 8:359, 1978
30. **Mackay B, Silva EG:** Diagnostic electron microscopy in oncology. Pathol Annu, 15(2):241, 1980

INDEX